L. Reznik, V. Kreinovich (Eds.)

Soft Computing in Measurement and Information Acquisition

Springer

Berlin
Heidelberg
New York
Hong Kong
London
Milano
Paris
Tokyo

Studies in Fuzziness and Soft Computing, Volume 127
http://www.springer.de/cgi-bin/search_book.pl?series=2941

Editor-in-chief
Prof. Janusz Kacprzyk
Systems Research Institute
Polish Academy of Sciences
ul. Newelska 6
01-447 Warsaw
Poland
E-mail: kacprzyk@ibspan.waw.pl

Further volumes of this series can be found on our homepage

Vol. 107. V. Dimitrov and B. Hodge
Social Fuzziology, 2002
ISBN 3-7908-1506-3

Vol. 108. L.M. Sztandera and C. Pastore (Eds.)
Soft Computing in Textile Sciences, 2003
ISBN 3-7908-1512-8

Vol. 109. R.J. Duro, J. Santos and M. Graña (Eds.)
Biologically Inspired Robot Behavior Engineering, 2003
ISBN 3-7908-1513-6

Vol. 110. E. Fink l. 112. Y. Jin
Advanced Fuzzy Systems Design and Applications, 2003
ISBN 3-7908-1523-3

Vol. 111. P.S. Szcepaniak, J. Segovia, J. Kacprzyk and L.A. Zadeh (Eds.)
Intelligent Exploration of the Web, 2003
ISBN 3-7908-1529-2

Vol. 112. Y. Jin
Advanced Fuzzy Systems Design and Applications, 2003
ISBN 3-7908-1537-3

Vol. 113. A. Abraham, L.C. Jain and J. Kacprzyk (Eds.)
Recent Advances in Intelligent Paradigms and Applications", 2003
ISBN 3-7908-1538-1

Vol. 114. M. Fitting and E. Orowska (Eds.)
Beyond Two: Theory and Applications of Multiple Valued Logic, 2003
ISBN 3-7908-1541-1

Vol. 115. J.J. Buckley
Fuzzy Probabilities, 2003
ISBN 3-7908-1542-X

Vol. 116. C. Zhou, D. Maravall and D. Ruan (Eds.)
Autonomous Robotic Systems, 2003
ISBN 3-7908-1546-2

Vol 117. O. Castillo, P. Melin
Soft Computing and Fractal Theory for Intelligent Manufacturing, 2003
ISBN 3-7908-1547-0

Vol. 118. M. Wygralak
Cardinalities of Fuzzy Sets, 2003
ISBN 3-540-00337-1

Vol. 119. Karmeshu (Ed.)
Entropy Measures, Maximum Entropy Principle and Emerging Applications, 2003
ISBN 3-540-00242-1

Vol. 120. H.M. Cartwright, L.M. Sztandera (Eds.)
Soft Computing Approaches in Chemistry, 2003
ISBN 3-540-00245-6

Vol. 121. J. Lee (Ed.)
Software Engineering with Computational Intelligence, 2003
ISBN 3-540-00472-6

Vol. 122. M. Nachtegael, D. Van der Weken, D. Van de Ville and E.E. Kerre (Eds.)
Fuzzy Filters for Image Processing, 2003
ISBN 3-540-00465-3

Vol. 123. V. Torra (Ed.)
Information Fusion in Data Mining, 2003
ISBN 3-540-00676-1

Vol. 125. M. Inuiguchi, S. Hirano and S. Tsumoto (Eds.)
Rough Set Theory and Granular Computing, 2003
ISBN 3-540-00574-9

Vol. 126. J.-L. Verdegay (Ed.)
Fuzzy Sets Based Heuristics for Optimization, 2003
ISBN 3-540-00551-X

L. Reznik
V. Kreinovich (Eds.)

Soft Computing in Measurement and Information Acquisition

Springer

Dr. Leon Reznik
Professor of Computer Science
Rochester Institute of Technology
102 Lomb Memorial Drive
Rochester, NY 14623-5608
USA
E-mail: lr@cs.rit.edu

Dr. Vladik Kreinovich
Department of Computer Science
University of Texas at El Paso
500 W. University
El Paso, TX 79968
USA
E-mail: vladik@cs.utep.edu

ISSN 1434-9922
ISBN 978-3-642-53509-3 ISBN 978-3-540-36216-6 (eBook)
DOI 10.1007/978-3-540-36216-6

Library of Congress Cataloging-in-Publication-Data applied for

A catalog record for this book is available from the Library of Congress.

Bibliographic information published by Die Deutsche Bibliothek
Die Deutsche Bibliothek lists this publication in the Deutsche Nationalbibliographie; detailed bibliographic data is available in the internet at <http://dnb.ddb.de>.

Springer-Verlag Berlin Heidelberg New York
a member of BertelsmannSpringer Science+Business Media GmbH
http://www.springer.de

Softcover reprint of the hardcover 1st edition 2003

Typesetting: camera-ready by editors
Cover design: E. Kirchner, Springer-Verlag, Heidelberg
Printed on acid free paper 62/3020/M - 5 4 3 2 1 0

Preface

This volume covers the fields of measurement and information acquisition. It contains a collection of papers representing the current research trends in these areas. What are those trends? The first one is the enormous growth in the amount of information and the amazing technologies, which make this information available anywhere and anytime. The second one is a substantial development of methods of the information presentation including, to name just a few, multimedia, virtual environment, computer animation. The third one is the all-time boosting demand for improving the quality of decisions made on the base of this information in various applications ranging from engineering to business. Nowadays information acquisition should not only provide more information but also provide it in such a way as to assure effective and efficient processing of this information. And here comes a relatively new methodology of soft computing. Application of soft computing in measurement and information acquisition is considered in this volume.

How can we assure that the acquired information leads to high quality decisions? Nowadays information delivery includes the engagement of complex information technology (IT) systems, including hardware, software, and the component called brainware (know-how and other applied knowledge). Availability of credible measurement and test methods is an important step toward assuring the quality of such systems. Many national and international organizations are now working toward this goal, including the US National Institute of Standards and Technology (NIST), national metrology institutes in the European Community and Japan, professional organizations such as IEEE and ACM, and industry groups such as Open Group (X/Open) and Underwriters Laboratories.

Since measurement is performed by more and more complex intelligent systems, it is necessary to use intelligent methodologies in the analysis of such systems. Some steps in this direction have been already made. For example, a fuzzy model can be easily extended to the domain of qualitative measurements as well as quantitative ones. It will allow to develop the common methodology for expressing and characterising uncertainty in both "machine-produced" and "human-performed" measurements resulting in a significant advancement towards the general measurement theory development. Of course, some important problems should be solved. How do humans measure values? What scale do they use? How to formalise their results and their methods in mathematical terms? Those questions are answered in the first part of the volume.

L. Mari and L. Reznik (Chapter 1) discuss philosophical and mathematical fundamentals of measurement uncertainty and offer linguistic scales as a replacement for the numerical scales in measurement. In this discussion they refer to ISO Guide to the expression of uncertainty in measurement, highlighting its merits and flaws.

H. T. Nguyen, V. Kreinovich, G. N. Solopchenko, and C. W. Tao (Chapter 2) apply the formalism of soft computing to statistical methods in measurement practice. Specifically, they provide a theoretical explanation for an empirical fact that among different "k sigma" intervals, the "two sigma" intervals are the most robust to possible changes in the probability distribution. Thus, they justify the choice of a parameter in establishing the width of the range so that receiving the measurement result beyond it is practically impossible.

A. Ryjov (Chapter 3) studies the mathematical aspects of a formalised description of "human-based" measurements. He develops a mathematical foundation for the construction of linguistic scales, which are applied to express the measurement results, and proves that his choice is the optimal one.

B. Pham and J. Zhang (Chapter 4) discuss one of the possible applications of linguistic scales and models in system design. They propose a shape specification based on fuzzy logic, which aims at facilitating a conceptual design by providing an opportunity to specify and work with rough models in a more intuitive fashion. This approach is more typical for a human design and would allow automation of human interaction of the initial design stages, when a human factor plays a major role.

A. Verkeyn, M. De Cock, D. Botteldooren, and E. E. Kerre (Chapter 5) develop a procedure of obtaining membership functions (which is necessary for linguistic scales modelling) from experimental data. The data are collected from the measurements obtained in psycholinguistic experiments. As a case study, they analyse the words that respondents use to describe the extent to which they are bothered by noise.

Vigorous development of Internet, computer graphics and animation and other information technologies have significantly expanded the volumes and sources of information available for decision making. On the other hand, in order to reach a high-quality decision in business, engineering and social applications, nowadays one has to fuse information of different kinds (e.g., numerical, statistical, textual, visual, audio) from a variety of sources (e.g. engineering measurement systems, expert's opinions, images, sound tracks). Due to their nature, these sources differ in reliability and uncertainty of the information produced. Uncertainty depends not only on the source but also on procedures and tools used to acquire and process information. To estimate uncertainty of expert information we might want to take into consideration such factors as psychological personal profiles and emotional behaviour, that immensely influence the results of information acquisition. Taking this psychological factors into consideration is difficult.

D. Lambert (Chapter 6) provides an overview of the Joint Directors of Laboratories (JDL) model of information fusion, one of the main data fusion techniques. He explains different aspects of object, situation, and impact fusions. This paper offers definitional, conceptual, and theoretical comments on components of the IDL model. Detailed algorithms for addressing specific fusion problems are presented in a joint paper by D. Lambert and M. Rutten (Chapter 8), which is devoted specifically to the situation assessment problem and its assessment.

S. Kaci, S. Benferhat, D. Dubois, and H. Prade (Chapter 7) develop mathematical background for soft computing and possibility theory applications in information fusion. The paper discusses the information fusion within the framework of the possibilistic logic, comparing a variety of merging operators at the semantic and syntactic levels, including conjunctive, disjunctive, reinforcement, adaptive and averaging operators.

D. Campbell and M. Lees (Chapter 9) consider a practical application of soft computing in measurement and information acquisition. Specifically, they describe a measurement system for a modern brewery factory. The food and beverage industries would benefit greatly from the ability to measure product parameters and evaluate its quality on-line and in real-time. To date, this has rarely been possible. Recent biotechnology developments have produced biosensors, which may allow real-time sensing. Would application of soft computing methodologies where some "human-based" models are used make bio-measurements faster and more reliable? And how to derive the integral characterisitic of the product quality from different qualitative and quantitative indicators?

L. Berrah, G. Mauris, and L. Foulloy (Chapter 10) consider the fuzzy mathematical foundations for the approach to integrating several qualitative and quantitative criteria into a single indicator. Although mathematical, the paper is strongly oriented towards real industrial applications. The paper studies the use of the Choquet fuzzy integral family for modelling different interactions and aggregating performance indicators.

Since its birth computational intelligence has been looking for inspiration and technical solutions in the operation of its natural counterpart. Modern advancement of both biological and computational sciences have made this interaction much more intense, with both sides aiming at achieving new results in this interdisciplinary region. This volume includes several chapters, which demonstrate a further movement towards closer collaboration.

S. Kameda and T. Yagi (Chapter 11) start with biology: based on recent physiological research results they model the architecture of the outer retina (which is a part of the central nervous system in the vertebrate) consisting of photoreceptors, horizontal and bipolar cells with an analog circuit. They analyze this model and apply this analysis to develop a VLSI circuit called silicon retina. This circuit performs a real time image processing in a natural illumination.

A. Chong, T.D. Gedeon and K.W. Wong (Chapter 12) report the results of developing a fuzzy inference system, that improves accuracy of predicting secondary structures of proteins. Proteins are very important for any biological creature. They come in many types or modifications. Over the years many new proteins have been identified by genome sequence projects and numerous methods of their structure prediction have been developed. Some of the most successful methods are based on neural networks. How to make the prediction more accurate? The authors demonstrate that a combination of fuzzy and neural techniques - implemented in ANFIS – substantially improves the accuracy of the structure prediction.

D. Tikk, L.T. Koczy and T.D. Gedeon (Chapter 13) study accuracy and other approximation characteristics of two soft computing techniques: radial basis function neural networks and fuzzy systems. Their paper concludes that certain fuzzy systems and neural networks could be considered as having similar input-output characteristics and hence capable of replacing and approximating each other. What sort of fuzzy systems could be replaced and which neural networks are to be applied? The paper offers a radial basis function approximation scheme.

M. Minami, J. Agbanhan and T. Asakura (Chapter 14) present a new method of scene recognition for a robot-manipulator. The method considers simultaneous shape recognition and determination of the position and orientation. How to do it? The hybrid genetic algorithm has been developed, employing the raw image to direct the search. The practical experiments with a hand-eye camera of a robot-manipulator reported in this chapter have proved reliability and robustness of the method.

T. Nanayakkara, K. Watanabe, K. Kiguchi and K. Izumi (Chapter 15) investigate a robot-manipulator dynamics under the condition of uncertainty in input joint torques or the gains of the servo controllers. The conventional methods application requires special hardware changes in an industrial manipulator. How to avoid it? The chapter introduces a method based on neural networks, and tests it on the Mitsubishi seven-link industrial robot PA-10. The sub-networks consist of radial basis function neural networks (which have become popular in modern research) optimised by an evolutionary algorithm.

The last cluster of chapters in this volume is directly devoted to the classification problems.

D.D. Coblentz, V. Kreinovich, B.S. Penn and S.A. Starks (Chapter 16) consider a practical problem of a geophysical area segmentation. Quite often different pieces of evidence lead experts to different sub-division schemes. How to reconcile them and develop a reliable clustering algorithm? The chapter proposes the use of extra topographical information and an application of interval and fuzzy analysis.

S. Abe (Chapter 17) introduces a new trainable fuzzy classifier with hyperbox regions. How to get a reliable classification which is fast in training? The proposed solution is to define a fuzzy rule for each class and tune them depending on the results.

A.C. Yeo, K.A. Smith, R.J. Willis and M. Brooks (Chapter 18) apply classification to a business problem. In the automobile insurance industry the expected value of claims is known with a high degree of uncertainty. To get a more accurate prediction of the claim value, the authors classify policy holders into groups with similar behaviour, and make different claim predictions for each group. Specifically, the chapter considers risk assessment and prediction of claim cost using k-means clustering, fuzzy c-means clustering and a heuristic method. The authors compare the results of these clustering techniques on real data.

S.E. Bedingfield and K.A. Smith (Chapter 19) consider credit scoring, which is another classification problem in business application. How to evaluate the risk of customers defaulting on granted credit when some data are missing or contain errors, and the cost of a mistake is high? The chapter presents a new approach to classification in which an evolutionary algorithm is used to generate and optimise classification rules.

And the last but definitely not the least, V. Dimitrov and B. Stewart (Chapter 20) describe the application of fuzzy logic and fuzzy measurement for modelling the behaviour and actions of government agencies.

El Paso, Texas, USA
November, 2002

L. Reznik and V. Kreinovich

VOLUME NAVIGATION HELP

To help the readers find the way through the volume, we provide a short manual describin possible navigation through this volume depending on the reader's interest.

Where to go? based on

1. Soft computing methodology applied

Fuzzy set theory (general)	Chapters 1,3,5,6,7,8,17,20
Fuzzy inference	Chapters 10,12,13,18
Fuzzy control	Chapters 13,15
Interval analysis	Chapters 2,16
Neural networks	Chapters 11,13,15
Evolutionary/genetics programming	Chapters 14,19

2. Problem being solved

Measurement	Chapters 1,3
Classification	Chapters 4,5,6,7,8,9,12,16,17,18,19
Prediction	Chapters 12,18,19
Image processing	Chapters 4,11,14,17
Robotics	Chapters 11,14,15

3. Results achieved and tested by

Mathematical analysis	Chapters 2,3,5,6,7,8,10,13,16,17,18
Computer analysis and simulation	Chapters 4,12,14,18,19
Electronics design	Chapters 11,15
Practical experiments	Chapters 5,11,12,14,15,18,19
Industrial application	Chapters 9,10,15,18,19

4. Research field represented

Science and engineering (general)	Chapters 11,16,17
Biotechnology and bioinformatics	Chapters 9,12
Computer science	Chapters 6,8,13,14,15,16
Human-computer interaction	Chapters 3,4,10
Mathematics	Chapters 2,3,7,16
Measurement science and practice	Chapters 1,2,5,9,14
Business and social science	Chapters 18,19,20

Contents

Uncertainty in Measurement: Some Thoughts about its Expressing and Processing

Luca Mari[1], Leon Reznik[2]
[1]Università C.Cattaneo - LIUC, C.so Matteotti 22, 21053 Castellanza (Va), Italy
email: lmari@liuc.it
[2]School of Communications and Informatics, Victoria University, PO Box 14428, Melbourne City MC, VIC 8001, Australia
email: Leon.Reznik@vu.edu.au

Summary: The paper discusses the concept of non-exactness of measurement results, and analyzes it by clearly distinguishing between: (i) the way the results are expressed to make their uncertainty explicit; (ii) the way the chosen expression is interpreted as a suitable combination of non-specificity and uncertainty; (iii) the way the interpreted results are formally dealt with. In this perspective the merits and flaws of the ISO Guide to the expression of uncertainty in measurement are highlighted.

Keywords: Foundations of Measurement, Measurement Uncertainty

1. Why Non-exactness is an Issue in Measurement

Measurement is a means of setting up a bridge between the empirical world (to which the measured thing belongs) and the linguistic/symbolic world (to which the measurement result belongs):

"things"
↓
"symbols"

The pragmatic aim of measurement is to enable symbolic processing of data drawn from the empirical world, so that any result obtained in data processing can be re-interpreted in terms of the measured things:

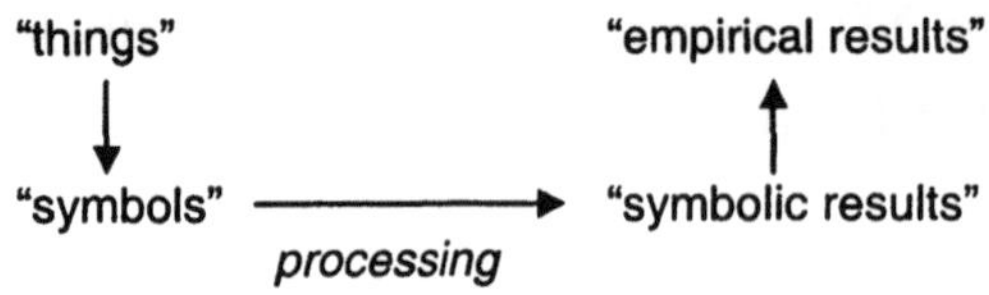

Crucial for the validity of this re-interpretation is therefore the *faithfulness* of the operation that associates symbols with things. In terms of the following diagram:

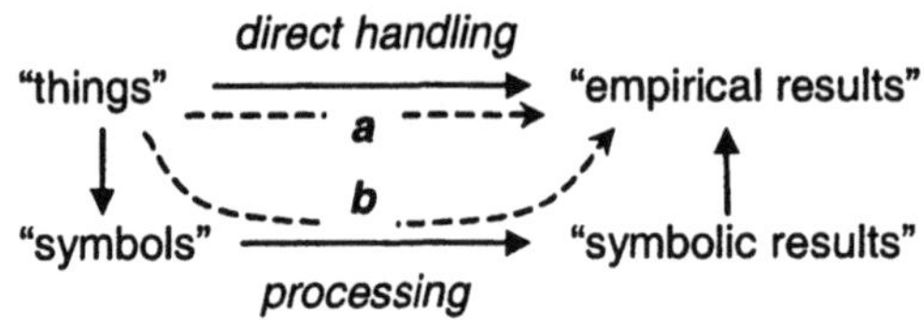

the issue is whether the procedures *a* and *b* would lead to the same result.

In the case of measurement such a notion of faithfulness is peculiar, since the result of any measurement is defined in reference to a given scale, playing the role of both conceptual and operational context (i.e.: model) in which the result is interpreted [Mari, 1999]. To be faithful, measurement is then required not only to keep a stable association among things and symbols, but also to preserve any structure (e.g., ordering) empirically observed among things.

The fact is that these two worlds, the one of empirical things and the one of symbols, are inherently different. According to Bridgman, «there are certain human activities which apparently have perfect sharpness. The realm of mathematics and of logic is such a realm, par excellence. Here we have yes-no sharpness. But this yes-no sharpness is found only in the realm of things we say, as distinguished from the realm of things we do. Nothing that happens in the laboratory corresponds to the statement that a given point is either on a given line or it is not» [Bridgman, 1959].

As Pattee pointed out [Pattee, 1989], the basic measurement problem is that semantic grounding of symbols by measurement is a controlled action by an observer that could not functionally be described by any laws. More accurately, if a measuring device, which is certainly a physical system obeying laws, is actually described by these laws by combining the device with the original system being measured, then the initial conditions are no longer separated, and additional new measuring devices are required to establish the initial conditions for this combined system. Therefore, syntactic description of measurement by laws destroys the semantic function of the measurement.

L.Zadeh classifies all the information commonly available into three groups:

- factual information which is numerical and measurement-based;
- pseudo-measurement based and pseudo-numerical (e.g. "checkout time is 11.00") information;
- perception based information which is mainly linguistic (e.g. "Robert is honest").

Those three groups are commonly assumed to differ in the degree of uncertainty, which the corresponding information has. However, does it mean that those groups have a fundamental, ontological differences as one traditionally assumes?

The classification follows the traditional division between quality and quantity, hard and soft sciences. This division could be illustrated with two quotations: "Quàlitative is nothing but poor quantitative" (Ernest Rutherford) and "A social scientist is a person who counts telephone poles" (Robert Hutchins).

For Rutherford everything we call a quality or a percept is expressible in terms of numerical magnitudes, without loss or distortion. Therefore, for him every quality can be quantified and hence measured and computed. For Rutherford science does not begin until quantification is made, until crude and inexact talk about quality is replaced by precise, exact, and completely equivalent talk about numbers. Here he follows another famous quotation by Dmitry Mendeleev: "Science begins where measurements are started".

Hutchins tacitly accepts Rutherford's equation of science with quantitative but for him this makes the phrase "social scientist" a contradiction in terms. For Hutchins, the features or qualities of a social structure which are of interest or importance are precisely those which are unquantifiable, and conversely anything that can be counted is trivial or irrelevant.

One of the not-so-commonly discussed consequence of such assumptions is that some classical distinctions, such as "linguistic" vs. "numerical", and "qualitative" vs. "quantitative", become rather ... fuzzy, in the proper sense that no clear-cut threshold can be drawn to define them, but by mere convention. Explicitly: even admitting that nominal scale measurement is inherently non-numerical [Mari 2000], ... what are "numbers"? On this subject some interesting lessons would come from the theory of algebraic structures (mainly focused on the concepts of relational systems - a generalization of universal algebras - and morphisms among them). In terms of scale types, the status of the so-called interval type could be considered, measuring quantities such as temperature (before Kelvin scale) and potential energy. According to a strong tradition, any kind of (integer, rational, real, complex) numbers derive from the natural sequence (for example, Kronecker said that natural numbers were given to manhood by God, while all other numbers have been created by us). Therefore the emphasis is on the existence of a unit, from which the whole sequence is inductively generated (as in the axiomatic approach by Peano). But the formal entities embedded in an interval scale do not empirically obey to any unit identification (consider the case of temperature: surely the unit degree is not an empirically primitive entity in the measurand definition). On the other hand, who would dare not to consider temperature degrees in Celsius scale as "numbers"? As a consequence, it seems that the very concept of number was implicitly extended, to embody "entities with a total order and a compatible metric on it". Is this the "final" extension? Or even "entities with a total order", i.e. in an ordinal scale, are "numbers"?

R. Rosen gives another example [Rosen, 1987]. It is a fact of experience that 2 sticks + 3 sticks = 5 sticks. On its face, this is a proposition about sticks. But it is not the same kind of proposition as "sticks burn" or "sticks float". It differs from them because it is something else besides sticks, and that "something else" according to Rosen is the mathematics. The mathematical world is embodied in percepts, but exists independent of them. "Truth" in the mathematical world is likewise manifested in, but independent of, any material embodiment, and is thus outside of conventional perceptual categories like space and time.

With a purely pragmatic position, one might see the distinction between "numbers" and "non-numbers" as purely conventional, and definitely useless. Furthermore, "numbers", whatever they are, are surely particular linguistic entities (characterized by specific algebraic properties), so that in any case the distinction should be "numbers" vs. "non-numbers", and not "numbers" vs. "linguistic entities".

The non-exactness (in the following the difference between non-exactness and uncertainty will be maintained and discussed) of measurement results accounts for such a distinction, although «by forcing the physical experience into the straight jacket of mathematics, with its yes-no sharpness, one is discarding an essential aspect of all physical experience and to that extent renouncing the possibility of exactly reproducing that experience. In this sense, the commitment of physics to the use of mathematics itself constitutes, paradoxically, a renunciation of the possibility of rigor» [Bridgman, 1959].

2. The Expression of Non-exact Measurement Results

Taking into account the linguistic side of the problem, the first decision to be made is related to the form a measurement result should be given to make its non-exactness explicit. According to the ISO *Guide to the expression of uncertainty in measurement* (GUM) ([ISO, 1993]; a useful synthesis of the Guide can be found in [Taylor, Kuyatt, 1997]), any measurement result must account for both the measurand value and its estimated uncertainty, and is therefore expressed as a couple:

measurement result = $<x,s(x)>$

where the first term, representing the measurand value, could be obtained as the average of the population of the instrument readings, and the second term, representing the uncertainty value, as its estimated standard deviation.

This position has the fundamental consequence that the concepts of measurement result and measurand value cannot be dealt with as equivalent, as instead, often implicitly, assumed in traditional approaches: measurement results express information not only on the measurand value, but also on its uncertainty. In other terms, "exact" measurement becomes the exception, not the rule.

If innovative in this *conceptual* assumption, the GUM maintains a strong link with the tradition in the *formal* side of its recommendations: indeed different, and somehow more general, representations could be chosen, for example subsets, or fuzzy subsets, or probability distributions. All of these are more widely applicable than the representation suggested by the GUM, that can be employed only for algebraically strong scales (although the generality is understood, because of the applicability of the Chebyshev's inequality, of expressing measurement results as couples $<x,s(x)>$, being x and $s(x)$ the average and standard deviation of the (unknown) assumed probability distribution).

3. Evaluation Methods and Related Meanings

The choice to express measurement results as $<x,s(x)>$ still leaves open the decision about the evaluation methods that can be adopted to obtain such results and the meaning to be attributed to both the terms of the couple. Once more, the GUM standpoint at this regard is a mix of innovation and tradition. With respect to the evaluation methods, the GUM embodies a recommendation issued by the CIPM in 1981 [CIPM, 1981] and admits both statistical (so-called "type A") and non-statistical ("type B") methods. The condition to make this pluralism operatively acceptable is that the suggested techniques to formally deal with the results are independent of the "type" of the evaluation method and therefore the same in both cases. From the conceptual point of view this position is an important step against the radical objectivism of some classical interpretations of measurement: some subjective information, in the form of "degrees of belief" (to quote the GUM) is present and required even in the case of an "objective" operation as measurement [Mari, Zingales, 1999].

On the other hand, not so pluralistic is the position of the GUM in reference to the meaning of the term $s(x)$. Its interpretation is meant to be statistical, as a standard deviation of x, thought of as a random variable in itself and therefore randomly extracted from a, typically unknown, probability distribution. The main application suggested by the GUM for this so-called "standard uncertainty" is to compute the "law of uncertainty propagation" (what is classically called "error propagation") through functional relations, thus essentially a problem of derived measurement. Only for specific applications a further interpretation is recognized as useful, in which $s(x)$ (and more precisely $ks(x)$, being k a proportionality factor usually in the range [1,3]), in this case called "expanded uncertainty", is considered to express the half-width of the interval of which x is the center point. This set-theoretic interpretation is however deemed as explicitly dependent on the statistical one and formally derived from it.

4. Two Categories of Applications

The position of the GUM is clearly conservative in this regard. A more general standpoint recognizes that the same result $<x,s(x)>$ could admit distinct interpretations in distinct applications. Given a set $\{<x_i,s(x_i)>\}$ of such measurement results, two basic areas of applications can be identified:

- "derived measurement": a quantity Y is known as analytically dependent of the quantities $X_1,\ldots,X_n$ through a function f, $Y=f(X_1,\ldots,X_n)$, and each $<x_i,s(x_i)>$ is the measurement result of a quantity X_i; the function f must be then somehow applied to the terms $<x_i,s(x_i)>$ to compute a measurement result $<y,s(y)>$ for Y;
- "measurement results comparison": all the $<x_i,s(x_i)>$ are repeated measurement results of the same quantity X, and must be compared to each other via a relation r, of which they are arguments, to establish whether such a relation holds among them or not.

Uncertainty propagation is clearly related to the first area of application, a case in which the statistical interpretation is plausibly the preferred one. The GUM suggests to compute y and $s(y)$ with distinct procedures, only in dependence on the terms x_i and $s(x_i)$ respectively:

$$y = f(x_1,\ldots x_n) \qquad\qquad s(y) = [c_i^2\, s^2(x_i)]^{1/2}$$

(in the case the quantities X_i are statistically correlated further terms should be introduced) having the factors c_i the role of "sensitivity coefficients", $c_i = \partial f/\partial x_i$, to weigh the corresponding standard uncertainties.

The most important example in the second area of application is the computation of what could be called "generalized equality", aimed at establishing whether two or more measurement results expressed as $<x_i,s(x_i)>$ can be considered undistinguishable with each other. The formal identity $<x_i,s(x_i)>=<x_j,s(x_j)>$, i.e.

$x_i = x_j$ and $s(x_i) = s(x_j)$, is clearly a "too exact" criterion in this case, and different, more general, principles have been proposed, typically based on the set-theoretical interpretation of the terms $<x_i, s(x_i)>$. For example, if the couple $<x_i, s(x_i)>$ is interpreted as the interval $[x_i - s(x_i), x_i + s(x_i)]$ then two results could be judged "compatible" with each other in the case their intersection is non-null, as the Italian standard [UNI, 1984] suggests.

It is worth to note that the GUM does not even mention this second category of applications. For important problems such as the definition of the procedures to compare national standards and express the results of the comparison an agreed position is still an open issue.

5. (Non-)Exactness as (Non-)Specificity and (Un)Certainty

The innovative conceptions of the GUM are surely of the greatest importance for both scientific advances and practical applications, and hopefully will finally remove some still widespread metaphysical assumptions such as the ones related to the concept of "true value". On the other hand, the GUM appears to be still very traditional in some of its groundings, and in particular in its assumption that the measurand value and its estimated uncertainty be definitely distinct values, both expressed as scalar and interpreted within the probabilistic framework.

On the contrary of the GUM viewpoints, we suggest that the non-exactness of a measurement result should be formalized as a combination of *two* distinct conceptual components, which can be called "uncertainty" and "non-specificity". An example is helpful to introduce the meaning of such concepts and their relations. Let us consider the following two statements:

A = "this is a 120-page book"
B = "this is a book"

aimed at expressing the knowledge of an observer on a given thing under examination. On the basis of the form of A and B, two conclusions can be immediately drawn:

- A entails B: if A is true then also B must be true (in set-theoretical terms, A is a subset of B); therefore A is more specific than B;
- regardless of the particular uncertainty assignment chosen, A is at most as certain as B, and plausibly more uncertain than it.

Hence the same formal expression, $<x, s(x)>$, admits two distinct, and actually opposite, meanings:

- the measurand value, the singleton x, is maximally specific but uncertain, with uncertainty $s(x)$;
- the measurand value, the interval $[\tilde{x}s(x),xs(x)]$, is not completely specific but considered certain.

The same empirical information, as obtained by the measurement, can be represented by suitably balancing the specificity and the certainty of the result (for example, if the statement A cannot considered certain on the basis of the available information, then:

C = "this is a more-than-100-page book"

could be adopted, less specific but more certain than A. In the case the object is factually a 120-page book, then A would not be "truer" than C, but more specific, and therefore more informative, than it).

Here one might see the relevance of Zadeh's classification referred to in section 1, which is considering both (non-)specificity and (un)certainty.

6. Conclusion

The traditional standpoint asserts that each quantity measured on a given thing has a "true" value, expressed as a singleton, i.e. a maximally specific value. A less-than-complete knowledge on that quantity would generate an uncertainty on such a true value, which, as a consequence, remains hidden to the observer. A clear distinction between the empirical world, to which the measured thing belongs, and the linguistic/symbolic world, to which the measurement result belongs, shows that this standpoint cannot be maintained against any operational analysis of the epistemic role of measurement and the way it is performed. In this view the ISO Guide to the expression of uncertainty in measurement is a mix of innovation and conservation. It has removed the metaphysical assumption of the existence of a "true" value, but it has substantially preserved the requirement that the measurement value be formalized as a singleton, coupled with an estimation of its uncertainty.

A more general position is based on a clear distinction between the evaluation of the measurement result and its expression. The empirical component of the measurement (the one usually performed by a sensor) is aimed at gathering information from the field on the measurand. Once this operation has been completed, the observer is able to evaluate such information, whose quantity depends on the quality of the measurement system and procedure adopted. The same information can be then formally expressed and interpreted in different ways, in view of specific applications. In particular, the *same* measurement result could be formalized as an uncertain singleton or an interval, or (if we allow to loose the compatibility with the GUM) a more general entity, such as a probability distribution or a fuzzy set. Because of this distinction, we can conclude that measurement results are neither specific nor non-specific, and neither certain nor uncertain. They are neither singletons-with-uncertainty-degrees nor intervals.

Entities such as singletons, intervals, probability distributions, fuzzy sets, ... belong to the linguistic/symbolic world, not to the empirical one: that is why the same empirical state (e.g. a measurand evaluated on a given thing) can be formally expressed in so different ways. In such a formalization the issue is not truth, but adequacy to given goals.

References

Bridgman P.W.(1959) How much rigor is possible in physics? In: Henkin et al. (eds.), The Axiomatic method, North-Holland.

CIPM, BIPM (1981) Proc. Com. Int. Poids et Mesures, 49, 8-9, 26 (in French) (also reported in the Introduction of the GUM).

ISO, et al., (1993) Guide to the expression of uncertainty in measurement (GUM). International Organization of Standardization, Geneva.

Mari L. (1999). Notes towards a qualitative analysis of information in measurement results, Measurement, 25, 3, 183-192.

Mari L.(2000) Metrological issues in weak scale type evaluation, Proc. IMEKO XVI.

Mari, L., Zingales, G.(2000) Uncertainty in measurement science. In: Measurement Science – A discussion, K.Karija, L.Finkelstein (eds.), Ohmsha, IOS Press, p. 135-147

Pattee H.H. (1989) The measurement problem in artificial world models, BioSystems, 23: p.281-290

Rosen R. (1987) On the scope of syntactics in mathematics and science: The machine metaphor. In: Casti J. and A.Karlqvist, Eds. Real brains, artificial minds, North Holland, Amsterdam, p. 1-23

Taylor N., Kuyatt C. (1997). Guidelines for evaluating and expressing the uncertainty of NIST measurement results. NIST Technical Note 1297 (also available in the web: physics.nist.gov/Pubs/guidelines/contents.html).

UNI, 1984. Misure e misurazioni. Termini e definizioni fondamentali. Italian Standard UNI 4546 (in Italian).

Why Two Sigma? A Theoretical Justification

Hung T. Nguyen[1], Vladik Kreinovich[2], Gennady N. Solopchenko[3], and Chin-Wang Tao[4]

[1] Department of Mathematical Sciences, New Mexico State University, Las Cruces, NM 88003, USA, email hunguyen@nmsu.edu
[2] Department of Computer Science, University of Texas at El Paso, El Paso, TX 79968, USA, email vladik@cs.utep.edu
[3] Information-Measurement Department, State Technical University of St. Petersburg, Politechnicheskaya 29, St. Petersburg, 195251 Russia, email sol@study.ssl.stu.neva.ru
[4] Department of Electrical Engineering, National I-Lan Institute of Technology, 260 I-Lan, Taiwan, email cwtao@mail.ilantech.edu.tw

Abstract. For a normal distribution, the probability density $\rho(x)$ is everywhere positive, so in principle, all real numbers are possible. In reality, the probability that a random variable is far away from the mean is so small that this possibility can be often safely ignored. Usually, a small real number k is picked (e.g., 2 or 3); then, with a probability $P_0(k) \approx 1$ (depending on k), the normally distributed random variable with mean a and standard deviation σ belongs to the interval $\mathbf{a} = [a - k \cdot \sigma, a + k \cdot \sigma]$.

The actual error distribution may be non-Gaussian; hence, the probability $P(k)$ that a random variable belongs to $\mathbf{a}$ differs from $P_0(k)$. It is desirable to select k for which the dependence of $P_0(k)$ on the distribution is the smallest possible. Empirically, this dependence is the smallest for $k \in [1.5, 2.5]$. In this paper, we give a theoretical explanation for this empirical result.

1 Formulation of the Problem

For many measuring instruments, the measurement error is normally distributed; see, e.g., [10,11]. This known empirical fact has a good theoretical explanation (see, e.g., [2–4,13]; see also [14] pp. 2.17, 6.5, 9.8, and references therein): Usually, the manufacturers of the measuring instruments have made their best effort to eliminate the major sources of measurement error. The resulting measurement error comes from a variety of small independent error sources, and thus, can be described as a sum of a large number of small independent random variables. According to the central limit theorem, such a sum, under reasonable conditions, converges to normal distribution. Thus, if there are sufficiently many small random components, the resulting error distribution is indeed close to normal.

For a normal distribution, the probability density $\rho(x)$ is positive for all x, so in principle, all real numbers are possible. In reality, however, the probability of a random variable to be far away from the mean is so small that in many practical applications, this possibility can be safely ignored. So, the values of a normally distributed random variable are located, with a reasonably high probability, within a finite interval. To implement this idea, in practice, usually, a small real number k is picked (typically, $k = 2$ or $k = 3$). Then, with a probability $P_0(k) \approx 1$ (depending on k), the values of a normally distributed random variable with mean a and standard deviation σ belong to the interval $[a - k \cdot \sigma, a + k \cdot \sigma]$. For a normal distribution, this probability does not depend on a and σ, only on k. For $k = 2$, we have $P_0(k) \approx 0.95$; for $k = 3$, we have $P_0(k) \approx 0.999$; for $k = 6$, we have $P_0(k) \approx 1 - 10^{-6}$, etc.

For a *normal* distribution, we can pick an arbitrary k and get the interval which contains all the values with the corresponding probability $P_0(k)$. In many real-life situations, however, the actual error distribution is close to Gaussian but not exactly normal [10,11]. The deviation from a Gaussian distribution can be characterized by one or several parameters ε (so that Gaussian distribution corresponds to $\varepsilon = 0$). For a non-Gaussian distribution characterized by a parameter ε, the probability $P(k, \varepsilon)$ that a random variable belongs to the interval $\mathbf{a} = [a - k \cdot \sigma, a + k \cdot \sigma]$ is, in general, different from $P_0(k)$. It is therefore desirable to select k for which the dependence of $P(k, \varepsilon)$ on ε is the smallest possible, i.e., for which we can guarantee that $P(\eta \in \mathbf{a}) \approx P_0(k)$ irrespective of whether η is normally distributed or not.

The empirical analysis of actual probability distributions of different measuring instruments show that the smallest possible dependence occurs when k is between 1.5 and 2.5 [10]. This empirical fact is an important part of measurement practice, but until now, it has not been theoretically explained – not because we get a difficult-to-solve precisely formulated statistical problem, but because, due to uncertainty, this informal problem is very difficult to formalize in precise terms.

In this paper, we show how this problem can be formalized, and we show that this formalization indeed justifies the empirical choice of $k \in [1.5, 2.5]$.

2 Selecting a Class of Non-Gaussian Probability Distributions

2.1 Selecting distributions can be reduced to selecting functions

In principle, there can be many different probability distributions which are close to Gaussian. For different possible deviations from the Gaussian distribution, different values of k may be the least sensitive to the corresponding deviations. Therefore, before we start looking for the optimal value of k, we must first select a reasonable class of such deviations.

In many practical situations, there is no reason why a positive error value x should be more probable or less probable than the corresponding negative value $-x$, so we can assume that these probabilities coincide, and the probability distribution is symmetric w.r.t. $x \to -x$ (i.e., the corresponding probability density function $\rho(x)$ is even).

In particular, we will consider symmetric Gaussian distributions, i.e., Gaussian distributions with zero mean. It is known that an arbitrary distribution of this type, with an arbitrary standard deviation σ, can be obtained from a "standard" Gaussian distribution (with zero mean and unit standard deviation) by a linear transformation $f(z) = \sigma \cdot z$. In other words, if ζ is a random variable which is distributed according to the standard Gaussian distribution, then the variable $\eta = f(\zeta) = \sigma \cdot \zeta$ is distributed according to the Gaussian law with zero mean and standard deviation σ.

One can show that non-Gaussian distributions can be obtained in a similar manner, but with possibly non-linear increasing functions $f(z)$. Indeed, an arbitrary probability distribution can be described by its cumulative distribution function (cdf) $F(x) = P(\eta \square x)$. Let $F_0(x) = P(\zeta \square x)$ denote a cdf which corresponds to the Gaussian distribution. Let us show that by choosing an appropriate function $f(z)$, we can make $\eta = f(\zeta)$ have the desired cdf $F(x)$. Indeed, since we are only considering increasing functions $f(z)$, the inequality $f(\zeta) \square x$ is equivalent to $\zeta \square f^{-1}(x)$, where $f^{-1}(x)$ is the function which is inverse to $f(x)$ (i.e., $f^{-1}(x) = z$ if and only if $f(z) = x$). Thus, to guarantee that $P(f(\zeta) \square x) = F(x)$ for all x, we must guarantee that $P(\zeta \square f^{-1}(x)) = F(x)$. Since ζ is distributed according to the standard Gaussian distribution, we have $P(\zeta \square f^{-1}(x)) = F_0(f^{-1}(x))$. Thus, we must guarantee that for every x, we have $F(x) = F_0(f^{-1}(x))$. If we denote $z = f^{-1}(x)$, then we have $x = f(z)$, and the desired equality takes the form $F(f(z)) = F_0(z)$, hence we can take $f(z) = F^{-1}(F_0(z))$. So, every distribution can indeed be described as $\eta = f(\zeta)$ for an appropriate function $f(z)$.

Since we only consider symmetric distribution, these increasing functions $f(z)$ have to be odd: $f(-z) = -f(z)$. Hence, $f(0) = 0$, $f(z) \geq 0$ for $z \geq 0$, and to reconstruct the entire function $f(z)$, it is sufficient to know its values for $z \geq 0$.

In view of this representation of a arbitrary probability distribution by a transformation function $f(z)$, instead of selecting a class of probability distributions, we can select a class of functions $f(z)$. Then, we will be able to use, as new random variables, combinations $\eta = f(\zeta)$, where ζ has a standard Gaussian distribution and $f(z)$ is one of the selected functions.

The question is: How to select the "best" (most appropriate) functions $f(z)$?

2.2 Best in what sense?

What do we mean by "the best"? It is not so difficult to come up with different criteria for choosing a functions $f(z)$:

- We may want to choose the function $f(z)$ for which the average distance $D(f)$ between the resulting probability distribution and the actual empirical distributions of measuring instruments is the smallest possible.

- We may also want to choose the function $f(z)$ for which the *average computation time* $C(f)$ of some statistical processing algorithms is the smallest (average in the same of some reasonable probability distribution on the set of all problems).

At first glance, the situation seems hopeless: it is difficult to feasibly estimate these numerical criteria even for a single function $f(z)$, so it may look like we therefore cannot undertake an even more ambitious task of finding the *optimal* function $f(z)$. Hopefully, the situation is not as hopeless as it may seem, because there is a symmetry-based formalism (actively used in the foundations of fuzzy, neural, genetic computations, see, e.g., [8]) which will enable us to find the optimal function $f(z)$.

2.3 We must choose a family of functions

If we simply replace the original measurement unit by a new unit which is C times smaller, then all the numerical values of the measurement error η get multiplied by C. Thus, if the function $f(z)$ (which describes the original probability distribution) is a reasonable transformation function, then the function $C \cdot f(z)$ which corresponds to the same distribution expressed in the new units is also reasonable. Thus, with every function $f(z)$, all the functions $C \cdot f(z)$ should be selected as well, the whole *family* of functions $\{C \cdot f(z)\}$ (characterized by a parameter $C > 0$) must be selected.

Thus, instead of selecting the "best" (more appropriate) *functions*, we should talk about selecting the best *families*.

In the following text, we will denote families of functions by caligraphic capital letters, such as $\mathcal{F}$, $\mathcal{F}_i$, $\mathcal{G}$, etc.

2.4 An optimality criterion can be non-numeric

Traditionally, optimality criteria are *numerical*, i.e., to every family $\mathcal{F}$, we assign some value $J(\mathcal{F})$ expressing its quality, and choose a family for which this value is minimal (i.e., when $J(\mathcal{F}) \square J(\mathcal{G})$ for every other alternative $\mathcal{G}$). However, it is not necessary to restrict ourselves to such numeric criteria only.

For example, if we have several different families $\mathcal{F}$ that have the same average distance $D(\mathcal{F})$, we can choose between them the one that has the minimal computational time $C(\mathcal{F})$. In this case, the actual criterion that we use to compare two families is not numeric, but more complicated: A family $\mathcal{F}_1$ is better than the family $\mathcal{F}_2$ if and only if either $D(\mathcal{F}_1) < D(\mathcal{F}_2)$, or $D(\mathcal{F}_1) = D(\mathcal{F}_2)$ and $C(\mathcal{F}_1) < C(\mathcal{F}_2)$.

The only thing that a criterion *must* do is to allow us, for every pair of families $(\mathcal{F}_1, \mathcal{F}_2)$, to make one of the following conclusions:

- the first family is better with respect to this criterion (we'll denote it by $\mathcal{F}_1 \succ \mathcal{F}_2$, or $\mathcal{F}_2 \prec \mathcal{F}_1$);
- with respect to the given criterion, the second family is better ($\mathcal{F}_2 \succ \mathcal{F}_1$);
- with respect to this criterion, the two families have the same quality (we'll denote it by $\mathcal{F}_1 \sim \mathcal{F}_2$);
- this criterion does not allow us to compare the two families.

Of course, it is necessary to demand that these choices be consistent. For example, if $\mathcal{F}_1 \succ \mathcal{F}_2$ and $\mathcal{F}_2 \succ \mathcal{F}_3$ then $\mathcal{F}_1 \succ \mathcal{F}_3$.

2.5 Optimality criterion must be final

A natural demand is that this criterion must choose a *unique* optimal family (i.e., a family that is better with respect to this criterion than any other family). The reason for this demand is very simple.

If a criterion *does not choose* any family at all, then it is of no use.

If *several* different families are the best according to this criterion, then we still have the problem of choosing the best among them. Therefore we need some additional criterion for that choice, like in the above example: If several families $\mathcal{F}_1, \mathcal{F}_2, \ldots$ turn out to have the same average distance ($D(\mathcal{F}_1) = D(\mathcal{F}_2) = \ldots$), we can choose among them a family with minimal computation time ($C(\mathcal{F}_i) \to \min$).

So what we actually do in this case is abandon that criterion for which there were several "best" families, and consider a new "composite" criterion instead: $\mathcal{F}_1$ is better than $\mathcal{F}_2$ according to this new criterion if either it was better according to the old criterion, or they had the same quality according to the old criterion and $\mathcal{F}_1$ is better than $\mathcal{F}_2$ according to the additional criterion.

In other words, if a criterion does not allow us to choose a unique best family, it means that this criterion is not final, we'll have to modify it until we come to a final criterion that will have that property.

2.6 The criterion must not change if we change the measuring unit corresponding to the original Gaussian distribution

The exact mathematical form of a function $f(z)$ depends on the exact choice of units for measuring the original normally distributed variable ζ. If we replace this unit by a new unit that is λ times larger, then the same physical value that was previously described by a numerical value ζ will now be described, in the new units, by a new numerical value $\widetilde{\zeta} = \zeta/\lambda$.

How will the expression for $f(z)$ change if we use the new units? In terms of $\widetilde{\zeta}$, we have $\zeta = \lambda \cdot \widetilde{\zeta}$. Thus, the variable η which was originally represented by a function $f(\zeta)$, will be described, in the new units, as $f\left(\lambda \cdot \widetilde{\zeta}\right)$, i.e., as $\widetilde{f}\left(\widetilde{\zeta}\right)$, where $\widetilde{f}(z) = f(\lambda \cdot z)$.

There is no reason why one choice of a unit should be preferable to another. Therefore, it is reasonable to assume that the relative quality of different families should not change if we simply change the units, i.e., if the family $\mathcal{F}$ is better than a family $\mathcal{G}$, then the transformed family $\widetilde{\mathcal{F}}$ should also be better than the family $\widetilde{\mathcal{G}}$.

We are now ready for the formal definitions.

2.7 Definitions and the main result

Definition 1. *Let $f(z)$ be a differentiable strictly increasing function from real numbers to non-negative real numbers. By a family that corresponds to this function $f(z)$, we mean a family of all functions of the type $\widetilde{f}(z) = C \cdot f(z)$, where $C > 0$ is an arbitrary positive real number. (Two families are considered equal if they coincide, i.e., consist of the same functions.)*

In the following text, we will denote the set of all possible families by Φ.

Definition 2. *By an optimality criterion, we mean a consistent pair $\langle \prec, \sim \rangle$ of relations on the set Φ of all alternatives which satisfies the following conditions, for every $\mathcal{F}, \mathcal{G}, \mathcal{H} \in \Phi$:*

1. *if $\mathcal{F} \prec \mathcal{G}$ and $\mathcal{G} \prec \mathcal{H}$ then $\mathcal{F} \prec \mathcal{H}$;*
2. *$\mathcal{F} \sim \mathcal{F}$;*
3. *if $\mathcal{F} \sim \mathcal{G}$ then $\mathcal{G} \sim \mathcal{F}$;*
4. *if $\mathcal{F} \sim \mathcal{G}$ and $\mathcal{G} \sim \mathcal{H}$ then $\mathcal{F} \sim \mathcal{H}$;*
5. *if $\mathcal{F} \prec \mathcal{G}$ and $\mathcal{G} \sim \mathcal{H}$ then $\mathcal{F} \prec \mathcal{H}$;*
6. *if $\mathcal{F} \sim \mathcal{G}$ and $\mathcal{G} \prec \mathcal{H}$ then $\mathcal{F} \prec \mathcal{H}$;*
7. *if $\mathcal{F} \prec \mathcal{G}$ then $\mathcal{G} \not\prec \mathcal{F}$ and $\mathcal{F} \not\sim \mathcal{G}$.*

Comment. The intended meaning of these relations is as follows:

- $\mathcal{F} \prec \mathcal{G}$ means that with respect to a given criterion, $\mathcal{G}$ is better than $\mathcal{F}$;
- $\mathcal{F} \sim \mathcal{G}$ means that with respect to a given criterion, $\mathcal{F}$ and $\mathcal{G}$ are of the same quality.

Under this interpretation, conditions *1.-7.* have simple intuitive meaning; e.g., the condition *1.* means that if $\mathcal{G}$ is better than $\mathcal{F}$, and $\mathcal{H}$ is better than $\mathcal{G}$, then $\mathcal{H}$ is better than $\mathcal{F}$.

Definition 3.

- *We say that an alternative $\mathcal{F}$ is optimal (or best) with respect to a criterion $\langle \prec, \sim \rangle$ if for every other alternative $\mathcal{G}$ either $\mathcal{F} \succ \mathcal{G}$ or $\mathcal{F} \sim \mathcal{G}$.*
- *We say that a criterion is final if there exists an optimal alternative, and this optimal alternative is unique.*

Definition 4. *Let $\lambda > 0$ be a positive real number.*

- *By a λ-rescaling of a function $f(x)$ we mean a function $\widetilde{f}(x) = f(\lambda \cdot x)$.*
- *By a λ-rescaling $R_\lambda(\mathcal{F})$ of a family of functions $\mathcal{F}$ we mean the family consisting of λ-rescalings of all functions from $\mathcal{F}$.*

Definition 5. *We say that an optimality criterion on Φ is unit-invariant if for every two families $\mathcal{F}$ and $\mathcal{G}$ and for every number $\lambda > 0$, the following two conditions are true:*

i) if $\mathcal{F}$ is better than $\mathcal{G}$ in the sense of this criterion (i.e., $\mathcal{F} \succ \mathcal{G}$), then $R_\lambda(\mathcal{F}) \succ R_\lambda(\mathcal{G})$;
ii) if $\mathcal{F}$ is equivalent to $\mathcal{G}$ in the sense of this criterion (i.e., $\mathcal{F} \sim \mathcal{G}$), then $R_\lambda(\mathcal{F}) \sim R_\lambda(\mathcal{G})$.

Theorem 1. *If a family $\mathcal{F}$ is optimal in the sense of some optimality criterion that is final and unit-invariant, then every function $f(z)$ from this family $\mathcal{F}$ has the form $C \cdot z^\alpha$ for some real numbers C and α.*

Comment. For the convenience of the readers, all the proofs are placed in the last section.

Since $f(z)$ is an odd function, we can therefore conclude that the corresponding random variable η can be described as $\eta = \mathrm{sign}(\zeta) \cdot |\zeta|^\alpha$, where ζ is a standard Gaussian random variable, i.e., a normally distributed random variable with zero mean and unit standard deviation. This is indeed a good description for empirical distributions of measurement error [10].

Gaussian variables correspond to $\alpha = 1$; so, since we are interested in distributions which are close to Gaussian, we should consider α close to 1, i.e., $\alpha = 1 + \varepsilon$ for some small ε.

3 Selecting the Optimal Value of k

Let us consider the class of probability distributions described in the previous section. We want to find k for which the dependence of $P(k, \varepsilon)$ on ε is the smallest. Since empirical distributions are close to normal, we have $\varepsilon \approx 0$. For $\varepsilon \approx 0$, we can neglect quadratic and higher order terms in the dependence of $P(k, \varepsilon)$ on ε, and conclude that $P(k, \varepsilon) \approx P_0(k) + \varepsilon \cdot P_1(k)$, where

$$P_1(k) = \frac{\partial P(k, \varepsilon)}{\partial \varepsilon}_{|\varepsilon=0}. \tag{1}$$

Thus, this dependence is the smallest if and only if the absolute value $|P_1(k)|$ of the coefficient $P_1(k)$ is the smallest possible. It turns out $|P_1(k)|$ achieves its smallest value $|P_1(k)| = 0$ for some k which is indeed close to the interval $[1.5, 2.5]$, thus justifying the above empirical fact:

Definition 6. *We say that the value k is the least sensitive to the possible non-Gaussian character of the probability distribution if for this k, the expression $|P_1(k)|$, where $P_1(k)$ is determined by the formula (1), attains the smallest possible value.*

The formulation of the result uses the Euler constant

$$\gamma = \lim_{n\to\infty}\left(1 + \frac{1}{2} + \ldots + \frac{1}{n} - \ln(n)\right) \approx 0.577.$$

Theorem 2. *The value* $k = \frac{e}{\sqrt{2}\cdot e^{\gamma/2}} \approx 1.44$ *is the least sensitive to the possible non-Gaussian character of the probability distribution, and for this k, we have* $|P_1(k)| = 0$.

Comment. For this value k, $P_0(k) \approx 0.85$, so at least 85% of the values of the random variable lie in the interval $[a - k\cdot\sigma, a + k\cdot\sigma]$. This value is in good accordance with common sense, namely, with the 20-80 "Pareto" law, according to which:

- 20% of the people drink 80% of the beer,
- 20% of the researchers write 80% of all papers etc.

This number is also in good accordance with the experimental fact that from each 25 rules typically discovered by a data mining system, approximately 20 (i.e., about 80%) are already known (see, e.g., [7]). It is probably worth mentioning that in [12], we give an alternative explanation of this same fact – by using fuzzy logic techniques (see, e.g., [6,9]) instead of probabilities.

4 Proofs

4.1 Proof of Theorem 1

This proof is based on the following lemma:

Lemma. *If an optimality criterion is final and unit-invariant, then the optimal family* $\mathcal{F}_{\text{opt}}$ *is also unit-invariant, i.e.,* $R_\lambda(\mathcal{F}_{\text{opt}}) = \mathcal{F}_{\text{opt}}$ *for every number* λ.

Proof of the Lemma. Since the optimality criterion is final, there exists a unique family $\mathcal{F}_{\text{opt}}$ that is optimal with respect to this criterion, i.e., for every other $\mathcal{F}$, either $\mathcal{F}_{\text{opt}} \succ \mathcal{F}$, or $\mathcal{F}_{\text{opt}} \sim \mathcal{F}$.

To prove that $\mathcal{F}_{\text{opt}} = R_\lambda(\mathcal{F}_{\text{opt}})$, we will first show that the re-scaled family $R_\lambda(\mathcal{F}_{\text{opt}})$ is also optimal, i.e., that for every family $\mathcal{F}$: either $R_\lambda(\mathcal{F}_{\text{opt}}) \succ \mathcal{F}$, or $R_\lambda(\mathcal{F}_{\text{opt}}) \sim \mathcal{F}$.

If we prove this optimality, then the desired equality will follow from the fact that our optimality criterion is final and therefore, there is only one optimal family (so, since the families $\mathcal{F}_{\text{opt}}$ and $R_\lambda(\mathcal{F}_{\text{opt}})$ are both optimal, they must be the same family).

Let us show that $R_\lambda(\mathcal{F}_{\text{opt}})$ is indeed optimal. How can we, e.g., prove that $R_\lambda(\mathcal{F}_{\text{opt}}) \succ \mathcal{F}$? Since the optimality criterion is unit-invariant, the desired relation is equivalent to $\mathcal{F}_{\text{opt}} \succ R_{\lambda^{-1}}(\mathcal{F})$. Similarly, the relation $R_\lambda(\mathcal{F}_{\text{opt}}) \sim \mathcal{F}$ is equivalent to $\mathcal{F}_{\text{opt}} \sim R_{\lambda^{-1}}(\mathcal{F})$.

These two equivalences allow us to complete the proof of the lemma. Indeed, since $\mathcal{F}_{\text{opt}}$ is optimal, we have one of the two possibilities: either $\mathcal{F}_{\text{opt}} \succ R_{\lambda^{-1}}(\mathcal{F})$, or $\mathcal{F}_{\text{opt}} \sim R_{\lambda^{-1}}(\mathcal{F})$. In the first case, we have $R_\lambda(\mathcal{F}_{\text{opt}}) \succ \mathcal{F}$; in the second case, we have $R_\lambda(\mathcal{F}_{\text{opt}}) \sim \mathcal{F}$.

Thus, whatever family $\mathcal{F}$ we take, we always have either $R_\lambda(\mathcal{F}_{\text{opt}}) \succ \mathcal{F}$, or $R_\lambda(\mathcal{F}_{\text{opt}}) \sim \mathcal{F}$. Hence, $R_\lambda(\mathcal{F}_{\text{opt}})$ is indeed optimal and thence, $R_\lambda(\mathcal{F}_{\text{opt}}) = \mathcal{F}_{\text{opt}}$. The lemma is proven.

Let us now prove the theorem. Since the criterion is final, there exists an optimal family $\mathcal{F}_{\text{opt}} = \{C \cdot f(z)\}$. Due to the lemma, the optimal family is unit-invariant.

¿From unit-invariance, it follows that for every λ, there exists a real number $A(\lambda)$ for which $f(\lambda \cdot z) = A(\lambda) \cdot f(z)$. Since the function $f(z)$ is differentiable, we can conclude that the ratio $A(\lambda) = f(\lambda \cdot z)/f(z)$ is differentiable as well. Thus, we can differentiate both sides of the above equation with respect to λ, and substitute $\lambda = 1$. As a result, we get the following differential equation for the unknown function $f(z)$:

$$z \cdot \frac{df}{dz} = \alpha \cdot f,$$

where by α, we denoted the value of the derivative $dA/d\lambda$ taken at $\lambda = 1$. Moving terms dz and z to the right-hand side and all the term containing f to the left-hand side, we conclude that

$$\frac{df}{f} = \alpha \cdot \frac{dz}{z}.$$

Integrating both sides of this equation, we conclude that $\ln(f) = \alpha \cdot \ln(z) + C$ for some constant C, and therefore, that $f(z) = \text{const} \cdot z^\alpha$. The theorem is proven.

4.2 Proof of Theorem 2

As we have shown in Section 2, for each ε, the corresponding random variable η can be described as $\eta = \text{sign}(\zeta) \cdot |\zeta|^{1+\varepsilon}$, where ζ is a standard Gaussian random variable, i.e., a normally distributed random variable with zero mean and unit standard deviation. For this random variable, the mean is equal to $E_\varepsilon(\eta) = 0$. Let $\sigma(\varepsilon) = \sqrt{E_\varepsilon(\eta^2)}$ denote its standard deviation. Then, the probability $P(k, \varepsilon)$ is equal to the probability that $\eta \in [-k \cdot \sigma(\varepsilon), k \cdot \sigma(\varepsilon)]$, i.e., to the probability that $|\eta| \,\square\, k \cdot \sigma(\varepsilon)$.

Since $|\eta| = |\zeta|^{1+\varepsilon}$, the probability $P(k, \varepsilon)$ is equal to the probability that $|\zeta|^{1+\varepsilon}$ □ $k \cdot \sigma(\varepsilon)$, i.e., that for a standard Gaussian random variable ζ, we have $|\zeta|$ □ $B(\varepsilon)$, where we denoted $B(\varepsilon) = (k \cdot \sigma(\varepsilon))^{1/(1+\varepsilon)}$. In other words, $P(k, \varepsilon) = F_{\text{erf}}(B(\varepsilon))$, where we denoted

$$F_{\text{erf}}(z) = \frac{1}{\sqrt{2\pi}} \cdot \int_{-z}^{z} \exp\left(-\frac{t^2}{2}\right) \mathrm{d}t. \tag{2}$$

Due to the chain rule, $P_1(k) = F'(B(0)) \cdot B'(0)$, where F'_{erf} and B' denote derivatives with respect to ε. The function $F_{\text{erf}}(z)$ is a strictly increasing function of z, with $F'_{\text{erf}}(z) > 0$ for all z. Hence, $P_1(k) = 0$ if and only if $B'(0) = 0$.

By definition, $B(\varepsilon) = (k \cdot \sigma(\varepsilon))^{1/(1+\varepsilon)} = \exp(b(\varepsilon))$, where we denoted

$$b(\varepsilon) = \frac{\ln(k) + \ln(\sigma(\varepsilon))}{1 + \varepsilon}. \tag{3}$$

Therefore, $B'(\varepsilon) = \exp(b(\varepsilon)) \cdot b'(\varepsilon)$. The first factor in this product is always positive, so $B'(0) = 0$ if and only if $b'(0) = 0$. Let use the equation $b'(0) = 0$ to determine the desired value k. Differentiating the above expression for $b(\varepsilon)$ and substituting $\varepsilon = 0$, we conclude that $(\ln(\sigma(\varepsilon))' - \ln(k) - \ln(\sigma(\varepsilon)) = 0$, i.e., $\sigma'(0)/\sigma(0) - \ln(k) - \ln(\sigma(0)) = 0$. For $\varepsilon = 0$, we have a standard Gaussian distribution, for which $\sigma(0) = 1$. Thus, the above equation takes the form $\sigma'(0) - \ln(k) = 0$, hence $\ln(k) = \sigma'(0)$ and

$$k = \exp(\sigma'(0)). \tag{4}$$

To complete our proof, let us find the explicit expression for $\sigma'(0)$. By definition,

$$\sigma^2(\varepsilon) = E_\varepsilon(\eta^2) = E_0(|\zeta|^{2+2\varepsilon}) = \frac{1}{\sqrt{2\pi}} \cdot \int_{-\infty}^{\infty} |\zeta|^{2+2\varepsilon} \cdot \exp\left(-\frac{\zeta^2}{2}\right) \mathrm{d}\zeta. \tag{5}$$

Since negative and positive values ζ lead to an equal contribution to this integral, we can conclude that

$$\sigma^2(\varepsilon) = \frac{2}{\sqrt{2\pi}} \cdot \int_0^{\infty} \zeta^{2+2\varepsilon} \cdot \exp\left(-\frac{\zeta^2}{2}\right) \mathrm{d}\zeta =$$

$$\sqrt{\frac{2}{\pi}} \cdot \int_0^{\infty} \zeta^{2+2\varepsilon} \cdot \exp\left(-\frac{\zeta^2}{2}\right) \mathrm{d}\zeta. \tag{6}$$

To simplify this integral, we introduce a new variable $z = \zeta^2/2$; then $\zeta = 2^{1/2} \cdot z^{1/2}$, $\mathrm{d}\zeta = (\sqrt{2}/2) \cdot z^{-1/2} \cdot \mathrm{d}z$, and hence,

$$\sigma^2(\varepsilon) = \frac{2^{1+\varepsilon}}{\sqrt{\pi}} \cdot \int_0^{\infty} z^{1/2+\varepsilon} \cdot \exp(-z)\, \mathrm{d}z. \tag{7}$$

By definition of a gamma function

$$\Gamma(n) = \int_0^\infty z^{n-1} e^{-z}\, dz \tag{8}$$

(see, e.g., [1], p. 350; [5], Appendix A, Table 17), we thus have

$$\sigma^2(\varepsilon) = \frac{1}{\sqrt{\pi}} \cdot 2^{1+\varepsilon} \cdot \Gamma\left(\frac{3}{2} + \varepsilon\right). \tag{9}$$

For $\varepsilon = 0$, we have $\sigma(0) = 1$ and $\Gamma(3/2) = \sqrt{\pi}/2$ (see [1]), so this equality clearly holds.

Differentiating both sides of the equality (9) with respect to ε, we conclude that

$$2\sigma(\varepsilon) \cdot \sigma'(\varepsilon) = \frac{1}{\sqrt{\pi}} \cdot \left(2^{1+\varepsilon} \cdot \ln(2) \cdot \Gamma\left(\frac{3}{2} + \varepsilon\right) + 2^{1+\varepsilon} \cdot \Gamma'\left(\frac{3}{2} + \varepsilon\right)\right). \tag{10}$$

Substituting $\varepsilon = 0$, taking into consideration that $\Gamma(3/2) = \sqrt{\pi}/2$, and dividing both sides of the resulting equality by 2, we conclude that

$$\sigma'(0) = \frac{\ln(2)}{2} + \frac{\Gamma'(3/2)}{\sqrt{\pi}}. \tag{11}$$

To compute $\Gamma'(3/2)$, we can use the following known equality (see, e.g., [5]):

$$\Gamma(z) \cdot \Gamma\left(z + \frac{1}{2}\right) = (2\pi)^{1/2} \cdot 2^{1/2-2z} \cdot \Gamma(2z); \tag{12}$$

hence,

$$\Gamma\left(z + \frac{1}{2}\right) = \frac{\Gamma(2z)}{\Gamma(z)} \cdot 2 \cdot \sqrt{\pi} \cdot 2^{-2z}. \tag{13}$$

In particular, for $z = 1 + \varepsilon$, we get

$$\Gamma\left(\frac{3}{2} + \varepsilon\right) = \frac{\Gamma(2 + 2\varepsilon)}{\Gamma(1 + \varepsilon)} \cdot \frac{\sqrt{\pi}}{2} \cdot 2^{-2\varepsilon}. \tag{14}$$

One of the main properties of a gamma function is that $\Gamma(n+1) = n \cdot \Gamma(n)$; hence $\Gamma(2 + 2\varepsilon) = (1 + 2\varepsilon) \cdot \Gamma(1 + 2\varepsilon)$, and the equation (13) takes the form:

$$\Gamma\left(\frac{3}{2} + \varepsilon\right) = \frac{\Gamma(1 + 2\varepsilon) \cdot (1 + 2\varepsilon)}{\Gamma(1 + \varepsilon)} \cdot \frac{\sqrt{\pi}}{2} \cdot 2^{-2\varepsilon}. \tag{15}$$

It is known [1] that $\Gamma'(1)$ is equal to $-\gamma$, where γ is the Euler's constant. Thus, for small ε, $\Gamma(1 + \varepsilon) = 1 - \gamma \cdot \varepsilon + o(\varepsilon)$, $\Gamma(1 + 2\varepsilon) = 1 - 2\gamma \cdot \varepsilon + o(\varepsilon)$, and $2^{-2\varepsilon} = e^{-2\varepsilon \cdot \ln(2)} = 1 - 2\ln(2) \cdot \varepsilon + o(\varepsilon)$. Hence, the equation (15) takes the form:

$$\Gamma\left(\frac{3}{2}+\varepsilon\right)=\frac{(1-2\gamma\cdot\varepsilon)\cdot(1+2\varepsilon)}{(1-\gamma\cdot\varepsilon)}\cdot\frac{\sqrt{\pi}}{2}\cdot(1-2\ln(2)\cdot\varepsilon)+o(\varepsilon)=$$

$$(1+\varepsilon\cdot(2-\gamma-2\ln(2)))\cdot\frac{\sqrt{\pi}}{2}+o(\varepsilon). \tag{16}$$

Thus,

$$\Gamma'\left(\frac{3}{2}\right)=\frac{\sqrt{\pi}}{2}\cdot(2-\gamma-2\ln(2))=\sqrt{\pi}\cdot\left(1-\frac{\gamma}{2}-\ln(2)\right). \tag{17}$$

Substituting (17) into (11), we conclude that

$$\sigma'(0)=1-\frac{\gamma}{2}-\frac{\ln(2)}{2}. \tag{18}$$

¿From the formula (4), we can now get the desired expression for k. The theorem is proven.

Acknowledgments

This work was supported in part by NASA under cooperative agreement NCC5-209, by NSF grants No. DUE-9750858 and CDA-9522207, by the United Space Alliance, grant No. NAS 9-20000 (PWO C0C67713A6), by the Future Aerospace Science and Technology Program (FAST) Center for Structural Integrity of Aerospace Systems, effort sponsored by the Air Force Office of Scientific Research, Air Force Materiel Command, USAF, under grant number F49620-95-1-0518, and by the National Security Agency under Grant No. MDA904-98-1-0561.

References

1. Beyer, W. H. (1991) CRC Standard Mathematical Tables and Formulae, CRC Press, Boca Raton, FL.
2. Clifford, A. A. (1973) Multivariate Error Analysis, Wiley, New York.
3. Fuller, W. A. (1987) Measurement Error Models, Wiley, New York.
4. H. G. Hecht, H. G. (1990) Mathematics in Chemistry. An Introduction to Modern Methods, Prentice Hall, Englewood Cliffs, NJ.
5. Itô, K., ed. (1993) Encyclopedic Dictionary of Mathematics, MIT Press, Cambridge, MA.
6. Klir, G., Yuan, B. (1995) Fuzzy Sets and Fuzzy Logic: Theory and Applications, Prentice Hall, Upper Saddle River, NJ.
7. Kruse, R., Borgelt, C., and Nauck, D. (1999) Fuzzy data analysis: challenges and perspectives, Proceedings of the 8th IEEE International Conference on Fuzzy Systems (FUZZ-IEEE'99), Seoul, Korea, August 22–25, 1999, **3**, 1211–1216.

8. Nguyen, H. T., Kreinovich, V. (1997) Applications of Continuous Mathematics to Computer Science, Kluwer, Dordrecht.
9. Nguyen, H. T., Walker, E. A. (1999) First Course in Fuzzy Logic, CRC Press, Boca Raton, FL.
10. Novitskii, P. V., Zograph, I. A. (1991) Estimating the Measurement Errors, Energoatomizdat, Leningrad (in Russian).
11. Orlov, A. I. (1991) How often are the observations normal? Industrial Laboratory **57**, No. 7, 770–772.
12. Osegueda, R. A., Ferregut, C., Kreinovich, V., Seetharami, S., Schulte, H. (2000) Fuzzy (granular) levels of quality, with applications to data mining and to structural integrity of aerospace structures, Proceedings of the 19th International Conference of the North American Fuzzy Information Society NAFIPS'2000, Atlanta, Georgia, July 13–15, 2000 (to appear).
13. Rabinovich, S. (1993) Measurement Errors: Theory and Practice, American Institute of Physics, New York.
14. Wadsworth, H. M. Jr., ed. (1990) Handbook of Statistical Methods for Engineers and Scientists, McGraw-Hill Publishing Co., New York.

Fuzzy Linguistic Scales: Definition, Properties and Applications

Alexander Ryjov
Chair of Mathematical Foundation of Intelligent Systems
Department of Mechanics and Mathematics
Lomonosov' Moscow State University
119899 Moscow
Russia

1. Problem Description

This paper studies the mathematical aspects of a formalized description of the humans acting as measuring devices. It is assumed that the person describes the real object properties in the form of linguistic values. The subjective degree of convenience of such a description depends on the selection and the composition of such linguistic values. Let us explain this on a model example.

Example 1. Let us evaluate the height of a man and consider two extreme situations:

Situation 1. It is allowed to use only two values: "small" and "high".
Situation 2. It is allowed to use many values: "very small", "not very high", ... , "not small and not high", ... , "very high".

Situation 1 is inconvenient. In fact, for many people both the permitted values may be unsuitable and, in applying them, they have to select between two "bad" values. Situation 2 is also inconvenient. In fact, in describing height of a man, several of the permitted values may be suitable. We again experience a problem but now due to the fact that we are forced to select between two or more "good" values. Could a set of linguistic values be optimal in this sense?

One object may be described by different experts (persons). Therefore it is desirable to have assurance that the different experts describe one and the same object in the most "uniform" way. On the basis of the above we may formulate the first problem as follows:

Problem 1. Is it possible, taking into account certain features of the human perception of real world objects and their description, to formulate a rule for selection of the optimal set of characteristics on the basis of which these objects may be described? Two optimal criteria are possible:

Criterion 1. We regard as optimum those sets of values through which use a human experiences the minimum uncertainty in describing objects.
Criterion 2. If the object is described by a certain number of experts, then we regard as optimum those sets of values, which provide the minimum degree of divergence of the descriptions.

This paper proves that we can formulate a method of selecting the optimum set of values of qualitative attributes. Moreover, it is shown that such a method is stable, i.e. the natural small errors that may occur in constructing the membership functions do not have a significant influence on the selection of the optimum set of values. The sets, which are optimal according to criteria 1 and 2, coincide. What gives us the optimal set of values of qualitative attributes for a solution of the practical tasks? To answer this question, let us assume that the human's description of an objects make a data base of some data management system. In this connection the following problem arises.

Problem 2. Is it possible to define the indices of quality of information retrieval in fuzzy (linguistic) databases and to formulate a rule for the selection of such a set of linguistic values, use of which would provide the maximum indices of quality of information retrieval?

It is shown that it is possible to introduce indices of the quality of information retrieval in fuzzy (linguistic) databases and to formalise them. It is shown that it is possible to formulate a method of selecting the optimum set of values of qualitative attributes, which provide the maximum quality indices of information retrieval. Moreover, it is shown that such a method is stable, i.e. the natural small errors in the construction of the membership functions do not have a significant effect on the selection of the optimum set of values.

2. Complete Orthogonal Fuzzy Linguistic Scales

The model of an estimating of real object's properties by a person as the procedure of measuring in fuzzy linguistic scale (FLS) is been analyzed at first time in [Ryjov, A. (1987), Ryjov, A. (1992)]. The set of scale values of some FLS is a collection of fuzzy sets defined on the same Universe.

Let us consider t fuzzy variables with the names a_1, a_2, ..., a_t, specified in one universal set (Fig. 1). We shall call such a set the *scale values set of a FLS.* Let us introduce a system of limitations for the membership functions of the fuzzy variables comprising s_t. For the sake of simplicity, we shall designate the membership function a_j as μ_j. We shall consider that:

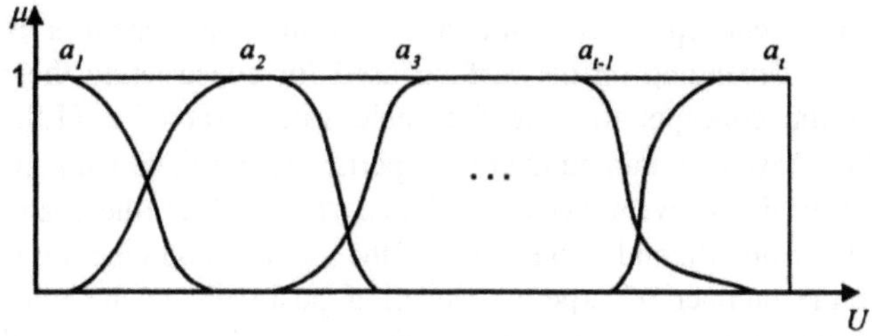

Fig.1.

1. $\forall\ \mu_j\ (1 \le j \le t)\ \exists\ U^1_j \ne \varnothing$, where $U^1_j = \{u \in U: \mu_j = 1\}$, U^1_j is an interval or a point;
2. $\forall\ j\ (1 \le j \le t)\ \mu_j$ does not decrease on the left of U^1_j and does not increase on the right of U^1_j (since, according to 1, U^1_j is an interval or a point, the concepts "on the left" and "on the right" are determined unambiguously).

Requirements 1 and 2 are quite natural for membership functions of concepts forming the scale values set of the FLS. In fact, the first one signifies that, for any concept used in the Universal set, there exists at least one object which is standard for the given concept. If there are many standards, they are positioned in a series and are not "scattered" around the Universe. The second requirement signifies that, if the objects are "similar" in the metrics sense in the Universal set, they are also "similar" in the sense of FLS.

Henceforth, we shall need to use the characteristic functions as well as the membership functions, and so we shall need to fulfil the following technical condition:

3. $j\ (1 \le j \le t)\ \mu_j$ has not more than two points of discontinuity of the first kind.

For simplicity let us designate the requirements 1-3 as L. Let us also introduce a system of limitations for the sets of membership functions of fuzzy variables comprising s_t. Thus, we may consider that:

4. $\forall\ u \in U\ \exists\ j\ (1 \le j \le t)$: $\mu_j(u) > 0$;
5. $\forall\ u \in U\ \sum_{j=1}^{t} \mu_j(u) = 1$.

Requirements 4 and 5 also have quite a natural interpretation. Requirement 4, designated the *completeness* requirement, signifies that for any object from the universal set there exists at least one concept of FLS to which it may belong. This means that in our scale values set there are no "holes". Requirement 5, designated the *orthogonality* requirement, signifies that we do not permit the use of semantically similar concepts or synonyms, and we require the sufficient distinction of the concepts used. Note also that this requirement is often fulfilled or not fulfilled depending on the method used for constructing the membership functions of the concepts forming the scale values set of a FLS. Thus, for example, if we have a certain number of experts, present them with an object $u \in U$ and permit only the answers "Yes, $u \in a_j$" and "No, $u \notin a_j$" (the answer "I do not know" is not permitted), and as the value of the membership function $\mu_j(u)$ we take the ratio of the number of experts answering positively to the total number of experts, this requirement is automatically fulfilled. Note also that all the results given below are justified with a some weakening of the orthogonality requirement [Ryjov, A. (1987)], but for its description it is necessary to introduce a series of additional concepts. Therefore, let us dwell on this requirement.

For simplicity we shall designate requirements 4 and 5 as G.

We shall determine the FLS with scale values set consisting of fuzzy variables, the membership functions of which satisfy the requirements 1-3, and their populations the requirements 4 and 5, a *complete orthogonal FLS* and denote it $G(L)$.

3. Choosing the Optimal Scale Value Set of the FLS

As one can see from Example 1 (Section 1), the different FLS have different degrees of internal uncertainty. Is it possible to measure this degree of uncertainty? For the complete orthogonal FLS the answer to this question is yes. To prove this fact and derive a corresponding formula, we need to introduce a series of additional concepts.

3.1. Degree of FLS uncertainty

Let there be a certain population of t membership functions $s_t \in G(L)$. Let

$$s_t = \{\mu_1, \mu_2, \ldots, \mu_t\}.$$

Let us designate the population of t characteristic functions

$$\hat{s}_t = \{h_1, h_2, \ldots, h_t\}$$

as the most similar population of characteristic functions, if

$$h_i(u) = \begin{cases} 1, & \text{if } \mu_i(u) = \max\limits_{1 \le j \le t} \mu_j(u) \\ 0, & \text{otherwise} \end{cases} \quad (1 \le i \le t). \tag{1}$$

It is not difficult to see that, if the complete orthogonal FLS consists not of membership functions but of characteristic functions, then no uncertainty will arise when describing objects in it. The expert unambiguously chooses the term a_j, if the object is in the corresponding region of the universal set. Some experts describe one and the same object with one and the same term. This situation may be illustrated as follows. Let us assume that we have scales of a certain accuracy and we have the opportunity to weigh a certain material. Moreover, we have agreed that, if the weight of the material falls within a certain range, it belongs to one of the categories. Then we shall have the situation accurately described. The problem lies in the fact that for our task there are no such scales nor we have the opportunity to weigh on them the objects of our interest.

However, we can assume that of the two FLSs, the one having the least uncertainty will be that which is most "similar" to the space consisting of the populations of characteristic functions. In mathematics, the distance can be a degree of similarity. Is it possible to introduce a distance among FLSs? For a complete orthogonal FLS it is possible. If we are considering complete orthogonal FLSs, the membership functions of the concepts employed satisfy the conditions L. This is a subset of the well-known space of the functions integrated over a certain interval U. It is well known that it is possible to introduce distance in this space, for example as follows:

$$d(f,g) = \int_U |f(u) - g(u)| du .$$

We can introduce distance in $G(L)$ with the aid of lemma 1.

Lemma 1. Let s_t, $s_t' \in G(L)$, $s_t = \{\mu_1, \mu_2, \ldots, \mu_t\}$, $s_t' = \{\mu_1', \mu_2', \ldots, \mu_t'\}$, $d(\mu, \nu)$ - be the metrics in L. Then

$$d(s_t, s_t') = \sum_{j=1}^{t} d(\mu_j, \mu_j')$$

is the metrics in $G(L)$.

The proof of this lemma is given in [Ryjov, A. (1987)].

The semantic statements formulated above may be formalized as follows.
Let $s_t \in G(L)$. For the measure of uncertainty of s_t we shall take the value of the functional $\xi(s_t)$, determined by the elements of $G(L)$ and assuming the values in [0,1] (i.e. $\xi: G(L) \to [0,1)$, satisfying the following conditions (axioms):

A1. $\xi(s_t) = 0$, if s_t is a set of characteristic functions;
A2. Let s_t, $s'_{t'} \in G(L)$, t and t' may be equal or not equal to each other. Then

$$\xi(s_t) \le \xi(s'_{t'}), \text{ if } d(s_t, \mathfrak{s}_t) \le d(s'_{t'}, \mathfrak{s}'_{t'})$$

(Let us recall that $\mathfrak{s}_t$ is the set of characteristic functions determined by (1) closest to s_t.)

Do such functionals exist? The answer to this question is given by the following theorem [Ryjov, A. (1987)].

Theorem 1 (of existence). Let $s_t \in G(L)$. Then the functional

$$\xi(s_t) = \frac{1}{|U|} \int_U f\left(\mu_{i_1^*}(u) - \mu_{i_2^*}(u)\right) du, \tag{2}$$

is a measure of uncertainty of s_t, i.e. satisfies the axioms A1 and A2, where

$$\mu_{i_1^*}(u) = \max_{1 \le j \le t} \mu_j(u), \; \mu_{i_2^*}(u) = \max_{\substack{1 \le j \le t, \\ j \ne i_1^*}} \mu_j(u), \tag{3}$$

f satisfies the following conditions:

F1. $f(0)=1, f(1)=0$;
F2. f does not increase -

There are many functionals satisfying the conditions of Theorem 1. They are described in sufficient detail in [Ryjov, A. (1987)]. The simplest of them is the functional in which the function f is linear. It is not difficult to see that conditions F1 and F2 are satisfied by the sole linear function $f(x) = 1 - x$. Substituting it in (2), we arrive at the following simplest measure of uncertainty of the complete orthogonal FLSs:

$$\xi(s_t) = \frac{1}{|U|} \int_U \left(1 - \left(\mu_{i_1^*}(u) - \mu_{i_2^*}(u)\right)\right) du, \tag{4}$$

where $\mu_{i_1^*}(u)$, $\mu_{i_2^*}(u)$ are determined by the relations (3).

Let us denote the sub-integral function in (4) by $\eta(s_t, u)$:

$$\eta(s_t, u) = 1 - \left(\mu_{i_1^*}(u) - \mu_{i_2^*}(u)\right). \tag{5}$$

Now we may adduce the following interpretation of the measure of uncertainty (4).

3.2. Interpretation of degree of uncertainty of FLS

Let us consider the process of describing objects in the framework of the FLS $s_3 \in G(L)$ (see Fig. 2).

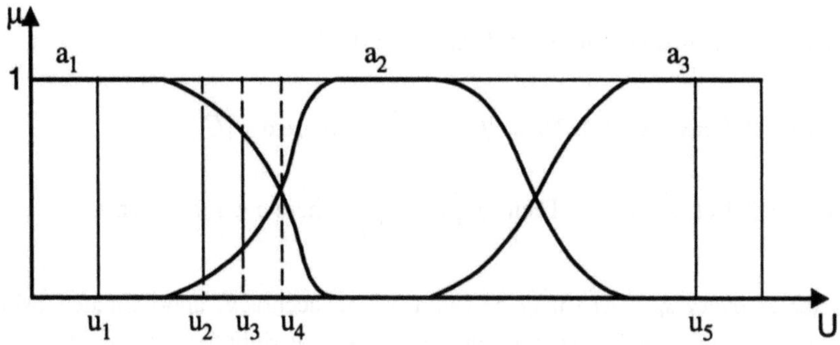

Fig.2.

For the objects u_1 and u_5, a human will without hesitation select one of the terms (a_1 and a_3 respectively). For the object u_2 the user starts selecting between the terms a_1 and a_2. This hesitation increases and attains its peak for the object u_4: at this point the terms a_1 and a_2 are indistinguishable. If we remember the procedure for constructing the membership functions described in the analysis of the orthogonality characteristic (Section 2), we can also confirm that all the experts will be unanimous in describing the objects u_1 and u_5, while in describing u_2 a certain divergence will arise which attains its peak for the object u_4.

Let us now consider formula (5). It is not difficult to see that

$$0 = \eta(s_t, u_5) = \eta(s_t, u_1) < \eta(s_t, u_2) < \eta(s_t, u_3) < \eta(s_t, u_4) = 1$$

($\eta(s_t, u_j)$ is equal to the length of the dotted line at point u_j).

Thus, $\eta(s_t, u)$ actually reflects the degree of uncertainty which a human experiences in describing objects in the framework of the corresponding semantic space or the degree of divergence of opinion of the experts in such a description. Then the degree of fuzziness of $\xi(s_t)$ (2) is an average measure of such uncertainty in describing all the objects of the Universal set.

3.3. Some properties of the FLS uncertainty degree

Let us define the following subset of function set L:

- $\overline{L}$ is a set of functions from L, which are part-linear and linear on

$$\overline{U} = \left\{ u \in U\colon \forall j (1 \le j \le t)\ 0 < \mu_j(u) < 1 \right\};$$

- $\hat{L}$ is a set of functions from L, which are part-linear on U (including $\overline{U}$).

The following theorems are hold [Ryjov, A. (1987)].

Theorem 2. Let $s_t \in G(\overline{L})$. Then $\xi(s_t) = \dfrac{d}{2|U|}$, where $d = |\overline{U}|$.

Theorem 3. Let $s_t \in G(\hat{L})$. Then $\xi(s_t) = c\dfrac{d}{|U|}$, where $c < 1$, $c = \text{Const}$.

As far as every $s_t \in G(L)$ may be (with a high accuracy) approximated using a system of sets $s_t \in G(\hat{L})$ then the same relationship is right for every $s_t \in G(L)$.

Let g be a one-to-one function, which is defined on U. This function is induced by the transformation of some complete orthogonal FLS $s_t \in G(L)$ on the Universe U to complete orthogonal FLS $g(s_t)$ on the Universe U', where

$$U' = g(U) = \{u' = g(u), u \in U\}.$$

The above induction is defined by following way: $g(s_t)$ is a set of membership functions $\mu_j'(u)$, where $\mu_j'(u') = \mu_j'(g(u)) = \mu_j(g^{-1}(u')) = \mu_j(u)$, $(\mu_j(u) \in s_t,\ 1 \le j \le t)$.

The following example illustrate this definition.

Example 2. Let $s_t \in G(L)$, U is the Universe s_t and g is expansion (compression) of the Universe U. In this case, $g(s_t)$ is a set of functions produced from s_t by the same expansion (compression).

Theorem 4. Let $s_t \in G(L)$, U is the Universe s_t, g is the linear one-to-one function defined on U and $\xi(s_t) \neq 0$. Then $\xi(s_t) = \xi(g(s_t))$.

This property means that a human describes different-type objects using some FLS with equal difficulties, if physical parameters of objects of one type may be produced from the parameters of objects of other type by a linear transformation. For example, using the scale value set {*high*, *medium*, *low*}, we describe people, trees, buildings and etc. with equal difficulties; using the scale value set {*very near*, *near*, *not near*, *far away*} we describe distances between molecules, distances between town street, distances between towns on a map and etc. with equal difficulties.

The fuzziness degree of a single fuzzy set induced by $\xi(s_t)$ is defined as fuzziness degree of the trivial complete orthogonal FLS, determined with the fuzzy set $\mu(u)$:

$$\xi(\mu) = \frac{1}{|U|} \int_U \left(1 - |2\mu(u) - 1|\right) du, \tag{6}$$

It is easy to prove that (6) satisfies all the axioms for the single set fuzziness degree [Ryjov, A. (1987)]. It can be shown that the introduced in the report more general notion $\xi(s_t)$ had been correctly defined.

3.4. The stability of our uncertainty degree

Here, we present the results of our model analysis, when the membership functions which are members of the given collection of fuzzy sets, are not determined with the absolute precision, but with some maximal inaccuracy δ (Fig. 3). Let us call this particular situation the δ-model and denote it by $G^{\delta}(L)$.

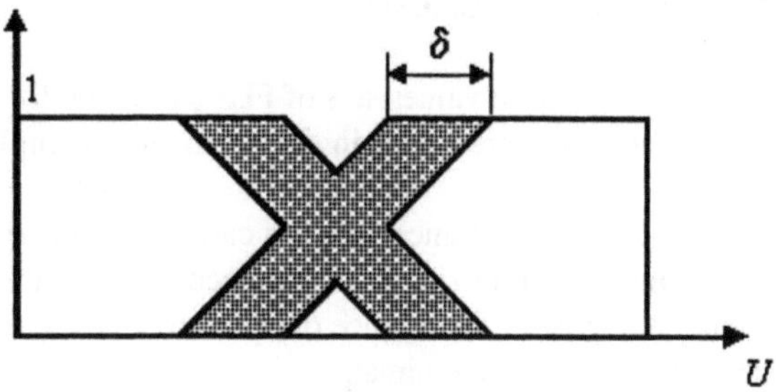

Fig. 3.

In this situation we can calculate the top ($\overline{\xi}(s_t)$) and the bottom ($\underline{\xi}(s_t)$) valuations of the degree of fuzziness. The collection of fuzzy sets with the minimum and maximum degree of fuzziness is given in Fig. 4a) and 4b) correspondingly.

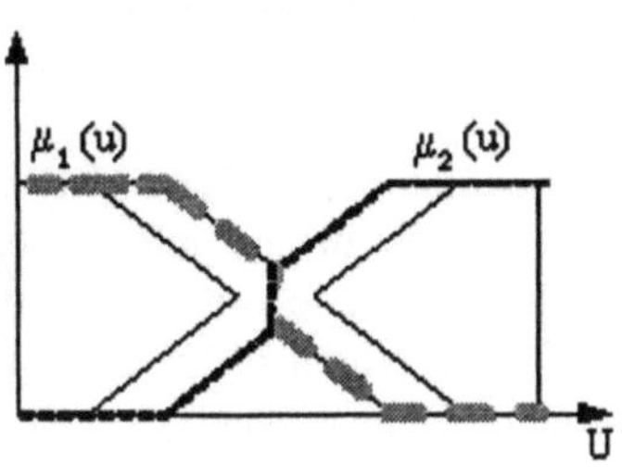

Fig. 4a

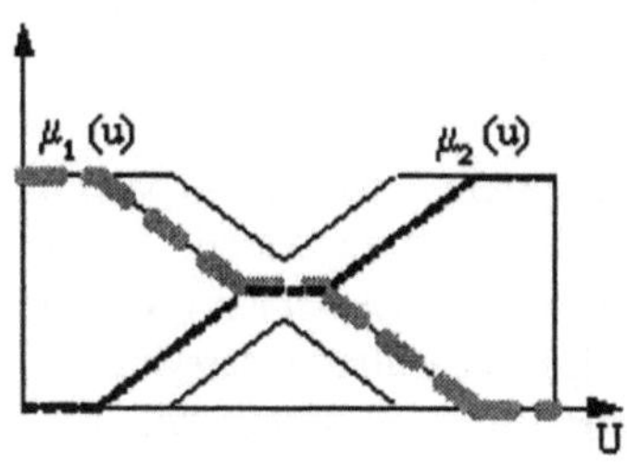

Fig. 4b

The following theorem is holds [Ryjov, A. (1998)].

Theorem 5. Let $s_t \in G^{\delta}(\overline{L})$. Then

$$\underline{\xi}(s_t) = \frac{d(1-\delta)^2}{2|U|}, \overline{\xi}(s_t) = \frac{d(1+2\delta)}{2|U|},$$

where d = $|\overline{U}|$.

By comparing the results of theorem 2 and theorem 5, one can see that for small δ values, the main laws of our model are preserved. Therefore, we can use our technique of estimation of the fuzziness degree in practical tasks, since we have shown it to be stable.

3.5. Technique of choosing of the FLS optimal scale value set

Based on results of sections 3.2 - 3.4, we can formulate the following technique of choosing of the FLS optimal scale value set:

Step 1. All the "reasonable" scale values sets of FLS should be formulated.
Step 2. Each of such sets is represented in the form of a complete orthogonal FLS.
Step 3. For each set the measure of uncertainty is calculated according to (2).
Step 4. As the optimum set minimizing both the uncertainty in the description of objects and the degree of divergence of opinions of experts we select one, the uncertainty of which is minimal.

Following this method, we may describe objects with the *minimum possible* uncertainty, i.e. the *guaranteed optimum* operation of the human-computer information/intelligent systems from this point of view.

4. The Model of Information Retrieval in Fuzzy (linguistic) Databases

The system database is a basis of a subject area information model. The quality of this basis is expressed, in particular, through parameters of the information retrieval. If the database containing the linguistic descriptions of objects of the subject area allows carrying out qualitative and effective search of the relevant information, the system will work also effectively and with a high quality.

4.1. Concepts of information loss and information noise in fuzzy data bases

As in section 3.1, we shall consider that the set of the linguistic meanings can be submitted as $G(L)$.

In our study of the process of information searches in data bases whose objects have a linguistic description, we introduced the concepts of loss of information ($\Pi_X(U)$) and of information noise ($H_X(U)$). These concepts apply to information searches in these databases, whose attributes have a set of significances X, which are modelled by the fuzzy sets in s_t. The meaning of these concepts can informally be described as follows. While interacting with the system, a user formulates his query for objects satisfying certain linguistic characteristics, and gets an answer according to his search request. If he knows the real (not linguistic) values of the characteristics, he would probably delete some of the objects returned by the system (information noise), and he would probably add some others from the database, not returned by the system (information loss). Information noise and information losses have their origin in the fuzziness of the linguistic descriptions of the characteristics.

These concepts can be formalized as follows.

Let's consider the case $t = 2$. Let's fix the number $u^* \in U$ and introduce following notations:

- $N(u^*)$ is the number of objects, the descriptions of which are stored in the database, that possesses a real (physical, not linguistic) significance equal to u^* ;
- N^{user} - the number of users in the system.

Then

- $N_{a_1}(u^*) = \mu_{a_1}(u^*)N(u^*)$ - the number of database descriptors, which have real meaning of some characteristic equal u^* and is expressed by the information source as a_1;
- $N_{a_2}(u^*) = \mu_{a_2}(u^*)N(u^*)$ - the number of the objects, which are described as a_2;
- $N_{a_1}^{user}(u^*) = \mu_{a_1}(u^*)N^{user}$ - the number of the system users who believe that u^* is a_1;
- $N_{a_2}^{user}(u^*) = \mu_{a_2}(u^*)N^{user}$ - the number of the users who believe that u^* is a_2.

Note that for $s_t \in G(L)$ taking into account the requirement of orthogonality, the following matches are right:

$$\mu_{a_1}(u^*)N(u^*) + \mu_{a_2}(u^*)N(u^*) = N(u^*),$$
$$\mu_{a_1}(u^*)N^{user} + \mu_{a_2}(u^*)N^{user} = N^{user}.$$

That's why under the request "To find all objects which have a meaning of an attribute, equal a_1" (let's designate it as $\langle I(O)= a_1\rangle$) the user gets $N_{a_1}(u^*)$ descriptions of objects with real meaning of search characteristic equal to u^*. Under these circumstances $N_{a_1}^{user}(u^*)$ users do not get $N_{a_2}(u^*)$ object descriptions (they carry losses). It delivers descriptions of objects, which have the meaning of characteristic equal u^*, but described by sources as a_2. By analogy the rest $N_{a_2}^{user}(u^*)$ users get noise ("unnecessary" descriptions in the volume of given $N_{a_1}(u^*)$ descriptions).

Average individual loses for users in the point u^* under the request are equal

$$\pi_{a_1}(u^*) = \frac{1}{N^{user}} N_{a_1}^{user}(u^*) \times N_{a_2}(u^*) = \mu_{a_1}(u^*)\mu_{a_2}(u^*)N(u^*), \tag{7}$$

By analogy average individual noises in the point u^*

$$h_{a_1}(u^*) = \frac{1}{N^{user}} N_{a_2}^{user}(u^*) \times N_{a_1}(u^*) = \mu_{a_1}(u^*)\mu_{a_2}(u^*)N(u^*), \tag{8}$$

Average individual information loses and noises, given under analysed request ($\Pi_{a_1}(U)$ and $\mathrm{H}_{a_1}(U)$ accordingly) are naturally defined as

$$\Pi_{a_1}(U) = \frac{1}{|U|}\int_U \pi_{a_1}(u)du,$$
$$\mathrm{H}_{a_1}(U) = \frac{1}{|U|}\int_U h_{a_1}(u)du.$$

It's obvious that

$$\Pi_{a_1}(U) = \mathrm{H}_{a_1}(U) = \frac{1}{|U|}\int_U \mu_{a_1}(u)\mu_{a_2}(u)N(u)du, \tag{9}$$

By analogy for the request $\langle I(O)= a_2\rangle$ or from symmetry considerations we can get that in this case average loses and noises are equal ($\Pi_{a_2}(U) = H_{a_2}(U)$) too and are equal the right part of (9). Under information loses and noises appearing during some actions with characteristic which has the set of significance $X = \{a_1, a_2\}$ (($\Pi_X(U)$ and $H_X(U)$) we naturally understand

$$\Pi_X(U) = p_1\Pi_{a_1}(U) + p_2\Pi_{a_2}(U),$$
$$H_X(U) = p_1 H_{a_1}(U) + p_2 H_{a_2}(U),$$

where p_i ($i = 1,2$) - the probability of some request offering in some i-meaning of the characteristic.

It's obvious that as $p_1 + p_2 = 1$, then

$$\Pi_X(U) = H_X(U) = \frac{1}{|U|}\int_U \mu_{a_1}(u)\mu_{a_2}(u)N(u)du, \tag{10}$$

Let's consider general case: t- meanings of the retrieval attribute. We can generalize the formula (10) in case of t meanings of the retrieval attribute the following way [Ryjov, A. (1996)]:

$$\Pi_X(U) = H_X(U) = \frac{1}{|U|}\sum_{j=1}^{t-1}\left(p_j + p_{j+1}\right)\int_U \mu_{a_j}(u)\mu_{a_{j+1}}(u)N(u)du, \tag{11}$$

where: $X = \{a_1, \ldots, a_t\}$, p_i ($i = 1,2, \ldots, t$) - the probability of the request offering in an i value of the characteristic.

4.2. Properties of information loss $\Pi_X(U)$ and of information noise $H_X(U)$

Theorem 6. Let $s_t \in G(\overline{L})$, $N(u) = N = Const$ and $p_j = \frac{1}{t}$ ($j=1, \ldots, t$). Then

$$\Pi_X(U) = H_X(U) = \frac{ND}{3t|U|}, \quad \text{where } D = |\overline{U}|.$$

Corollary 1. Let the restrictions of the theorem 6 be valid. Then

$$\Pi_X(U) = H_X(U) = \frac{2N}{3t}\xi(s_t).$$

To prove the Corollary one can compare theorems 2 and 6.

We can generalize corollary 1 for $s_t \in G(L)$. The following theorem is hold.
Theorem 7. Let $s_t \in G(L)$, $N(u) = N = Const$ and $p_j = \frac{1}{t}$ ($j=1, \ldots, t$). Then

$$\Pi_X(U) = \mathrm{H}_X(U) = \frac{c}{t}\xi(s_t),$$

where c is a constant with depends on N only.

Theorems 6 and 7 are proved in [Ryjov, A. (1996)]. These theorems show that the volumes of information loss and information noise, arising by search of the information in a linguistic (fuzzy) databases are coordinated with a degree of the object description uncertainty. It means that describing objects by the optimal way (see section 3.5) we provide also the optimal search of the information in databases.

4.3. The stability of information loss $\Pi_X(U)$ and of information noise $\mathrm{H}_X(U)$

By analogy with section 3.4 we can construct the top ($\overline{\Pi}_X(U), \overline{\mathrm{H}}_X(U)$) and bottom ($\underline{\Pi}_X(U), \underline{\mathrm{H}}_X(U)$) valuations of the $\Pi_X(U)$ and $\mathrm{H}_X(U)$.

The following theorems and corollaries are hold [Ryjov, A. (1996)].

Theorem 8. Let $X = \{a_1, \ldots, a_t\}$, $s_t \in G^{\delta}(\overline{L})$, $N(u)=N=Const$ and $p_j = \frac{1}{t}$ ($j=1, \ldots, t$). Then

$$\underline{\Pi}_X(U) = \underline{\mathrm{H}}_X(U) = \frac{ND(1-\delta)^3}{3t|U|}, \qquad \text{where } D = |\overline{U}|. \qquad (12)$$

Corollary 2. Let the restrictions of the theorem 8 are true. Then

$$\underline{\Pi}_X(U) = \underline{\mathrm{H}}_X(U) = \frac{2N}{3t}(1-\delta)\underline{\xi}(s_t).$$

Theorem 9. Let $X = \{a_1, \ldots, a_t\}$, $s_t \in G^{\delta}(\overline{L})$, $N(u)=N=Const$ and $p_j = \frac{1}{t}$ ($j=1, \ldots, t$). Then

$$\overline{\Pi}x(U)=\overline{\text{H}}x(U)=\frac{ND(1-\delta)^3}{3t|U|}+\frac{2ND\delta}{t|U|}\text{, where } D=|\overline{U}| \tag{13}$$

Corollary 3. Let the restrictions of the theorem 9 are true. Then

$$\overline{\Pi}x(U)=\overline{\text{H}}x(U)=\frac{2N}{t(1+2\delta)}\left[\frac{(1-\delta)^3}{3}+2\delta\right]\xi(s_t),$$

By comparing the results of the theorem 6 and theorems 8 and 9 or the corollary 1 and corollaries 2 and 3, one can see that for small δ values, the main laws of our information retrieval model are preserved. Therefore, we can use our technique of estimating the uncertainty degree and our model of information retrieval in fuzzy (linguistic) databases in practical tasks, since we have shown it to be stable.

5. Conclusions

Many characteristics of real objects and processes have measurement instruments developed to measure them. However, quite a few characteristics of many objects and processes in medicine, psychology, sociology and other social sciences have the only measurement instrument left, which is a human being. Many problems in management, planning, information search apply human estimates in their solving. Often these are the problems where information technologies are used. This is why modelling of characteristics of a human being acting as a measurement instrument should be considered an important problem with a wide spectrum of practical applications.

Exactly this problem of a human as a measurement instrument is studied in this chapter. Fuzzy set theory provides a natural framework for its solving. The problem of the scale labels optimal set choice when a human acts as a measurement instrument is formalised. The fuzzy sets aggregate given on the same Universe serves as the model of such a set. This problem is solved under some natural conditions limiting the scale labels set. The concept of fuzziness degree for these structures is introduced, its properties are studied. The choice methodology of the optimal set is formulated. The stability problem, which is very important in practical applications, is considered.

In order to study practical applications the information search model in fuzzy (linguistic) databases is proposed. It is demonstrated that the more un-fuzzy information in the database, the better information search characteristics. This intuitively understandable property is proved formally. By this way, even before the design of the search system, one is able to "calculate" the limitation values of the search characteristics and based on the specifications given one can determine if these specifications could be satisfied for this subject area.

The limited size of this chapter does not allow to include other results on relationship fuzziness degree measurement and fuzzy classification system quality estimates. One may note that here we are able to determine the classification system quality depending on the initial information quality (fuzziness degree) as well. The problems, which have not been solved until now, include a fuzziness degree determination under weaker limitations as well as the relationship between the quality of the systems of different types and the initial information quality. Solving these problems in the future will allow to calculate the limitation characteristics of the fuzzy system quality by the similar way as one calculates rocket trajectories, or reactor regimes.

The author wishes to take this opportunity to express his appreciation to Prof. Lotfi Zadeh (University of California at Berkeley, USA) for his interest to this work and support, Gert De Coman (Gent University, Belgium) for many discussion and very valuable comments and Leonid Reznik (Victoria University, Australia) for his assistance in this paper preparation and comments significantly improving its quality.

Bibliography

Pfanzgal (1976): The Measurement Theory. Mir, Moscow (in Russian)

Ryjov, A. (1987): The Degree of Fuzziness of Fuzzy Descriptions. In: Krushinsky, L.V., Yablonsky, S.V., Lupanov, O.B. (Eds). Mathematical Cybernetics and its Application to Biology. Moscow University Publishing, Moscow,.60-77 (in Russian)

Ryjov, A. (1992): The Axiomatic Definition of a Linguistic Scale Fuzziness Degree, its Major Properties and Applications. North American Fuzzy Logic Proceeding Society (NAFIPS'92). Proceedings of a Conference held in PUERTO VALLARTA, MEXICO, 21 - 29

Ryjov A. (1993): The Information Retrieval in Fuzzy Data Bases. Proceedings of the Fifth International Fuzzy Systems Association World Congress'93, Seoul, Korea, Vol.1, 477 - 480

Ryjov, A. (1996): On degree of fuzziness and its application to intelligent information systems. Intelligent Systems, V.1, 205 - 216 (in Russian).

Ryjov, A. (1998): The Principles of Fuzzy Set Theory and Measurement of Fuzziness. Dialog-MSU, Moscow (in Russian)

Ryjov, A., Belenki, A., Hooper, R., Pouchkarev, V., Fattah, A. and Zadeh, L.A. (1998): Development of an Intelligent System for Monitoring and Evaluation of Peaceful Nuclear Activities (DISNA). IAEA, STR-310, Vienna.

A Fuzzy Shape Specification System to Support Design for Aesthetics

Binh Pham and Jinglan Zhang
Faculty of Information Technology
Queensland University of Technology
GPO Box 2434
Brisbane Q 4001
Australia

Summary: A shape specification system based on fuzzy logic is proposed, which aims to facilitate designers in conceptual design stage by allowing them to specify and work with rough models in a more intuitive fashion. The system uses a 3D object representation called 'parametric geons' which provide both qualitative and quantitative descriptions, in addition to global shape constraints, via eight shape parameters. Shape descriptors and aesthetic descriptors are introduced and their fuzzy membership functions are constructed using results from subjective experiments on shape descriptions. Inference rules are then devised to link these descriptors with shape parameters. The system which models after designers' behaviour, allows designers' own aesthetic terms and interpretation of shapes to be added, and a library of successful designs with appropriate fuzzy descriptions to be constructed for later use.

Keywords: fuzzy logic, conceptual design, shape specification, aesthetics

1 Introduction

Design is generally defined as a process of creating a description of an artificial objects that satisfies certain constraints. These constraints can be functional, manufacturing, economical, efficiency and technical requirements, and aesthetic factors. Aesthetic constraints which induce the degree of satisfaction and emotional responses of users, have emerged as increasingly more important for marketing purposes in many products such as cars, household and office products. The degree of satisfaction means in practice a level of users' feeling with respect to some aesthetic factors, for example, being in balance and harmony with the design's surroundings. Generally, experienced designers can subconsciously create a design which not only meets the constraints but also has a higher degree of satisfaction. However, such knowledge and understanding which are only acquired after many successful designs, are not available to design novices.

Geometric modelling systems may be viewed as consisting of two essential components - geometric models and design methods - which in turn determine the types of systems. Underlying geometric models may be first chosen and a system is then constructed using design techniques that are appropriate for these geometric models. Alternatively, a system may be built based on some specified design techniques which would limit the choice of available geometric models used for representing objects. Both approaches suffer some drawbacks. Ideally, geometric models should be separated from design methods. In addition, designers should be relieved of the need to specify a geometric representation at exact low-level details, and to have access to a number of flexible and intuitive design tools. In particular, they wish to have an initial model by specifying some fuzzy criteria which are more in tune with the fuzziness in human thought process and subjective perception. This need for fuzziness also arises from our inability to acquire and process adequate information about a complex system. For example, it is difficult to extract exact relationships between what humans have in mind for objects' shape and what geometric techniques can offer due to the complexity of rules and underlying principles, viewed from both perceptual and technical perspectives.

Fuzzy logic which was introduced by Zadeh [14] in 1965, has been used extensively in many areas, especially in control systems and in social sciences and engineering (e.g. [6,11]). Very few attempts have been made to apply fuzzy logic to geometric modelling. For some applications, fuzzy systems often perform better than traditional systems because of their capability to deal with non-linearity and uncertainty. In such cases, while traditional systems make precise decisions at every stage, fuzzy systems retain the information about uncertainty as long as possible and only draw a crisp decision at the last stage. Another advantage is that linguistic rules used in fuzzy systems facilitate the design and intuitive understanding of the systems.

In previous papers [7-10], Pham proposed a systematic approach for exploring the interactions of aesthetic properties and design variables, by integrating knowledge from other fields such as philosophy, psychology and arts. She also analysed the way these principles influence product characteristics and based on this knowledge, she has extracted concrete and computable properties of products that may be varied to induce different aesthetic judgements and responses. In this paper, we design and construct a shape specification system which is based on fuzzy linguistic descriptions of the geometric and aesthetic characteristics of the shapes. We derive the membership functions for these descriptors from the results of a series of experiments which capture how people describe a particular shape. Parametric geons [12,13] which provide both qualitative and quantitative representation of a shape, are used for object representation in this system. Although the system was initially constructed based on the parametric geons, its essence is independent of the geometric model and hence the method can be directly applied to other geometric models with appropriate parameters or parametric function.

Section 2 discusses the components of the proposed fuzzy shape specification, while Section 3 justifies the use of parametric geons as object representation in conceptual design stage. Section 4 explains how the fuzzy membership functions for shape descriptors and inference rules are constructed, while Section 5 briefly discusses our other related work and plan to extend the concept of fuzzy shapes to support conceptual design and rapid prototyping.

2 Components of a Fuzzy Shape Specification System

Mathematical models are based on algebraic operations (e.g. equations, integrals), while *logic models* rely on logic-type connectives (and, or, if-then), often with linguistic parameters which give rise to rule-based and knowledge-based systems. *Fuzzy logic models* can combine both of these types of modelling via the fuzzification of algebraic and logical operations. There are three common classes of fuzzy logic models: information processing, decision and control models [4]. An information processing model describes probabilistic relationships between sets of inputs and outputs (using Shanon measure of information), while a decision model: models human behaviour by incorporating subjective knowledge and needs using decision variables. A control model, on the other hand, controls the operations of systems which are governed by many fuzzy parameters (e.g. flight control, air conditioning systems). These parameters may be linear or non-linear. Our shape specification system possesses properties shared by all three classes.

A typical fuzzy system has three main components: input fuzzification, inference engine and decision defuzzification. The type of such a system depend on the rules used to determine the exact meaning of the operand AND, the implication function THEN and the defuzzification method. A fairly large number of these rules exist in literature [2]. The most commonly used implication functions are Larsen's product function and Mamdani's minimum operation. A typical fuzzy system requires:

- A set of rules to represent the heuristic strategies of designers (e.g. to achieve shape characteristics using appropriate geometric functions);
- A set of input data (e.g. shape charateristics specified by designers) prior to actual decisions of system (e.g. calling appropriate geometric functions);
- A method for evaluating any proposed action (e.g. geometric functions) in terms of its conformity to expressed rules, given input data;
- A method for generating promising actions and for determining when to stop searching for optimal ones;

- A method for evaluating the degree of fulfilment of these rules for a generated output.

To achieve the above, we need to address the following questions:

- Which shape representation schemes are appropriate for integrating both *qualitative* and *quantitative* information as well as *local* and *global* shape constraints?
- What shape characteristics (in terms of both geometric and aesthetic) that designers normally require?
- What geometric parameters or functions can be used to achieve these characteristics?
- Can we use this knowledge to construct a set of rules to represent the heuristic strategies of designers?
- Can we incorporate designers' needs by providing customised models that satisfy certain subjective aesthetic constraints?
- Can we devise rules to generate promising output, given certain fuzzy geometric specifications or fuzzy subjective aesthetic constraints?
- Can we evaluate the degree of fulfilment of these rules for a generated output, given certain fuzzy geometric specifications or fuzzy subjective aesthetic constraints?

These questions are addressed in the next two sections while describing the process of constructing the fuzzy shape specification system. This system consists of three main components. A *shape database* from which alternative designs can be retrieved for a given set of constraints or specific requirements. A *fuzzy knowledge base* which stores knowledge extracted from the shape database. The knowledge conveys to designers the relationship between the aesthetic factors and the geometric properties in terms of fuzzy rules. A *graphical user interface* which provides facility for shape specification and display.

The following section discusses the reasons for our choice of parametric geons as the basic shape representation scheme for our work.

3 Parametric Geons as Object Representation for Conceptual Design

A few attempts have been made to explore the use of qualitative shape models for object recognition. Biederman [1] proposed that an object could be represented as a hierarchy of geometrical ions, called *geons*. This model which gives a qualitative description of the whole object and its topology, is based on the fact that our visual system uses co-linearity, curvature, symmetry and co-termination to discriminate and recognise shapes. Geons form a restricted class of generalised cylinders whose cross-sections, axes and sweep properties are arbitrary functions. As the geon representation describes larger structures of the object, it avoids some restrictions imposed by other models that are based on local information. However, this model suffers a drawback in not being able to provide sufficient quantitative information about the object. Such information is required at later stages for object refinement and manipulation.

This drawback was addressed by Dickinson and Metaxa [3], then later by Wu and Levine [12,13], who proposed a set of parametric geons which are based on simple, regular and symmetric shapes represented by super-ellipsoids and their deformations. These super-ellipsoids may be expressed by an implicit equation:

$$\left(\left| \frac{x}{a_1} \right|^{2/\varepsilon_2} + \left| \frac{y}{a_2} \right|^{2/\varepsilon_2} \right)^{\varepsilon_2/\varepsilon_1} + \left| \frac{z}{a_3} \right|^{2/\varepsilon_1} = 1$$

where, $\varepsilon_1, \varepsilon_2$ are shape parameters which control the surface in north-south direction and west-east direction respectively; and a_1, a_2, a_3 are scalar parameters which represent the length of a shape along x, y, z axis respectively.

The seven basic parametric geons are: ellipsoid, cylinder, cuboid, tapered cylinder, tapered cuboid, curved cylinder and curved cuboid. These parametric geons have distinct volumetric characteristics such as roundness, squareness, ellipsoidness, tapering and bending, which can be controlled by manipulating a set of parameters. While the geons are described in strictly qualitative and local attributes, the parametric geons provide both *qualitative* and *quantitative* characteristics, as well as *global shape constraints*.

The following transformations produce the global deformations along z-axis required to create the parametric geons:

(i) Tapering:

$$\begin{aligned} X &= \left(\frac{K_x}{a_3} z + 1\right) x \\ Y &= \left(\frac{K_y}{a_3} z + 1\right) y \\ Z &= z \end{aligned}$$

where K_x and K_y are tapering parameters in the x and y coordinates. To permit downward tapering only and avoid invalid tapering, the constraints

$0 \leq K_x \leq 1$ and $0 \leq K_y \leq 1$ may be imposed.

(ii) Bending:

$$\begin{cases} X = k^{-1} - \cos\theta\,(k^{-1} - x) \\ Y = y \\ Z = (k^{-1} - x)\sin\theta \end{cases}$$

where $\theta = k * z$ is the bending angle.

(iii) Twisting:

$$\begin{cases} X = x\cos\theta - y\sin\theta \\ Y = x\sin\theta + y\cos\theta \\ Z = z \end{cases}$$

where $\theta = t * (\frac{z}{a_3} + 1) * \frac{\pi}{2}$ is the twisting angle.

Fig. 1 shows some sample shapes that may be created using this representation.

Since the shape of an object does not relate to its size, we let $a_1 = 1$ and a_2, a_3 represent the ratio of a_2/a_1, a_3/a_1 and use 8 shape parameters $\{\varepsilon_1, \varepsilon_2, a_2, a_3, K_x, K_y, k, t\}$ to represent the shape characteristics, where:

ε_1, ε_2 : control the shape roundness/squareness.
K_x, K_y : control the tapering property
k : controls the bending property.
t : controls the twisting property.
a_2, a_3 : control the relative dimension along the axes.

However, to use such parametric geons for specification and representation of initial rough shapes is still a tedious task because mathematical insights and trial-and error experiments are needed to obtain appropriate values for these parameters in order to generate a desired shape. The main reason is that the variation of these parameters has non-linear and sometimes unpredictable effects on the shape. In addition, designers often have some fuzzy global geometric or aesthetic characteristics about the shape in mind and wish to obtain such a shape quickly. They might also wish to re-use some designs of similar global characteristics. A viable shape specification system for conceptual design purposes needs to be based on these fuzzy shape descriptors (e.g. roundness, oblongness, squareness, bendiness, taperness, etc.).

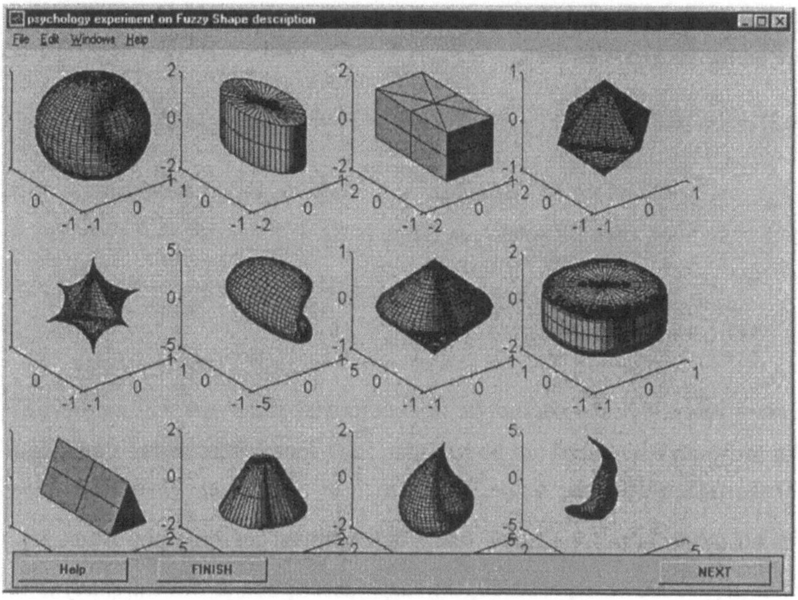

Fig.1. Examples of parametric geons

The following section describes a set of subjective experiments which has been carried out to investigate how people perceive and describe shape characteristics. The intention is to collect relevant data in order to construct fuzzy membership functions for these shape descriptors and the inference rules that link these descriptors to the values of the shape parameters.

4 Fuzzy Membership Functions and Inference Rules

To obtain fuzzy membership functions for each of these shape descriptors, we carried out experiments to build up knowledge on how people perceive and describe shapes. We categorise shapes using 10 shape descriptors: *roundish, squarish, cylindrical, ellipsoidish, pinched, bevel, oblong, tapered, bent* and *twisted.* We allow 3 fuzzy linguistic partitions for each shape descriptor: *slightly, moderately* and *extremely.*

Each parameter is partititioned into equally-spaced intervals and is varied one at a time. The corresponding shape is displayed and the subject is asked to select the most suitable shape descriptor. Since some ranges of the shape parameters do not produce valid 3D shapes or valid shape descriptors, the following constraints are imposed on the shape parameters:

$\varepsilon_1 \in [0,3]$, $\varepsilon_2 \in [0,3]$,

when ε_1, ε_2 are beyond of the range, the shape is either invalid (e.g. self-intersect) or is not a real 3D solid (e.g. some parts of the shape will converge to lines).

$a_2 \in [1,1000]$, $a_3 \in (0,1000]$,

the ratio value a_2, a_3 can not be zero otherwise the shape will not be a 3D solid shape and they need not be less than zero because the minus sign affects only the direction of the shape. When a_2, a_3 are greater than 1000, there exist no more shapes with new fuzzy descriptions, for example, when $a_3 >$ 500 or $a_3 > 1000$ or $a_3 > 10000$, the shape description is still *extremely long*. Furthermore, as a_2 is the ratio of two axis lengths of the shape cross-section, we can assume $a_2 \geq 1$.

$K_x, K_y \in [0,1]$,

when $K_x = 0, K_y = 0$, the shape has no tapering. When $K_x = 1, K_y = 1$, the shape has converged to one point at one end, hence it can not be tapered any more.

$k \geq 0$ and $t \geq 0$,

when $k = 0$ or $t = 0$, the shape has no bending or twisting.

When these parameters change, different shapes will be obtained. The deformation parameters K_x, K_y, k, t work independently, that is, the tapering property does not affect the bending or twisting property. We therefore need to test only one of them at a time. For the *oblong* property, we consider a_3 parameter only. The other parameters will be fixed at a point where the shape is a cube, that is, $\varepsilon_1 = 0, \varepsilon_2 = 0, a_2 = 1, k_x = k_y = k = t = 0$. When a_3 changes from 1 to 1000, the shape changes from a square to a very thin block.

ε_1 and ε_2 control the shape in north-south and east-west direction respectively from square to round to bevel, and finally to pinched. ε_1 and ε_2 works independently. But when testing whether a shape is round or square, we treat ε_1 and ε_2 in the same way since nobody would consider a shape is square or round if its one side is square and the other side is round or pinched. The same is true with bevel and pinch properties. Therefore, when testing the *squarish, roundish, bevel* and *pinched* property, we assume $\varepsilon_1 = \varepsilon_2$. In this case, we cover only one line in the ($\varepsilon_1, \varepsilon_2$) space in the experiment. The squarish, roundish properties are also affected by a_2 and a_3. We can fix $\varepsilon_1 = \varepsilon_2$ while changing a_2 and a_3 to see the effects on the squarish and roundish cognition.

For the cylindrical testing, we consider a small region where $\varepsilon_1 \approx 0$ and $\varepsilon_2 \approx 1$. We fix ε_1 and change ε_2 to examine how ε_2 affects the cylindrical property and then fix ε_2 and change ε_1.

Ellipsoidal testing is more time consuming. ε_1 and ε_2 are changed in the region of $\varepsilon_1 \approx 1$, $\varepsilon_2 \approx 1$ while a_2, a_3 in their valid range [1,1000] to examine how they affect the cognition of ellipsoid property.

We assume all the deformation parameters are zero while testing for squarish, roundish, cylindrical, ellipsoidal, bevel and pinched property. For the deformation property, we assume that all other parameters are fixed and their values do not affect the cognition on the deformation property.

By using eight shape descriptions with specific parameter ranges, only a small set of superquadric shapes are used in our system. Many shapes which superquadrics can represent are thus not contained in our fuzzy system. However, these constraints do not exert adverse effects since our main intention is to explore how the linguistic shape descriptions can be related to geometric parameters through a fuzzy system.

Data collected from the experiment is analysed using statistical methods and the membership function for each shape descriptor is constructed based on its frequency distribution. Four types of membership functions are used for curve fitting: triangle, trapezoid, bell and S curve. It is worthwhile to note that some resulting curves for different shape discriptors are very similar in shape and corresponding range of parameter values. For example, *extremely squarish* and *extremely cylindrical* are similar with respect to ε_1 ; while *extremely roundish* and *extremely cylindrical* are similar with respect to ε_2.

As one of our aims is to find as many shapes as possible, which satisfy specified constraints, the Mamdani implication method is most appropriate since it produces the most tolerant result. Similarly, the union operation is chosen as the most appropriate aggregation method. The fuzzy set of shapes resulted from such aggregated implication may be viewed as a set of design alternatives. We therefore wish to retain the fuzziness as long as possible. Commonly used defuzzification methods such as Centroid, Center of Maximum or Mean of Maximum may be used for defuzzification to obtain one shape candidate which satisfies the given shape specification. However, such a crisp solution is of a limited value for creative design stage where the priority is to explore alternative shapes. A better use of this fuzzy set of shape, therefore, is to deploy it as a search space for further shape optimisation. Given extra criteria for a specific design, evolutionary programming techniques such as genetic algorithms can be used to find an optimal shape within this set.

Data on each shape consists of two parts: geometric information in terms of values of shape parameters, and descriptive information in terms of shape descriptors. These two types of information are linked by appropriate inference rules. The shape specification system is implemented on a PC using MATLAB environment. A user can specify shape descriptors by selecting appropriate options on the graphical user interface. Instead of obtaining and displaying only a shape corresponding to a defuzzified result, the aggregated set of shapes resulted from the inference process is retained and saved into a database. In other words, a fuzzy set of shape parameters are retained as equally spaced sample discrete data sets. Each tuple of these parameter values corresponds to a potential design alternative whose shape can be displayed on the screen for inspection. The user has a number of options. If a particular shape is judged to be attractive, s/he can save it to a library of selected initial designs and then proceed with a later stage of design. If no shape has a satisfactory appeal, s/he can save all the fuzzy set of shapes for further manipulation. In addition, the user has the option to add his or her own aesthetic description to a chosen shape (e.g. sensual, aggressive) and save this description for future reference. Thus, the shape descriptors can be adapted to each user and his or her aesthetic sense. This capability is very important, given that the value of a design often relies strongly on its distinctive and unique style.

5 Conclusion

The incorporation of fuzzy set theory to a shape specification system has produced results which agree with human intuition and satisfy the fuzzy description and specification of shapes required in conceptual design stage. Such a system allows designers to explore alternative designs and examine the effects of variations of geometric attributes on aesthetic characteristics of a design. It can therefore be used for both supporting design for aesthetics and training inexperienced designers. Although parametric geons have been used as the underlying geometric representations, the methods discussed in this paper can be readily extended to other types of geometric models as long as they have identifiable shape parameters which can be used for shape control.

In related work, we have explored this concept further by designing and constructing a fuzzy shape database to support shape design by integrating fuzzy data processing and fuzzy querying function into an existing conventional database. We use a possibility-based relational database model and deal with issues arisen from fuzzy indexing, fuzzy querying and retrieval [15]. In addition, we have investigated the concept of shape aesthetic measures [10] with the intention to use them as criteria for searching for optimal shapes among the fuzzy set of shapes obtained from this fuzzy shape specification system. We plan to extend this work further to shape composition, classification and evolution, which are important tasks to support design and rapid prototyping.

References

1. Biederman I., Human image understanding and: Recent research and a theory, *Computer Vision, Graphics and Image Processing,* 32, 1985, 29-73.
2. Berkan R.C. and Trubatch S.L., *Fuzzy System Design Principles*, IEEE Press,NY, 1997.
3. Dickinson S.J. and Metaxa D., Integrating qualitative and quantitative shape discovery, *Inter. Jour. Comp. Vision* 13 (3), 1994, 311-330.
4. Farwowski W. and Mital A. (Eds.), *Applications of fuzzy set theory in human factors,* 1986.
5. McNeill D., Freioberger P., *Fuzzy Logic*, Simon & Scuster, NY, 1993.
6. Negotia C.V., Ralescu D., *Simulation Knowledge-based Computing and Fuzzy Statistics*, Van Nostrand Reinhold, NY, 1987.
7. Pham B., Fuzzy Logic Applications in CAD, in *Fuzzy System Design: Social and Engineering*, Eds. Reznik L., Dimitrov V., Kacprzyk J., Physica-Verlag, Heidelberg-New York, 1998, 73-85.
8. Pham B., A hybrid representation for aesthetic factors in design, *International Jour. On Machine Graphics & Vision 6 (2),* 1997, 237-246.
9. Pham B., Design for aesthetics: Interactions of design variables and aesthetic properties, *Proceedings of SPIE IS&T / SPIE 11th Annual Symposium - Electronic Imaging '99*, Jan 1999, San Jose, USA, 364-373.
10. Pham B., Aesthetic measures and their potential uses, *EIS'2000 – Second International Symposium on Engineering of Intelligent Systems*, June 2000, Paisley, Scotland, Paper 1303-093.pdf.

11. Reznik L., Dimitrov V., Kacprzyk J (Eds), *Fuzzy System Design: Social and Engineering*, Physica-Verlag, Heidelberg-New York, 1998.
12. Wu K., and Levine M.D., Parametric Geons: A Discrete Set of Shapes with Parameterised Attributes, Proceed. SPIE Inter. Symposium on Optical Engineering and Photonics in Aerospace Sensing: Visual Info. Process. V2239, 14-26, Orlando Fl., 1994.
13. Wu K., and Levine M.D., 3D Object Representation Using Parametric Geons, *Technical Report TR-CIM-93-13*, Centre for Intelligent Machines, McGill University, Montreal, Canada, 1993.
14. Zadeh L., Fuzzy Sets, *Information and Control* 8(3), 1965, 338-353.
15. Zhang J, Pham B., Chen P., Construction of a Fuzzy Shape Database, *Second International Discourse on Fuzzy Logic in the New Millennium,* Great Barrier Reef, Australia, September 2000.

Generating Membership Functions for a Noise Annoyance Model from Experimental Data

A. Verkeyn *, M. De Cock **, D. Botteldooren *, E. E. Kerre **

* Ghent University
Department of Information Technology
Sint-Pietersnieuwstraat 41, 9000 Gent, Belgium

** Ghent University
Department of Applied Mathematics and Computer Science
Krijgslaan 281 (S9), 9000 Gent, Belgium

Summary: The success of fuzzy expert systems could be mainly attributed to the inclusion of linguistic terms into their reasoning scheme. This allows reasoning about complex issues within a certain (tolerated) degree of imprecision. Hence, an important issue in the development of such systems is the choice of the membership functions that model the linguistic terms involved in the application. In this chapter we will describe several methods for the construction of these membership functions (which represent information) from measurements obtained in psycholinguistic experiments. Special attention will be paid to the inclusive and the non-inclusive interpretation of linguistic terms. Secondly, these techniques are applied to data gathered in an International Annoyance Scaling Study, where the relationship between more than 20 different linguistic terms and their corresponding noise annoyance level was under survey.

Keywords: fuzzy expert system, linguistic term, membership function, inclusive and non-inclusive interpretation, noise annoyance

1 Introduction

People tend to express real-life information by means of natural language. It allows them to reason about everyday issues within a certain (tolerated) degree of imprecision. It is, therefore, not surprising that the introduction of the fuzzy set theory as a framework for the mathematical representation of linguistic concepts has given a rise to an important evolution in the field of computer science. The fuzzy expert systems that emerged in this context have proved to be a useful tool in many real-life applications. A fundamental issue in the development of such systems, is the design of the membership functions that model the linguistic terms involved in the application. In this chapter we will describe several methods for the construction of those functions (which represent information) from measurement results obtained in psycholinguistic experiments.

The chapter is structured as follows: after presenting the problem of noise annoyance modelling (Section 2) and describing the form of the experimental data at hand (Section 3), we will briefly recall how linguistic terms are represented in fuzzy expert systems, thereby stressing the difference between an inclusive and a non-inclusive interpretation (Section 4). The main part of this chapter is an overview of well-known, improved and new methods for the construction of membership functions; we explain in detail how they can be applied for noise annoyance terms (Sections 5, 6, 7, 8).

2 Noise Annoyance Modelling

As an environmental factor, noise has several adverse effects on man. Annoyance or disturbance is commonly used as an impact indicator for these effects. Noise surveys can be used as a measurement tool. One of the most important questions in such a survey sounds out about the level in which someone is annoyed by the noise, namely "Thinking about the last 12 months, when you were here at home, how much did noise bother, disturb, or annoy you?" If the survey is conducted by telephone, the subjects are given a set of linguistic terms to choose from: e.g. *not at all annoyed*, *slightly annoyed*, *very annoyed*,... Throughout this chapter we will refer to the linguistic terms that are generated by applying an adverb to the base term *annoyed* as **annoyance terms**. In postal or face-to-face surveys, a numerical or graphical scale can be used. Even in this case the question must be asked whether the mark on a numerical or graphical scale is not a forced expression of a feeling that is more easily expressed using natural language.

The goal of an annoyance model is to predict the outcome of annoyance surveys. Such models can be used by noise policy makers to make strategic decisions. For instance, they can help to choose the "best" route for a new railway by comparing the predicted level of annoyance that the population living along two different possible routes will experience. Fuzzy techniques are very well suited for this modelling purpose: it is far more easy for humans to express annoyance by means of a linguistic term - which is intrinsically vague - than by some crisp number.

The output of a linguistic noise annoyance model will be a noise annoyance term. The input can be either crisp or vague facts. Examples of crisp facts that may influence noise annoyance are age, number of children under 18, average daytime noise exposure level, average nighttime noise exposure level, etc. An example of a vague fact is a person's sensitivity to noise or the state of the environment in general. Rules used in the model are always fuzzy. For more details on a fuzzy system to model noise annoyance we refer to [2].

3 Experimental Data: An International Scaling Study

To get a better understanding of what people really mean when they use some annoyance term, an International Annoyance Scaling Study was conducted (see [5]). In 9 different languages for 21 annoyance terms people were asked to indicate with a mark on a continuous line what each term meant to them, with the most left-hand side being no annoyance at all and the most right-hand side being the most possible degree of annoyance one can imagine. While processing the data, each mark on the line was converted into the distance (expressed in centimetres) from the left-hand side on the 10 centimetres long line. This resulted in a continuous numerical domain ranging from 0 to 10 and a dataset containing 21 values for each subject. In the same study, people were also asked to select five annoyance terms dividing the annoyance level scale in equal intervals. With these data and classical statistical analysis, five terms for each language were selected to globally represent five equidistant portions of the annoyance scale. For the English language they are: *not at all, slightly, moderately, very* and *extremely annoyed.* These five terms are the ones which will be used throughout this chapter. We will refer to them as A_1, A_2, A_3, A_4 and A_5 respectively. Finally, it should be stressed that these data were not gathered with the purpose of membership function construction in mind.

4 Representing Annoyance Terms

Representing terms

In fuzzy set theoretical contexts, a linguistic term is usually represented by a fuzzy set on the suitable universe X. This fuzzy set is characterized by a $X \rightarrow [0,1]$ mapping A which is called the membership function. For each x in X, A(x) is called the membership degree of x in the fuzzy set A. The class of all fuzzy sets on X will be denoted **F**(X). Throughout this chapter we will use the same symbol (e.g. A) to denote the term being modelled, the fuzzy set and the membership function. Recall that for A and B fuzzy sets on X, the inclusion is defined as follows: $A \subseteq B$ iff $(\forall x \in X)(A(x) \leq B(x))$.

The height of A is defined by hgt $A = \sup\{A(x) \mid x \in X\}$. If hgt A = 1 then A is called normalized. For the representation of annoyance terms, we will use the universe X = [0,10] which corresponds to the real interval denoted by the 10 centimetres long line used in the scaling study as described in the previous section. If A and B are two (annoyance) terms then the terms "A *and* B" and "*not* A" are usually represented by the fuzzy sets $A \cap B$ and co(A) respectively, defined by $(A \cap B)(x) = \min(A(x),B(x))$ and $co(A)(x) = 1-A(x)$, for all x in X.

Representing modified terms

In the literature, adverbs such as *slightly, very,...* are often interpreted as linguistic modifiers that alter the meaning of a linguistic term. Therefore, they are represented by a fuzzy modifier on X, i.e. a **F**(X)-**F**(X) mapping (for an overview we refer to [11]). For instance, in *very young*, the adverb *very* is represented by an operator that transforms the fuzzy set *young* into another fuzzy set which is then interpreted as *very young*. In the main part of this chapter however, we will consider annoyance terms as linguistic terms "as such": e.g. *very annoyed* will be considered as one term rather than the term *annoyed* being modified by *very*. In Section 8 this approach is compared to an approach based on modifiers.

General Shape Functions

For the representation of linguistic terms by fuzzy sets on a numerical universe, usually a general shape function is chosen. This allows for the fuzzy set to be fully described by only a small number of parameters. In this chapter we will use sigmoidal and bell-shaped functions, characterized by real parameters μ, σ, and δ, and defined by

$$SIGM(\mu,\sigma,x) = \frac{1}{1+\exp(\mu-\sigma x)}$$

$$BELL(\mu,\sigma,\delta,x) = \begin{cases} \exp\left(\dfrac{-(x-\mu)^2}{2\sigma^2}\right) & \text{if } x \leq \mu \\[2ex] \exp\left(\dfrac{-(x-\mu)^2}{2\delta^2}\right) & \text{if } x > \mu \end{cases}$$

for all x in **R**. For positive σ, a sigmoidal function is increasing. A bell-shaped function is partly increasing and partly decreasing: μ corresponds to the "top" of the bell, while σ and δ determine the width of the left and the right flank of the bell respectively. The shape of these functions is considered to be acceptable from a psycholinguistic point of view [7] for the representation of linguistic terms.

Inclusive and Non-Inclusive Interpretation

Linguistic models found in literature are usually based on one of two different interpretations of the linguistic terms involved, each characterized by a typical shape for the corresponding membership functions.

In the **inclusive interpretation**, we assume that the membership function for *not at all annoyed* is decreasing, while the membership functions for the four other terms are increasing. Furthermore

extremely annoyed $\subseteq$ *very annoyed* $\subseteq$ *moderately annoyed* $\subseteq$ *slightly annoyed*

The underlying semantics is that e.g. everybody who is *very annoyed* is also *moderately annoyed.* In this interpretation the membership degree of x in A clearly corresponds to the degree to which x **satisfies** the term modelled by A: indeed the degree to which somebody is *moderately annoyed* is always larger than or equal to the degree to which he or she is *very annoyed.* To distinguish this interpretation more clearly, the membership functions could be labelled with a preceding *at least.*

In the **non-inclusive interpretation,** *not at all annoyed* is decreasing and *extremely annoyed* is increasing, but the membership functions for the other three terms are partly increasing, partly decreasing (e.g. bell-shaped). They denote neither subsets nor supersets of each other, but different, possibly overlapping categories. In this interpretation the membership degree of x in A corresponds to the degree to which x is **representative** for the term modelled by A.

For a more detailed discussion about the inclusive and the non-inclusive interpretation we refer to [7,11,16]. In fuzzy control applications the non-inclusive interpretation is the most popular one (see e.g. [9]), although the inclusive interpretation is sometimes also used (see [15]).

5 Overview of Existing Methods

In this section, we will briefly describe the two broad categories of methods for the construction of membership functions that are commonly found in the literature. The first category constructs membership functions based on inquiries, solely done for this purpose. A second, fundamentally different approach, obtains the membership functions as a by-product of fuzzy clustering techniques.

5.1 Inquiry-driven methods

The bottomline of the inquiry is always the same: for a certain linguistic term A, we want for each element x of the universe, the degree of compatibility A(x) between the element and the term ("compatibility" may both refer to "satisfiability" or "representativity" as mentioned in the previous section, depending how the question is posed). However, the way in which the questions are formulated has a serious impact on the size and the position of the resulting membership function (see [12] for a more detailed discussion).

1. **Direct rating.** One possibility is to directly ask one or more experts, or the group in which the experiment is conducted, for the membership degree of some elements, for example: "How A is x?".
2. **Reverse rating.** In this approach, the questions are formulated in reverse form: "Which element x has a given degree A(x) of membership in A?".
3. **Polling.** In the two methods described above, people are very directly asked to assign a certain degree of membership to some elements. This is for most domains very difficult and at the same time somewhat arbitrary. Hersh and Caramazza [7] overcome this, by asking only yes/no questions of the form: "Does x belong to A?". Afterwards, A(x) is calculated as the total number of "yes" responses for x divided by the total number of responses for x (yes and no together).
4. **Indirect rating.** These methods also try to replace the direct assignments of degrees with simpler tasks, for instance with pairwise comparisons which are generally easier to estimate. One such method is the analytic hierarchy process (AHP) (see [14]), where questions are asked as "To what degree does x_1 imply A in comparison with x_2 ?". If the cardinality of the universe is an integer n, then all those answers, for instance on a scale from 1 till 10, result in a square matrix P of order n with $P_{ij} = 1/P_{ji}$ for all i=1,...,n and j=1,...,n. After column-wise normalisation, $A(x_i)$ is then calculated as the (if desirable, again normalized) row average of the i-th row of P.

Final construction phase

Whatever method above used, we always end up with couples (x, A(x)). Finally, the resulting membership function can then be constructed using some well-known curve fitting method, for instance Lagrange interpolation and least-square error method. Other techniques such as learning through neural networks can be also applied for this purpose (see [12]).

Annoyance terms

In the case of annoyance the universe is quite abstract: the scale from 0 to 10 does not correspond to any quantity people are used to experience. Therefore, it is not possible to use the techniques described above in their direct form.

5.2 Methods based on fuzzy clustering

Fuzzy clustering

The primary goal of fuzzy clustering algorithms, such as the fuzzy c-means (FCM) algorithm (see [8]), is to partition a given set of data or objects into fuzzy subsets called clusters such that objects strongly belonging to the same cluster (the membership degree of both objects to that cluster are close to 1) are as "similar" as possible and objects that belong to different clusters (the membership degree of one of those objects to that cluster is close to 0) are as "different" as possible. The notions "similar" and "different" are defined by a user given dissimilarity measure d (for instance, the Euclidean distance in a metric space). The aim of the clustering procedure is then to globally minimise this dissimilariy between elements belonging to the same cluster. In a two-dimensional universe X x Y, clustering is often used to extract the relationship between a variable **u** on X and a variable **v** on Y. Each cluster C (which is a fuzzy set on X x Y) gives rise to a fuzzy rule of the form: "IF **u** is A THEN **v** is B", in which A and B are fuzzy sets on X and Y respectively, obtained by "projecting" C on X and Y respectively. A common practice in the field of fuzzy clustering is to assign "ad hoc" linguistic terms to the obtained fuzzy sets A and B. This way, the resulting rules are fully linguistical and easy to understand by domain experts (not necessarily having much knowledge of fuzzy logic).

Annoyance terms

Suppose, we have data from an annoyance survey that contains the noise level the subject is exposed to (expressed in decibels) as well as the experienced annoyance level (expressed on a continuous 0-10 scale). We would then be able to use fuzzy clustering (where the variable **u** is the exposure and **v** is the annoyance), generate fuzzy rules describing that relationship and derive the necessary membership functions used in the rules. If five membership functions for the consequent variable **v** would emerge from the clustering process, we could consider to label them with the five annoyance terms introduced in Section 3. However, it can be expected that the generated rules and membership functions are optimally suited for the training data, but will not necessarily behave well in the general case. The membership functions associated with the annoyance terms would not really represent the meaning that people give to that term. Therefore, fuzzy clustering is less suitable for a real-world annoyance model with linguistic rules.

6 A Probabilistic Approach

The construction of membership functions presented in this section is based on a pure probabilistic approach, which resembles to the polling method. The underpinnings of these methods can be found in the statement by Zadeh that the probability of a fuzzy event is equal to the expected grade of membership of the event itself (see [17]). In the polling method as presented in [7], for each linguistic term A, 13 elements x_j of the universe were evaluated. Hence 13 pairs (x_j, <answer>) were obtained (j=1,..,13 and <answer> either "yes" or "no"). In the annoyance scaling study for each term A_m, (m=1,..,5) the informant placed a mark at a point x_m on the 0-10 scale. This corresponds to the pair (x_m,yes) for A_m. For all other x in $[0,10]\setminus\{x_m\}$, however, the scaling study does not give any information about the second part of the pair (x,<answer>). We can, however, make some intuitively justifiable assumptions about this, thereby forcing either the inclusive or the non-inclusive interpretation.

6.1 Inclusive interpretation

Assume that for all $x < x_m$ (m=2,..,5) the informant would have meant "no," and for all $x \geq x_m$ he would have answered "yes." In other words, he placed a mark on the line to indicate the annoyance level corresponding to *at least* A_m. Computing the average number of "yes"-responses for each x-value taken over all informants then comes down to constructing the cumulative distribution function. The inclusive membership functions is then created by fitting a sigmoidal S-function $SIGM(\mu,\sigma,.)$ on the cumulative distribution function.

Table 1: Numerical parameters for the fitted curves (probabilistic)

	Non-inclusive		Inclusive	
	μ	σ	μ	σ
not at all	-0.0128	0.0723	4.0456	49.0496
slightly	1.2928	0.6662	2.7888	1.9846
moderately	4.9634	0.1742	7.9807	1.7820
very	7.9206	0.7561	12.2168	1.5988
extremely	10.2227	0.3276	27.3687	2.8424

This approach works well for all the terms, except for A_1 = *not at all annoyed.* Indeed for this term, the informant would have answered "yes" for all $x \leq x_5$ and "no" for all $x > x_5$, thereby indicating with his mark the annoyance level corresponding to *at most* A. To obtain the curve of *not at all annoyed*, therefore the reversed cumulative distribution function (thus moving from 10 down to 0 in our domain) was used and fitted on the reflected sigmoidal function $co(SIGM(\mu, \sigma, .))$.

The generated membership functions are depicted in Figure 1b; the values of the parameters are given in Table 1. Please note, that not every fitted function A is normalized. If necessary, this can be solved by dividing all membership degrees of A by hgt A.

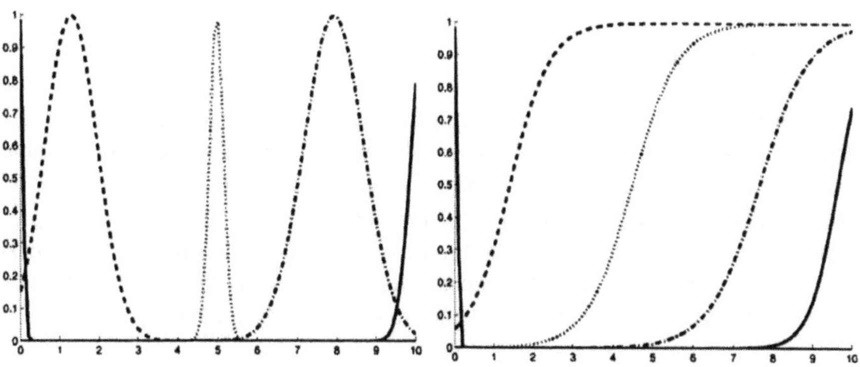

Figure 1: a) Non-inclusive (left) and b) inclusive (right) membership functions for *not at all, slightly, moderately, very* and *extremely annoyed* (probabilistic)

6.2 Non-inclusive interpretation

Another possibility is that the informant wanted to indicate with his mark on x_m (m=1,..,5): "this (and the surrounding) annoyance level(s) I call A_m, but the other ones not." This means that the answer is "yes" for x_m (and perhaps for the levels in a small interval containing x_m) but "no" for the others. Computing the average number of "yes" answers for every x taken over all informants now corresponds to constructing the (normalized) probabilistic histogram. The non-inclusive membership function is then derived from the histogram by fitting an exponential function $BELL(\mu, \sigma, \sigma, .)$ on it with the least-square error method. See Figure 1a for the membership functions and Table 1 for the obtained parameters of those functions.

7 Aggregation of Individual Curve Methods

The probabilistic method does not use any information about how far the marks for all linguistic terms placed on the 0-10 scale by an informant, lie from each other. Each term is modelled as such, without using the relationships between the terms. A method taking this relationship into account is explained in [3] for the non-inclusive interpretation. For every term first an individual curve for each informant is constructed. Then all these curves are combined into the final membership function for that term. The curves generated with the approach in [3] however tend to overlap each other a lot, which makes them less suitable for practical purposes. In this section we will present a variant to this technique, in which the overlap is intrinsically smaller and can even be controlled by a parameter. Furthermore we explain how the same approach can be used for the inclusive interpretation as well.

7.1 Non-inclusive interpretation

The general underlying idea is that the mark x_m (m=1,..5) placed by an informant for a term A_m can be considered as a value with some uncertainty. In the non-inclusive interpretation this uncertainty is proportional to the distance from x_m to the previous mark x_{m-1} (for term A_{m-1}) and the next mark x_{m+1} (for term A_{m+1}). Since there is no previous mark for x_1, we can only take the distance to the next mark x_2 into account to construct a curve for A_1. Likewise there is no next mark for x_5; hence we will only use the distance to the previous mark x_4.

Individual curves

First a fixed membership degree α is chosen. The individual curves will be constructed such that the height of the intersection of two succeeding curves A_m and A_{m+1} is α. For each informant and for each term A_m (m = 2,..,4) an individual bell-shaped function $A_m = BELL(x_m, \sigma_m, \delta_m.)$ is constructed such that for $\Delta x_m = (x_{m+1} - x_m)/2$

$$A_m(x_m+\Delta x_m) = \alpha = A_{m+1}(x_{m+1}-\Delta x_m)$$

The top of the bell for A_m corresponds to the mark x_m placed by the informant, and the width of the flanks is determined by the distance of the mark x_m to the previous mark x_{m-1} and the next mark x_{m+1}, as well as by the parameter α. Solving this equation results in the value for δ_m and σ_{m+1}, namely:

$$\delta_m = \sigma_{m+1} = \left(\frac{1}{\sqrt{-2\ln(\alpha)}}\right)\Delta_m .$$

For the left-most and the right-most terms A_1 and A_5:we will use the functions $co(SIGM(x_1, \delta_1,.))$ and $SIGM(x_5, \sigma_5,.)$ respectively.

Aggregation

After all the individual curves for a linguistic term A_m (m = 1,..,5) are calculated, they are numerically added and normalized. Finally, the curve is fitted to the appropriate shape function to produce the final membership function.

Remark. It can be observed that this method comes down to the probabilistic histogram method for the limit value 0 of parameter α, where $\lim_{\alpha \to 0} \frac{1}{\sqrt{-2\ln(\alpha)}} = 0$, which means that the width of all flanks is 0. Stated otherwise: no flanks are added at all, hence only the given crisp points are summed, which is exactly the same as the histogram approach.

The results of this method, fitted on the same exponential and sigmoidal functions, for parameter α=0.1 are shown in Figure 2a. The exact parameters of the fitted functions are given in Table 2.

Compared to Figure 1a, the curves in Figure 2a cover the whole universe in a uniform manner. This covering is induced by taking the distance between the marks into account when constructing the membership functions, and make them far more suitable for practical purposes such as an annoyance model. For if we would use the curves in Figure 1a, what would happen for an annoyance level of e.g. 4? It does not belong to any fuzzy set to a degree greater than 0, so how can we call it, how should we treat it?

On the other hand the functions in Figure 1a are more desirable from a linguistic point of view. The curves might in fact indicate that the number of terms taken into account is too small to cover the whole universe. Perhaps two more terms should be taken into consideration, namely one "concentrated" around 4, and one around 6.

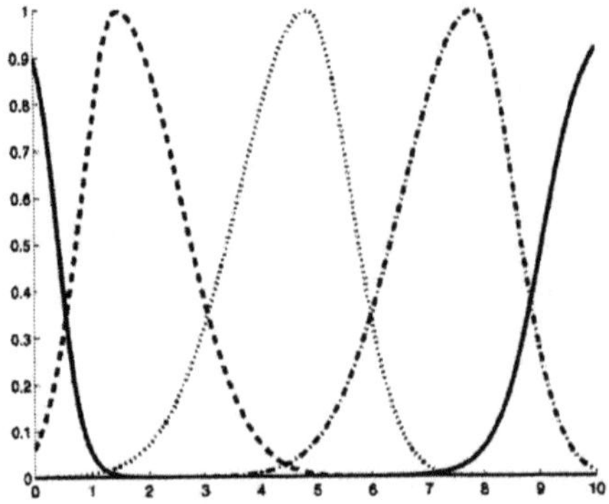

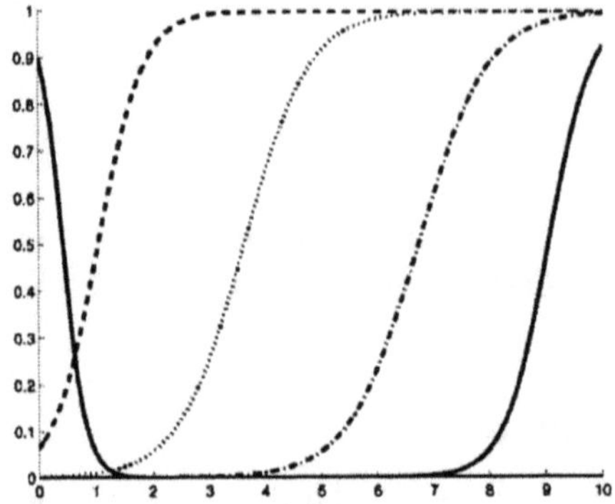

Figure 2: a) Non-inclusive (left) and b) inclusive (right) membership functions for *not at all, slightly, moderately, very* and *extremely annoyed* (aggregation)

Table 2: Numerical parameters for the fitted curves (aggregation)

	Non-inclusive			Inclusive	
	μ	σ	δ	μ	σ
Not at all	2,1832	5,0436	-	2,1832	5,0436
Slightly	1.4777	0.6261	1.0959	2,6666	2,5325
Moderately	4.8681	1.2452	0.7677	6,2700	1,7293
Very	7.8198	1.2588	0.7340	10,9270	1,6229
Extremely	23.1984	2.5647	-	23,6321	2,6111

7.2 Inclusive interpretation

Like in the previous section for the inclusive interpretation we will assume that, by placing a mark x_m (m=2,..,5), the informant indicates that all levels greater than that mark are surely A_m (A_m to degree 1). Hence the only uncertainty is situated left from the mark x_m. As a membership function for A_m we will therefore use a sigmoidal curve $SIGM(x_m, \sigma_m,.)$, in which σ_m is determined by the distance between x_m and the previous mark x_{m-1}, as well as by a parameter α. For the term A_1 on the other hand we will use a reflected sigmoidal curve, with a flank determined by the distance between x_1 and x_2 and the parameter α.

Those results for the same parameter value α=0.1, again after fitting to a sigmoidal function, are shown in Figure 2b and given in Table 2.

8 Using Fuzzy Logical Operators and Modifiers

8.1 Logical operators

From inclusive to non-inclusive

Please note that the membership functions for A_1 = not at all annoyed and A_5 = extremely annoyed in the inclusive and the non-inclusive interpretation are practically interchangeable. It is worth mentioning that also very acceptable membership functions for the terms A_2, A_3, and A_4 in the non-inclusive interpretation can be derived from those in the inclusive interpretation, using the fuzzy logical operations *not* and *but* (=*and*). E.g. the term *very annoyed* in the non-inclusive interpretation corresponds to *very annoyed but not extremely annoyed* in the inclusive interpretation. Formally - for m=2,..,4

$$(A_m)_{NI} = (A_m)_I \cap co((A_{m+1})_I)$$

in which NI stands for the non-inclusive and I for the inclusive interpretation.

Note that in the inclusive interpretation co(*slightly annoyed*) is also a possible membership function for *not at all annoyed.*

8.2 Fuzzy modifiers

As briefly mentioned in Section 4, fuzzy modifiers (mappings from **F**(X) to **F**(X)) can be used to derive the membership function for "<adverb> A" from that of A. In the five terms of the scaling study, the base term *annoyed* is lacking. Nevertheless it is interesting to consider A_3 = *moderately annoyed* as base term instead, and to try to derive the membership functions for the terms A_2, A_4 and A_5 from A_3 (generated with a method described in Section 6 or 7), using fuzzy modifiers.

Powering modifiers

The oldest and most popular modifiers used in the inclusive interpretation are the powering modifiers P_α originally developed by Zadeh [18] and defined by (for α in $[0,+\infty[$): $P_\alpha(A)(x) = (A(x))^\alpha$, for all A in **F**(X), x in X. The most important shortcoming is that for all x in X, $P_\alpha(A)(x) = 1$ iff $A(x) = 1$ and that $P_\alpha(A)(x) = 0$ iff $A(x) = 0$. Looking at the Figures 1b and 2b one immediately sees that the intervals in which the inclusive membership functions are 0 (1 respectively) are usually different. Powering modifiers are therefore not really suitable.

Shifting modifiers

The shifting modifiers S_α, informally suggested by Lakoff [13] and more formally developed by Hellendoorn [6], Bouchon-Meunier [1] and Kerre [10], are defined by (for α in **R**): $S_\alpha(A)(x) = A(x-\alpha)$, for all A in **F**(**R**), x in **R**. They simply shift the original membership function of A to the left or the right (for a positive and negative α respectively) and can be used in both the inclusive and the non-inclusive interpretation. In Figures 3a and 3b the membership functions for A_2, A_3, A_4 and A_5 from Figures 1a and 1b are repeated. The dashed curves correspond to shifted versions of A_3 that are most suitable to represent the other terms. Although the shape of all curves in Figure 1b is not really the same, the functions obtained by shifting A_3 in Figure 3b are very good approximations. The same holds for the membership functions in Figures 2a and 2b. In Figure 3a however, we are not able to derive from the small curve for A_3 the wide functions for *slightly annoyed* and *very annoyed* simply by using a shifting operator.

Modifiers based on fuzzy relations

In [4] a new class of fuzzy modifiers is introduced. They are based on fuzzy relations R on X, i.e. fuzzy sets on X x X. For y in X, the R-foreset of y is denoted Ry and defined by Ry(x) = R(x,y), for all x in X. Furthermore the concepts of *degree of inclusion* and *degree of overlap* are used. For A and B two fuzzy sets on X are defined by

$$\mathrm{INCL}(A,B) = \inf\{\min(1-A(x)+B(x),1) \mid x \in X\}$$

$$\mathrm{OVERL}(A,B) = \sup\{\max(A(x)+B(x)-1,0) \mid x \in X\}$$

Inclusive interpretation

For the inclusive interpretation a resemblance relation E_1 is used, i.e. for all x and y in X, E_1 (x,y) is the degree to which x and y resemble to each other. Hence E_1y is the fuzzy set of objects resembling to y. The general idea is that an object y can be called *slightly* A if it resembles to an object that can be called A; in other words if the set of objects resembling to y overlaps with A (cfr. a man can be called *slightly old* if he resembles to somebody who is *old*). On the other hand y can be called *very* A if every object resembling to y can be called A; in other words if the set of objects resembling y is included in A (cfr. a man can be called *very old* if everybody whom he resembles to is *old*). Formally and fuzzy:

$$\textit{slightly}\ A(y) = \mathrm{OVERL}(E_1y, A)$$

$$very\ A(y) = INCL(E_1 y, A)$$

Extremely A is modelled in a similar, but with a looser resemblance relation E_2 (i.e. $E_1 \subseteq E_2$):

$$extremely\ A(y) = INCL(E_2 y, A)$$

Following this scheme and using the resemblance relations

$$E_1(x,y) = \min(1, BELL(x,1.5,1.5,y)*10)$$
$$E_2(x,y) = \min(1, BELL(x,2.5,2.5,y)*10)$$

the dotted membership functions in Figure 3b were obtained.

Non-inclusive interpretation

As stated above, shifting modifiers work well for Figure 1b, 2a, and 2b. It is explained in [4] that S_α is actually a fuzzy modifier based on the fuzzy relation G_α defined by

$$G_\alpha(x,y) \begin{matrix} =1 & \text{if } x = y - \alpha \\ =0 & \text{otherwise} \end{matrix}$$

namely $S_\alpha(A)(y) = OVERL(G_\alpha y, A)$. G_α is based on the crisp equality between x and y - α. If this equality is fuzzified by means of a resemblance relation E, a more general kind of relation F_α arises, defined by $F_\alpha(x,y) = E(x, y - \alpha)$. The fuzzy modifier built on this relation, namely $WS_\alpha(A)(y) = OVERL(F_\alpha y, A)$ does not only have a shifting but also a widening effect on the membership function of A. Hence it can be applied to A_3 to obtain approximations for the curves in Figure 1a. Using the resemblance relation $E(x,y) = BELL(x,0.7,0.7,y)$ the dotted curves in Figure 3a were obtained. The α-values were chosen as –3.7, 3, and 5.2 respectively. Note that the obtained membership function for A_2 even coincides in the picture with the original membership function generated with the probabilistic approach; therefore the dotted curve for A_2 is not visible.

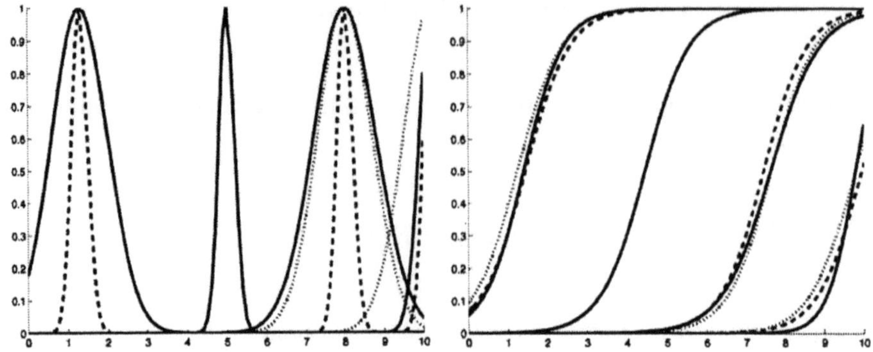

Figure 3: a) Non-inclusive (left) and b) inclusive (right) membership functions (fuzzy modifiers)

Acknowledgements

The authors would like to thank Dr. James Fields for the data of the International Scaling Study. M. De Cock would like to thank the Fund For Scientific Research Flanders (FWO) for funding the research reported on in this chapter.

Bibliography

[1] Bouchon-Meunier, B. (1993): La Logique Floue, Que sais-je?, Paris, vol. 2702

[2] Verkeyn, A., Botteldooren, D., De Tre, G., De Caluwe, R. (2001): Fuzzy modeling of traffic noise annoyance. In: Proceedings of the joint 9th IFSA World Congress and the 20th NAFIPS International Conference, Vancouver (Canada) (to be published)

[3] Cleeren, R. (1992): The Modelling of Linguistic Expressions using Fuzzy SetTheory, (in Dutch), Graduate thesis, Ghent University

[4] De Cock, M., Kerre, E.E (2000): A New Class of Fuzzy Modifiers, In: Proceedings ISMVL 2000, IEEE Computer Society, 121-126

[5] Fields, J.M., De Jong, R.G., Gjestland, T., Flindell, L.H., Job, R.F.S., Kurra, S., Lercher, P., Vallet, M., Yano, T., Guski, R., Felscher-Suhr, U., Schuemer, R. (2001): Standardized general-purpose noise reaction questions for

community noise surveys: research and a recommendation, Journal of Sound and Vibration, 242, 641-679

[6] Hellendoorn, H. (1990): Reasoning with Fuzzy Logic, PhD Thesis, T.U. Delft, Delft

[7] Hersh, H. M., Caramazza, A. (1976): A Fuzzy Set Approach to Modifiers and Vagueness in Natural Language, Journal of Experimental Psychology, 105 (3), 254-276

[8] Höppner, F., Klawonn, F., Kruse, R., Runkler, T. (1999): Fuzzy Cluster Analysis: Methods for classification, data analysis and image recognition, John Wiley & Sons, Chichester

[9] Kandel A., Langholz G. (1994): Fuzzy Control Systems, CRC Press, Inc., Boca Raton

[10] Kerre, E. E. (1993): Introduction to the Basic Principles of Fuzzy Set Theory and Some of its Applications, Communication and Cognition, Gent

[11] Kerre, E. E., De Cock, M. (1999): Linguistic Modifiers; an Overview. In : Fuzzy Logic and Soft Computing (G. Chen, M. Ying, K.-Y. Cai, eds.), Kluwer Academic Publishers, 69-85

[12] Klir, G., Yuan, B. (1995): Fuzzy Sets and Fuzzy Logic: Theory and Applications, Prentice Hall

[13] Lakoff, G. (1973): Hedges: a Study in Meaning Criteria and the Logic of Fuzzy Concepts, Journal of Philosophical Logic, 2, 458-508

[14] Mason, A. K., Korkmaz, S. (1998): Obtaining Fuzzy Membership Functions and Rule Sets in Fuzzy Expert Systems with AHP, IEEE Int. Conf. on Systems, Man and Cybernetics, 2315-2319

[15] Novak V., Kovár J. (2000): Linguistic if-then rules in large scale application of fuzzy control. In: Fuzzy If-Then Rules in Computational Intelligence : Theory and Applications (D. Ruan, E. E. Kerre, eds.), Kluwer Academic Publishers, 223-241.

[16] Van den Eynde C. (1996): A Very Difficult Problem: Modelling Modification using Very. A Semantic-Pragmatic Approach (in Dutch), FKFO-project, private communication.

[17] Zadeh, L. A. (1968): Probability measures of fuzzy events. Journal of Mathematical analysis and applications, 23, 421-427

[18] Zadeh, L. A. (1972): A Fuzzy-Set-Theoretic Interpretation of Linguistic Hedges. Journal of Cybernetics, 2(3):4-34

Possibilistic Logic: A Theoretical Framework for Multiple Source Information Fusion

Souhila Kaci, Salem Benferhat, Didier Dubois, and Henri Prade

Institut de Recherche en Informatique de Toulouse (I.R.I.T.) – C.N.R.S.
Université Paul Sabatier, 118 route de Narbonne 31062 TOULOUSE Cedex 4, FRANCE
E-mail:{kaci, benferhat, dubois, prade}@irit.fr

Abstract. The problem of merging or combining multiple sources information is central in many information processing areas such as databases integrating problems, expert opinion pooling, preference aggregation, etc. Possibilistic logic offers a qualitative framework for representing pieces of information associated with levels of uncertainty or priority. This paper discusses the fusion of multiple sources information in this setting. Different classes of merging operators are considered, at the semantic and the syntactic level, including conjunctive, disjunctive, reinforcement, adaptive and averaging operators. This framework appears to be the syntactic counterpart of the pointwise aggregation of possibility distributions or fuzzy sets.

1 Introduction

Possibility theory and fuzzy set-based approaches have been recently advocated for handling data fusion problems where information is poor or qualitative [7,10,11]. These approaches try to cope with some of the difficulties faced by the probabilistic approach on the problem of pooling imprecise information coming from heterogeneous sources. Their main features are: faithfulness of the representation of subjective data, no need for a priori knowledge, and a variety of fusion methods whose choice depends about the reliability of experts or sources and the level of conflict between their opinions. However, the possibilistic approach is not built as opposed to the probabilistic one; both approaches can be related, complementary, and shed light on each other.
An important issue when combining information is to clarify the nature of the items that are supposed to be merged. In particular, a major discussion is to be made between preference aggregation and information fusion. In the preference aggregation problem (e.g. [13,14]), it makes sense to find the opinion of the "average man" in an homogeneous group of individuals, and to look for trade-offs between preferences expressing different points of views. On the contrary, the information merging problem is a matter of truth and reliability: what is required is to make the best of the available information by discarding the wrong data when possible and keeping the right information. Logical combinations are natural candidates as fusion operators. Conjunctive

combinations apply when all the sources are reliable, while disjunctive combinations deal with the case of unreliable sources hidden in a group of other reliable ones. Obviously, prioritized logical combinations may be considered in particular when the sources are not equally reliable. Averaging operations in information aggregation can be justified when the set of sources can be viewed as a single random source producing different inputs. In that case, indeed, the set of data to be fused can be interpreted as standard statistics. For instance, several successive measurements from a single sensor can be viewed as the result of a random experiment. Then, the discrepancies between the sources can be explained in terms of random variability and the fusion comes close to a filtering problem. However in the case of unique measurements issued from heterogeneous sensors, or in the case of expert knowledge or databases, it is not clear that averaging combination modes make sense.

The way the information fusion problem is tackled depends on the way the information is represented. On the one hand, pieces of information pertaining to numerical parameters are usually represented by distribution functions (in the sense of some uncertainty theory), while preference are represented by means of value or utility functions. These distributions are directly combined by means of operations, which are in agreement with the uncertainty theory or the multiple criteria aggregation used. Their combination yields a new distribution on the set of the possible values of the considered parameter (here possibility is either a matter of plausibility or of satisfactoriness). On the other hand, information may be also naturally expressed in logical terms, especially in case of (symbolic) information pertaining to properties, which may be, however, pervaded with uncertainty. In this case, some uncertainty weights are attached to the logical formulas. Although similar issues are raised in the two representation frameworks, like the handling of conflicting information, the two lines of research in numerical data fusion (e.g. [1,12]) and in symbolic information combination (e.g. [2,4,8,15–20]) have been investigated independently. The fuzzy set framework enables us to reconcile the two views.

Indeed, a fuzzy set F, or a possibility distribution π $(= \mu_F)$, whose membership or possibility degrees $\mu_F(u)$ or $\pi(u)$ are ranging on a *finite* scale $\{\alpha_0 = 0 < \alpha_1 < \cdots < \alpha_n = 1\}$, can be equivalently viewed as a finite set of formulas in possibilitic logic. Each formula corresponds to a α-level cut F_α of the fuzzy set $(F_\alpha = \{u, \mu_F(u) \geq \alpha)$ associated with its necessity measure $(N(F_{\alpha_i}) = 1 - \alpha_{i-1}$ for $i > 1)$. In a previous work [5], a syntactic counterpart of fuzzy set aggregation operations has been proposed where the combination is directly performed on the possibilistic logic formulas. Depending on the situation a possibility distribution may represent a state of uncertain and imprecise knowledge or a preference profile in terms of levels of satisfaction; then the syntactic counterpart corresponds to piece of information with their level of certainty, or to goals with their level of priority respectively. This

format can be convenient for expressing knowledge or preference is a qualitative, granular and compact form.
The paper is organized in the following way. First a background on possibilistic logic and its semantic counterpart is provided, including a discussion and examples of syntactic combinations. Then in a second part, different families of merging operators are discussed in this framework.

2 Background on possibilistic logic

This introduction to possibilistic logic [9] emphasizes the semantical aspects first. The convention representations are given and the notion of entailment is defined in terms of possibility distributions, before being restated in the syntactic form of possibilistic logic, and exhibiting the correspondence between the syntactic and semantic representations.

2.1 Notations

We consider a propositional langage $\mathcal{L}$ over a finite alphabet $\mathcal{P}$ of atoms. We denote the atoms by $p, q, \cdots$, and the formulas by $\phi, \psi, \cdots$. A literal is an atom or a negation of an atom. Ω denotes the set of all interpretations. Logical equivalence is denoted by $\equiv$. Classical disjunction and conjunction are respectively represented by $\vee, \wedge$. $\top$ and $\bot$ are the top and bottom elements and denote the tautology and the contradiction respectively. The symbol $\vdash$ denotes the classical syntactic inference. Let ψ be a formula of $\mathcal{L}$, $Mod(\psi)$ denotes the set of all models of ψ. If $\omega \in Mod(\psi)$, then we write $\omega \models \psi$.

2.2 Possibility distributions

The basic element of the semantics of possibilistic logic is the notion of *a possibility distribution*, denoted by π, which is a mapping from Ω to a totally ordered set. Generally, this latter is the closed interval $[0, 1]$. $\pi(\omega)$ represents the degree of compatibility of the interpretation ω with the available beliefs on the real world if we are representing uncertain pieces of knowledge (or the degree of satisfaction of reaching state ω if we are modelling preferences). By convention, $\pi(\omega) = 1$ means that nothing prevents ω for being the real world (or that ω is fully satisfactory), $1 > \pi(\omega) > 0$ means that ω is only somewhat possible (or satisfactory), while $\pi(\omega) = 0$ means that ω is certainly not the real world (or not satisfactory at all). If $\pi(\omega) > \pi(\omega')$, then ω is more plausible situation than ω'.
π is said to be consistent or normalized if $\pi(\omega) = 1$ for some ω i.e., there is at least one interpretation which is completely possible in Ω. π is partially inconsistent or subnormalized if $max_{\omega \in \Omega} \pi(\omega) < 1$.

We now give some definitions which will be useful later.

Definition 1 *Let π be a possibility distribution. We call the kernel of π, denoted by $\mathcal{K}_\pi$, the set of interpretations having the highest possibility degree, i.e., $\mathcal{K}_\pi = \{\omega : \pi(\omega) = max_{\omega' \in \Omega}\pi(\omega')\}$.*
We define the support of π, denoted by S_π, the set of possible worlds w.r.t. π, i.e., $S_\pi = \{\omega : \pi(\omega) > 0\}$.

If π is normalized, then $\mathcal{K}_\pi = \{\omega : \pi(\omega) = 1\}$ and is called core of π. We now define the α-cut on possibility distributions. It simply consists of interpretations having the possibility degree higher than α.

Definition 2 *Let π be a possibility distribution. The α-cut (resp. strict α-cut) of π, denoted by $[\pi]_{\geq\alpha}$ (resp. $[\pi]_{>\alpha}$), is the set of interpretations having the possibility degree at least equal to (resp. strictly greater than) α.*

Definition 3 *Let π be a possibility distribution. We define the consistency level of π, denoted by $h(\pi)$, as follows:*

$$h(\pi) = max_{\omega \in \Omega}\pi(\omega).$$

Definition 4 is the usual way for normalizing possibility distributions in a qualitative setting (where division of π by $h(\pi)$ cannot make sense):

Definition 4 *Let π be a subnormalized possibility distribution. Then, the normalization of π, denoted by π_N, is defined by:*

$$\pi_N(\omega) = \begin{cases} 1 \; if \pi(\omega) = h(\pi) \\ \pi(\omega) \; otherwise. \end{cases}$$

Example 1. Let s, d, o be three propositional symbols, which respectively stand for *SQL*, *Datalog* and O_2 that a teacher can teach. Let π be the possibility distribution expressing the preferences of a given student.
Let $\Omega = \{\omega_0, \omega_1, \omega_2, \omega_3, \omega_4, \omega_5, \omega_6, \omega_7\}$ be the set of interpretations such that $\omega_0 = \neg s \neg d \neg o$, $\omega_1 = \neg s \neg do$, $\omega_2 = \neg sd \neg o$, $\omega_3 = \neg sdo$, $\omega_4 = s \neg d \neg o$, $\omega_5 = s \neg do$, $\omega_6 = sd \neg o$ and $\omega_7 = sdo$. Let π be defined as follows:

Table 1.

ω	$\pi(\omega)$
ω_0	.2
ω_1	1
ω_2	.2
ω_3	.4
ω_4	1
ω_5	1
ω_6	.4
ω_7	.4

Following Table 1, the more preferred situations to this student are ω_1, ω_4 and ω_5. The next ones are ω_3, ω_6 and ω_7, and the least preferred situations are ω_0 and ω_2.

Then, $\mathcal{K}_\pi = \{\omega_1, \omega_4, \omega_5\}$ and $S_\pi = \Omega$.
Let $\alpha = .4$. Then, $\pi_{\geq\alpha} = \{\omega_1, \omega_3, \omega_4, \omega_5, \omega_6, \omega_7\}$ and $\pi_{>\alpha} = \{\omega_1, \omega_4, \omega_5\}$.

We shall see in the following how π can be also induced from some granular specifications.

2.3 Possibility and necessity measures

A possibility distribution π induces two mappings grading respectively the consistency and the certainty degree of a formula ϕ (which is equivalently represented by a subset of Ω containing models of ϕ):

- the consistency (or possibility) degree of a formula ϕ is defined by $\Pi(\phi) = max\{\pi(\omega) : \omega \models \phi\}$, and evaluates to what extent ϕ is consistent with the available beliefs expressed by π. $\Pi(\phi) = 1$ means that ϕ is totally consistent with the available beliefs, while $\Pi(\phi) = 0$ means that ϕ is totaly inconsistent with them. The function Π satisfies the following conditions:
  i. $\Pi(\bot) = 0$,
  ii. $\forall\phi, \forall\psi, \Pi(\phi \vee \psi) = max(\Pi(\phi), \Pi(\psi))$.
- the certainty (or necessity) degree is defined by duality as $N(\phi) = 1 - \Pi(\neg\phi)$, and evaluates to what extent ϕ is entailed from the available beliefs. It satisfies the following conditions:
  i. $N(\top) = 1$,
  ii. $\forall\phi, \forall\psi, N(\phi \wedge \psi) = min(N(\phi), N(\psi))$.

In the rest of the paper $a, b, c, \cdots$ denote the possibility degrees of the interpretations.

2.4 Entailment between possibility distributions

In this subsection, we define the related notions of *informativeness*, and *entailment* between normalized possibility distributions.

Definition 5 *Let π and π' be two normalized possibility distributions. Then, π' is said to be a consequence of π, denoted by $\pi \models \pi'$, if π is more specific (informative) than π', namely*

$$\forall\omega,\ \pi(\omega) \leq \pi'(\omega).$$

Table 2.

ω	$\pi'(\omega)$
ω_0	1
ω_1	1
ω_2	.5
ω_3	.5
ω_4	1
ω_5	1
ω_6	.5
ω_7	.5

Example 2. Let us consider again the possibility distribution π of Example 1.
Let π' be another possibility distribution given in Table 2:
We have $\forall\omega$, $\pi(\omega) \leq \pi'(\omega)$. Then, π is more informative than π'.

We generalize this definition in the case where π is subnormalized (π' still normalized).

Definition 6 *Let π and π' be two possibility distributions such that π is* **subnormalized** *and π' is normalized. Then, π entails π', denoted also by $\pi \models \pi'$, if the normalization of π is more specific than π', namely $\pi_N \leq \pi'$.*

It is easy to check that:

Proposition 1 *Let π be a subnormalized possibility distribution, and π' be a normalized possibility distribution. Then, π entails π' iff*

- *π is more specific than π', and*
- *$\mathcal{K}_\pi \subseteq \mathcal{K}_{\pi'}$.*

Example 3. Let π and π' be two possibility distributions given in Table 3:
We have $\mathcal{K}_\pi = \{\omega_3\}$ and $\mathcal{K}_{\pi'} = \{\omega_0, \omega_2\}$.

Table 3.

ω	$\pi(\omega)$	$\pi'(\omega)$
ω_0	.4	1
ω_1	.4	.6
ω_2	.2	1
ω_3	.6	.6

We have well $\pi \leq \pi'$ but $\pi \not\models \pi'$ since $\mathcal{K}_\pi \not\subseteq \mathcal{K}_{\pi'}$.

2.5 Syntactic representation of ordered information

Possibilistic knowledge bases: At the syntactic level, uncertain information is represented by means of weighted formulas of the form $B = \{(\phi_i, \alpha_i) : i \in I\}$, where ϕ_i is a classical formula, and α_i is a real number belonging to $(0, 1]$. (ϕ_i, α_i) means that the certainty degree of ϕ_i is at least equal to α_i (i.e. $N(\phi_i) \geq \alpha_i$). The value 0 is excluded from the interval $[0, 1]$ because we want to represent only explicit beliefs.
We denote by B^*, the classical base corresponding to B obtained from B by forgetting the weights. Namely, $B^* = \{\phi_i : (\phi_i, \alpha_i) \in B\}$. In the following $\alpha, \beta, \gamma, \cdots$ denote the necessity degrees associated with formulas.

We now give some definitions which are used later:

Definition 7 *Let B be a possibilistic knowledge base, and $\alpha \in [0, 1]$. We call the α-cut (resp. strict α-cut) of B, denoted by $B_{\geq\alpha}$ (resp. $B_{>\alpha}$), the set of classical formulas in B having a certainty degree at least equal to α (resp. strictly greater than α).*

Definition 8 *Two possibilistic bases B and B' are said to be equivalent, denoted by $B \equiv_s B'$, iff $\forall \alpha \in [0, 1]$, $B_{\geq\alpha} \equiv B'_{\geq\alpha}$,
where $\equiv$ is the classical equivalence.*

We denote by $Inc(B) = max\{\alpha_i : B_{\geq\alpha_i}\ \textit{is inconsistent}\}$ the inconsistency degree of B.
When B is consistent, we have $Inc(B) = 0$. This is equivalent to the classical consistency of B^*. In the case of inconsistency, we define the useful consistent part of B, denoted by $\rho(B)$, as the base made of the classical formulas whose weights are above the inconsistency level, namely: $\rho(B) = \{\phi_i : (\phi_i, \alpha_i) \in B\ and\ \alpha_i > Inc(B)\}$.

2.6 Entailment between possibilistic knowledge bases

Let us now introduce the notion of *entailment* between possibilistic knowledge bases. Let B_1 and B_2 be two possibilistic knowledge bases. We say that B_1 entails B_2 if and only if it entails each formula belonging to B_2. We distinguish two different ways to define the entailment between possibilistic knowledge bases depending on whether priorities attached to conclusions are involved or not.

In the following definition, we define the plausible entailment of a piece of information from a possibilistic knowledge base.

Definition 9 *Let B be a possibilistic base. Then, a classical formula ϕ is said to be a plausible consequence of B, denoted by $B \vdash_P \phi$, if and only if*

$$\rho(B) \vdash \phi.$$

Example 4. Let B be a possibilistic base defined by $B = \{(s \vee o, .8); (\neg s, .6)\}$. Obviously $\rho(B) = \{s \vee o, \neg s\} \vdash o$. Then, o is a plausible consequence of B.

We now define the inference process taking into account the certainty or priority levels attached with the formulas.

Definition 10 *Let B be a possibilistic base. Let (ϕ, α) be a piece of information. (ϕ, α) is said to be a possibilistic consequence of B, denoted by $B \vdash_\pi (\phi, \alpha)$, if and only if*

$$\exists \gamma \geq \alpha \text{ such that } B_{\geq\gamma} \text{ is consistent and } B_{\geq\gamma} \vdash \phi.$$

Example 5. Let us consider again the base B of Example 4.
Let $(o, .7)$ be a formula.
There is no γ such that $\gamma \geq .7$, $B_{\geq\gamma}$ is consistent and $B_{\geq\gamma} \vdash o$. Then, $(o, .7)$ is not a possibilistic consequence of B.
Let us now consider the formula $(o, .5)$. Then, for $\gamma = .6$ we have $\gamma \geq .5$; $B_{\geq\gamma} = \{s \vee o, \neg s\}$ is consistent and $B_{\geq\gamma} \vdash o$. Then, $(o, .5)$ is a possibilistic consequence of B.

Definitions 9 and 10 can be easily generalized to the case of entailment between possibilistic knowledge bases. Indeed, we have $B_1 \vdash_P B_2$ iff $\forall(\psi, \beta) \in B_2, B_1 \vdash_P \psi$, and $B_1 \vdash_\pi B_2$ iff $\forall(\psi, \beta) \in B_2, B_1 \vdash_\pi (\psi, \beta)$.

2.7 From syntactic representation to semantic representation

Possibilistic knowledge bases can be viewed as a compact representation of possibility distributions. Namely, given a knowledge base B, we can generate a unique possibility distribution by associating with each interpretation, its level of compatibility with the beliefs or goals in B. The interpretations satisfying all the formulas in B will have the highest weight, namely 1 and the other interpretations will be ranked w.r.t. the highest weighted formulas that they falsify.
Let us first consider a knowledge base B with one formula, $B = \{(\phi, \alpha)\}$. Then, all interpretations satisfying ϕ will have the weight 1 i.e., $\pi(\omega) = 1$ and the others will have the weight $1 - \alpha$ i.e., $\pi(\omega) = 1 - \alpha$. We define the possibility distribution associated with B and denoted by $\pi_{(\phi,\alpha)}$ as follows: $\forall \omega \in \Omega$,

$$\pi_{(\phi,\alpha)}(\omega) = \begin{cases} 1 & if \omega \models \phi \\ 1 - \alpha & otherwise. \end{cases}$$

Let us now consider B a general knowledge base. Let ω and ω' be two interpretations. ω is said to be preferred to ω' if the highest weight falsified by ω is less than the one falsified by ω'. Hence, we define the possibility distribution associated with B, denoted by π_B, as follows [9]:

Definition 11 $\forall \omega \in \Omega$,

$$\pi_B(\omega) = \begin{cases} 1 & \forall (\phi_i, \alpha_i) \in B, \omega \models \phi_i \\ 1 - max\{\alpha_i : (\phi_i, \alpha_i) \in B \ and \ \omega \not\models \phi_i\} & otherwise. \end{cases}$$

Example 6. Let B be a possibilistic base defined by $\{(s \vee o, .8); (\neg d, .6)\}$. Then, if we compute π_B, the possibility distribution associated with B we recover the possibility distribution π of Example 1.
The interpretations ω_1, ω_4 and ω_5 are the most preferred since they are compatible with all the beliefs of B. The interpretations ω_3, ω_6 and ω_7 are more preferred than ω_0 and ω_2 since the highest belief in B that they falsify (i.e., $(\neg d, .6)$) is less certain than the highest belief in B falsified by ω_0 and ω_2.

When all the information in the knowledge base are completely certain (i.e. $\forall i, \alpha_i = 1$), we get a particular possibility distribution called binary possibility distribution, where each interpretation is associated with 1 or 0. The interpretations having the possibility degree equal to 1 are the models in the sense of the classical logic, and those having the possibility degree equal to 0 are the counter-models.

We now relate the inconsistency degree of a possibilistic knowledge base B and the one of π_B its associated possibility distribution.

Corollary 1 *Let B be a possibilistic knowledge base.*
Let π_B be the possibility distribution associated to B. Then,

$$Inc(B) = 1 - h(\pi_B).$$

Corollary 2 gives the link between the α-cut on a possibilistic knowledge base and the one on its associated possibility distribution.

Corollary 2 *Let B be a possibilistic knowledge base, and π_B be its associated possibility distribution.*
Then,

$$Mod(B_{\geq\alpha}) = \pi_{B_{>1-\alpha}} \ and \ Mod(B_{>\alpha}) = \pi_{B_{\geq 1-\alpha}}.$$

The following proposition gives the correspondences between the semantics and syntactic consequence relations:

Proposition 2 *Let B be a knowledge base, and π_B be its possibility distribution. Let (ϕ, α) be a formula. Then,*

1. $B \vdash_P \phi$ *iff* $\mathcal{K}_{\pi_B} \subseteq \mathcal{K}_{\pi_{(\phi,\alpha)}}$
2. $B \vdash_\pi (\phi, \alpha)$ *iff* $\pi_B \models \pi_{(\phi,\alpha)}$.

These correspondences mean that plausible inferences only refer to the most normal situations in the possibility distribution while possibilistic inferences consider all the possibility distributions.

3 Fusion of possibilistic knowledge bases

In this section, we recall the semantic and syntactic fusion of possibilistic knowledge bases developed in [6] in the framework of possibility theory. In this setting, the authors have defined several merging operators in order to fuse prioritized information, depending if the sources are independent or not, and if they are conflicting or not. In the following subsection, we give both semantic and syntactic definitions of these merging operators.

3.1 Possibilistic merging operators

Let B_1 and B_2 be two possibilistic knowledge bases, and π_1, π_2 be their associated possibility distributions respectively (using Definition 11). A possibilistic merging operator is defined as follows:

Definition 12 *A possibilistic merging operator, denoted by $\oplus$, is a two function whose domain is $[0,1] \times [0,1]$, used for aggregating $\pi_1(\omega)$ and $\pi_2(\omega)$, and which satisfies the following requirements:*
i. $1 \oplus 1 = 1$,
ii. If $a \geq c$ and $b \geq d$, then $a \oplus b \geq c \oplus d$ (monotonicity).

The first one acknowledges the fact that if two sources agree that ω is fully possible (or satisfactory), then the result of the combination should confirm it. The second property expresses that a degree resulting from a combination cannot decrease if the combined degrees increase.
In [6], it has been shown that the syntactic counterpart of the fusion of π_1 and π_2 is the following possibilistic base, denoted by $\mathcal{B}_\oplus$ (and sometimes by $B_1 \oplus B_2$), and which is made of the union of:

- *the initial bases* with new weights defined by:
$\{(\phi_i, 1-(1-\alpha_i)\oplus 1) : (\phi_i,\alpha_i) \in B_1\} \cup \{(\psi_j, 1-1\oplus(1-\beta_j)) : (\psi_j,\beta_j) \in B_2\}$
- and *the knowledge common* to B_1 and B_2 defined by:
$\{(\phi_i \vee \psi_j, 1-(1-\alpha_i)\oplus(1-\beta_j)) : (\phi_i,\alpha_i) \in B_1$ *and* $(\psi_j,\beta_j) \in B_2\}$.

In [6], it has been shown that $\pi_{\mathcal{B}_\oplus}(\omega) = \pi_1(\omega) \oplus \pi_2(\omega)$ where $\pi_{\mathcal{B}_\oplus}$ is the possibility distribution associated to $\mathcal{B}_\oplus$ using Definition 11, thus expressing the complete agreement between the syntactic and the semantic views.

Remarkable cases of $\oplus$ are the *minimum* (for short *min*), *maximum* (for short *max*), *Product* (for short *Pro*) and *weighted sum* (for short *ws*). The first one is meaningful when the sources are consistent and may be not independent, the second one is appropriate when the sources are highly conflicting, the third one deals with independent sources and the fourth deals with not equally reliable sources.

In the case of n sources, the syntactic computation of the resulting base can be easily applied when $\oplus$ is associative. Note that it is also possible to provide the syntactic counterpart for non-associative merging operator. In this case $\oplus$ is no longer a binary operator, but a n-ary operator applied to n-tuples of possibility distributions. The syntactic counterpart is as folllows: Let $\mathcal{B} = (B_1, ..., B_n)$ be a vector of possibilistic bases. Let $(\pi_1, ..., \pi_n)$ be their associated possibility distributions and $\pi_{\mathcal{B}_\oplus}$ be the result of combining $(\pi_1, ..., \pi_n)$ with $\oplus$. Then, the base associated to $\pi_{\mathcal{B}_\oplus}$ is:

$$\mathcal{B}_\oplus = \{(D_j, 1 - x_1 \oplus \cdots \oplus x_n) : j = 1, n\},$$

where D_j are disjunctions of size j between formulas taken from different B_i's $(i = 1, n)$. x_i is either equal to $1 - \alpha_i$ or to 1 depending if ϕ_i belongs to D_j or not.

Example 7. Let $\mathcal{B} = (B_1, B_2, B_3)$ be such that $B_1 = \{(s \vee o, .8); (\neg d, .6)\}$; $B_2 = \{(\neg s, .7); (d \vee o, .5); (\neg d \vee \neg o, .4)\}$ and $B_3 = \{(\neg d, .5)\}$.
Let us compute $\mathcal{B}_\oplus$ for $\oplus = ws$.
Let $(.2, .6, .2)$ be the vector of the weights associated to $\mathcal{B}$ (where $\lambda_1 + \lambda_2 + \lambda_3 = 1$).
Then, $\mathcal{B}_{ws} = \{(s \vee o, .16); (\neg d, .12)\} \cup \{(\neg s, .42); (d \vee o, .3); (\neg d \vee \neg o, .24)\} \cup \{(\neg d, .1)\} \cup \{(s \vee d \vee o, .46); (\neg s \vee \neg d, .54); (\neg d \vee \neg o, .36)\} \cup \{(s \vee \neg d \vee o, .26); (\neg d, .22)\} \cup \{(\neg s \vee \neg d, .52); (\neg d \vee \neg o, .34)\} \cup \{(\neg s \vee \neg d, .64), (\neg d \vee \neg o, .46)\}$.

3.2 Classification of possibilistic merging operators

In this section, we give a general classification of merging operators. The possibility distributions to be combined are assumed here to be normalized. We denote by $\pi_\oplus$ the result of combining two normalized possibility distributions π_1 and π_2 with $\oplus$.
Let B_1, B_2 and $\mathcal{B}_\oplus$ be the possibilistic bases associated with π_1, π_2 and $\pi_\oplus$ respectively.

We first need a further definition which will be useful in this section.

Definition 13 *Let π and π' be two normalized possibility distributions. Then, π and π' is consistent if and only if $\mathcal{K}_\pi \cap \mathcal{K}_{\pi'} \neq \emptyset$.*

This definition says that two normalized possibility distributions are together consistent if there is at least one interpretation which is completely possible in both distributions.
Throughout this section, we use the following example to illustrate the propositions:

Example 8. Let us extend Example 1 to five students denoted respectively by B_1, B_2, B_3, B_4 and B_5. The students give the following information described by π_1, π_2, π_3, π_4 and π_5 respectively.
Let $\Omega = \{\omega_0, \omega_1, \omega_2, \omega_3, \omega_4, \omega_5, \omega_6, \omega_7\}$ such that $\omega_0 = \neg s \neg d \neg o$; $\omega_1 = \neg s \neg d o$; $\omega_2 = \neg s d \neg o$; $\omega_3 = \neg s d o$; $\omega_4 = s \neg d \neg o$; $\omega_5 = s \neg d o$; $\omega_6 = s d \neg o$ and $\omega_7 = s d o$.

Table 4.

ω	$\pi_1(\omega)$	$\pi_2(\omega)$	$\pi_3(\omega)$	$\pi_4(\omega)$	$\pi_5(\omega)$
ω_0	.2	.5	1	.6	.9
ω_1	1	1	1	.6	1
ω_2	.2	1	.5	.8	.5
ω_3	.4	.6	.5	1	.5
ω_4	1	.3	1	.4	.1
ω_5	1	.3	1	.4	.1
ω_6	.4	.3	.5	.4	.1
ω_7	.4	.3	.5	.4	.1

The above possibility distributions give the level of preference of the students w.r.t. each situation. These preferences can be described in a compact way using weighted formulas. For example, the possibility distribution π_1 can be represented syntactically by $\{(s \vee o, .8); (\neg d, .6)\}$ where the weights .8 and .6 represent the levels of priority of $s \vee o$ and $\neg d$ respectively. Then, the situations $\neg s \neg d o$, $s \neg d \neg o$ and $s \neg d o$ are the preferred ones since they satisfy both $s \vee o$ and $\neg d$. The next preferred situations $\neg s d o$, $s d \neg o$ and $s d o$ falsify $\neg d$ which is less preferred than $s \vee o$. Lastly, the less preferred situations $\neg s \neg d \neg o$ and $\neg s d \neg o$ are those which falsify $s \vee o$, the more preferred formula.
Also, the possibility distribution π_3 contains only two levels which can be represented syntactically by $\{(\neg d, .5)\}$. Then, the preferred situations are those which satisfy $\neg d$.

Conjunctive operators
One of the important aims in merging uncertain information is to exploit complementarities between the sources in order to get a more complete and precise global point of view. Since we deal with prioritized information, two kinds of complementarities can be considered depending on whether we refer to the most normal situations, or to the whole possibility distribution. In this subsection, we introduce conjunctive operators which exploit the symbolic complementarities between sources. More formally,

Definition 14 *$\oplus$ is said to be a conjunctive operator if*
$$\forall a \in [0,1], a \oplus 1 = 1 \oplus a = a.$$

Note that this definition is less restrictive than the one usually used in the fuzzy set literature.

The following proposition shows indeed that conjunctive operators, in case of consistent sources of information, exploit their complementarities by recovering all the symbolic information.

Proposition 3 *Let π_1 and π_2 be two possibility distributions which are together consistent.*

Let $\oplus$ be a conjunctive operator. Then,

$$\mathcal{K}_{\pi_{\oplus}} = \mathcal{K}_{\pi_1} \cap \mathcal{K}_{\pi_2}.$$

One aim with conjunctive merging operators is, in case of consistency, to recover all the information provided by the two sources. Namely, $\pi_{\oplus}$ should be more informative than π_1 and π_2 i.e., $\pi_{\oplus} \leq \pi_1$ and $\pi_{\oplus} \leq \pi_2$. Moreover, since $1 \oplus a = a \oplus 1 = a$, we have $\forall \omega, \pi_{\oplus}(\omega) \geq \pi_*(\omega)$ where π_* is called *the drastic product* and defined as follows:

$$\pi_*(\omega) = \begin{cases} 1 & \text{if } \pi_1(\omega) = 1 \text{ and } \pi_2(\omega) = 1 \\ \pi_1(\omega) & \text{if } \pi_2(\omega) = 1 \\ \pi_2(\omega) & \text{if } \pi_1(\omega) = 1 \\ 0 & \text{otherwise.} \end{cases}$$

As pictured on Figure 1, the set of these solutions is defined as follows:
$\Gamma = \{\pi_{\oplus} : \pi_{\oplus}(\omega) = 1 \; if \; \pi_1(\omega) = 1 \; and \; \pi_2(\omega) = 1, \; and \; \pi_*(\omega) \leq \pi_{\oplus}(\omega) \leq min(\pi_1(\omega), \pi_2(\omega)) \; otherwise\}$.
We can easily check that the least specific possibility distribution in Γ is obtained only if $\oplus$ is the *min* operator.

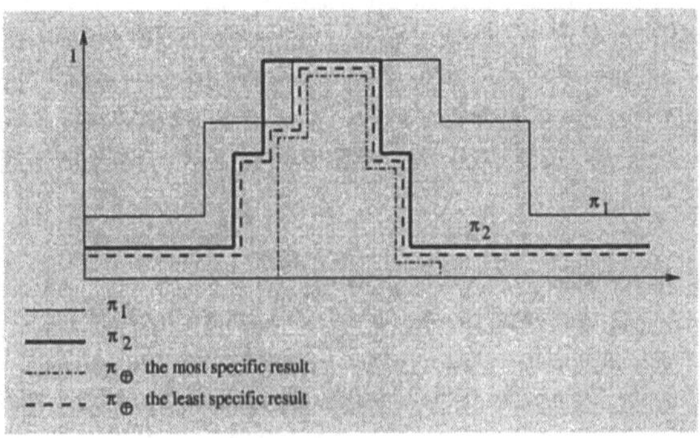

Fig. 1. $\oplus$ conjunctive, $\pi_{\oplus} \leq \pi_1$ and $\pi_{\oplus} \leq \pi_2$

Example 9. Let us consider the possibility distributions π_1 and π_2 of the above example.
We have $\mathcal{K}_{\pi_1} = \{\omega_1, \omega_4, \omega_5\}$ and $\mathcal{K}_{\pi_2} = \{\omega_1, \omega_2\}$.
$\mathcal{K}_{\pi_1} \cap \mathcal{K}_{\pi_2} \neq \emptyset$, then π_1 and π_2 is consistent.
Let $\oplus = min$ which is a conjunctive operator.
Then, $\pi_{min}(\omega_0) = \pi_{min}(\omega_2) = .2$; $\pi_{min}(\omega_4) = \pi_{min}(\omega_5) = \pi_{min}(\omega_6) = \pi_{min}(\omega_7) = .3$ and $\pi_{min}(\omega_1) = 1$.
We have $\mathcal{K}_{\pi_{min}} = \{\omega_1\} = \mathcal{K}_{\pi_1} \cap \mathcal{K}_{\pi_2}$.

An important feature of a conjunctive operator is its ability to give preference to more specific information. Namely, if a source of information S_1 contains all the information provided by S_2, then combining S_1 and S_2 with a conjunctive operator leads simply to S_1. More formally,

Proposition 4 *Let π_1 and π_2 be two possibility distributions such that $\pi_1 \leq \pi_2$. Let $\oplus$ be a conjunctive operator. Then,*

$$\mathcal{K}_{\pi_\oplus} = \mathcal{K}_{\pi_1}.$$

Now, when π_1 and π_2 are the possibility distributions associated to B_1 and B_2 respectively, then the possibilistic base associated with $\pi_\oplus$, denoted by $\mathcal{B}_\oplus$, is:

$$B_1 \cup B_2 \cup \{(\phi_i \vee \psi_j, 1-(1-\alpha_i)\oplus(1-\beta_j)) : (\phi_i,\alpha_i) \in B_1 \; and \; (\psi_j,\beta_j) \in B_2\}$$

Then, Proposition 3 means that if B_1 and B_2 is consistent, then $\mathcal{B}^*_\oplus \equiv B^*_1 \wedge B^*_2$, and Proposition 4 says that if B_1 entails B_2, then $\mathcal{B}^*_\oplus \equiv B^*_1$.

Disjunctive operators

Another important issue in information fusion is how to deal with conflicts. When all the sources are equally reliable and conflicting, we should not introduce any preference between the sources, and we should not infer from the result of the combination all the information provided by one of the sources. Such a behaviour cannot be captured by any conjunctive operator. This requirement is captured by the disjunctive operators defined by:

Definition 15 *$\oplus$ is said to be a disjunctive operator if*

$$\forall a \in [0,1], a \oplus 1 = 1 \oplus a = 1.$$

Since $\oplus$ is a disjunctive operator and π_1, π_2 are normalized then $\pi_\oplus(\omega)$ is necessarily normalized. So, as it is pictured on Fig. 2, the kernel of $\pi_\oplus$ is at least equal to the union of $\mathcal{K}_{\pi_1}$ and $\mathcal{K}_{\pi_2}$. Then, it is not included in any one which means that the result does not entail any source.
Then, we have:

Proposition 5 *Let π_1 and π_2 be two possibility distributions which are together inconsistent.*
Let $\oplus$ be a disjunctive operator. Then,

$$\mathcal{K}_{\pi_\oplus} \not\subseteq \mathcal{K}_{\pi_1} \; and \; \mathcal{K}_{\pi_\oplus} \not\subseteq \mathcal{K}_{\pi_2}.$$

Namely, the most normal situation in $\pi_\oplus$ should not be included in the most normal situations of π_1 and π_2.

A natural requirement when the sources are conflicting is to constrain the result $\pi_\oplus$ to be less specific than both π_1 and π_2, i.e. $\pi_\oplus \geq \pi_1$ and $\pi_\oplus \geq \pi_2$. Fig 2. explicits this requirement.

Now, if we take for $\pi_\oplus$ the least specific possibility distribution satisfying the two above constraints, we get the situation of total ignorance (i.e., $\pi_{ls}(\omega) = 1$, $\forall \omega \in \Omega$), which can be also obtained using the "vacuous" disjunctive operator defined by: $\forall a, \forall b, a \oplus b = 1$. Hence, we need to define a new family of disjunctive operators, called "regular disjunctive" as follows:

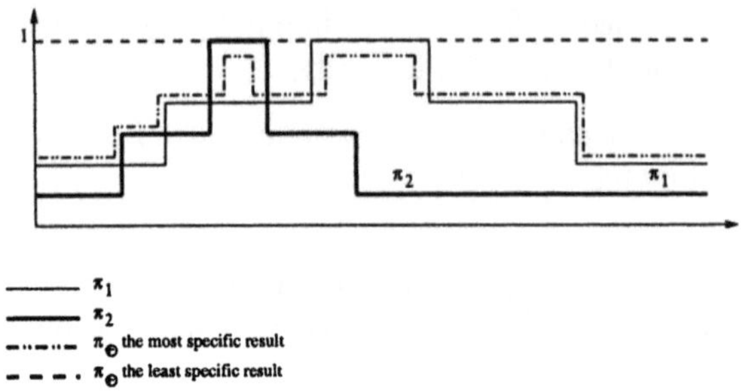

Fig. 2. $\oplus$ disjunctive, $\pi_\oplus \geq \pi_1$ and $\pi_\oplus \geq \pi_2$

Definition 16 *$\oplus$ is said to be a regular disjunctive operator if:*

- $\forall a, a \oplus 1 = 1 \oplus a = 1$,
- $\forall a \neq 1, \forall b \neq 1, a \oplus b \neq 1$.

Examples of regular disjunctive operators are the *max*, the so-called *"probabilistic sum"* defined by: $a \oplus b = a + b - ab$, and the dual of the geometric average defined by $a \oplus b = 1 - \sqrt{(1-a)(1-b)}$. Note that this definition is not the usual one in the fuzzy set literature where disjunctive operators should be usually greater or equal to maximum.

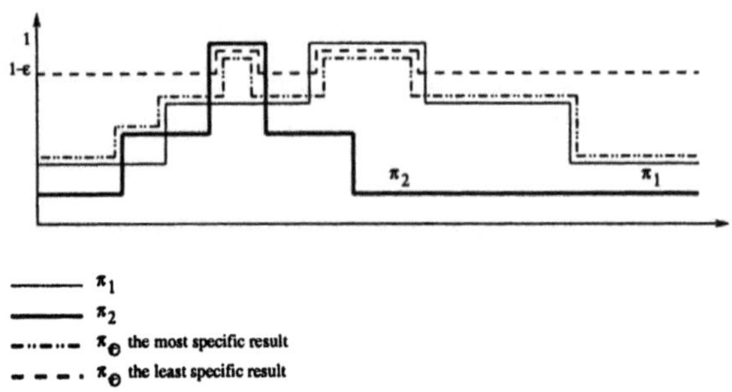

Fig. 3. $\oplus$ regular disjunctive, $\pi_\oplus \geq \pi_1$ and $\pi_\oplus \geq \pi_2$

We can easily check that the most specific possibility distribution $\pi_\oplus$ such that $\pi_\oplus \geq \pi_1$ and $\pi_\oplus \geq \pi_2$ is obtained only using the *max* operator.

Hence, we have the following proposition:

Proposition 6 *Let π_1 and π_2 be two possibility distributions, and $\oplus$ be a regular disjunctive operator. Then,*

$$\mathcal{K}_{\pi_\oplus} = \mathcal{K}_{\pi_1} \cup \mathcal{K}_{\pi_2}.$$

Lastly, note that regular disjunctive operators are not appropriate in the case of consistent sources since they only retain the common part of the information.

Besides, we have the following result:

Proposition 7 *Let π_1 and π_2 be two possibility distributions such that $\pi_1 \leq \pi_2$. Let $\oplus$ be a regular disjunctive operator. Then,*

$$\mathcal{K}_{\pi_\oplus} = \mathcal{K}_{\pi_2}.$$

Example 10. Let us consider π_1 and π_3, and $\oplus = max$.
Then, $\pi_{max}(\omega_2) = \pi_{max}(\omega_3) = \pi_{max}(\omega_6) = \pi_{max}(\omega_7) = .5$ and $\pi_{max}(\omega_0) = \pi_{max}(\omega_1) = \pi_{max}(\omega_4) = \pi_{max}(\omega_5) = 1$.
We have indeed $\pi_1 \oplus \pi_3 = \pi_3$.

When π_1 and π_2 are the possibility distributions associated to syntactic bases B_1 and B_2 respectively, then the possibilistic base associated with $\pi_\oplus$ and denoted by $\mathcal{B}_\oplus$ is equal to:

$$\{(\phi_i \vee \psi_j, 1 - (1 - \alpha_i) \oplus (1 - \beta_j)) : (\phi_i, \alpha_i) \in B_1 \textit{ and } (\psi_j, \beta_j) \in B_2\}.$$

Then, Proposition 5 says that if B_1 and B_2 is inconsistent, then the disjunctive operator does not give preference to any base i.e., $\mathcal{B}_\oplus$ does not entail neither B_1 nor B_2 and Proposition 7 means that if B_1 entails B_2 then $\mathcal{B}^*_\oplus \equiv B^*_2$.

Idempotent operators
Another important issue in fusing multiple sources information is how to deal with redundant information. There are two different situations: either we ignore the redundancies, which is suitable when the sources are not independent, or we view redundancy as a confirmation of the same information provided by independent sources. Idempotent operations aim at ignoring redundancies and are defined by:

Definition 17 *$\oplus$ is said to be an idempotent operator if*

$$\forall a \in [0,1], a \oplus a = a.$$

An important property of idempotence property is the fact for each interpretation ω, we have:

$$\forall \omega,\ min\{\pi_1(\omega), \pi_2(\omega)\} \leq \pi_1(\omega) \oplus \pi_2(\omega) \leq max\{\pi_1(\omega), \pi_2(\omega)\}.$$

Indeed, $\pi_1(\omega) \oplus \pi_2(\omega) \geq \pi_1(\omega) \oplus \pi_1(\omega)$ if $\pi_2(\omega) \geq \pi_1(\omega)$
$\Rightarrow \pi_1(\omega) \oplus \pi_2(\omega) \geq \pi_1(\omega)$
$\Rightarrow \pi_1(\omega) \oplus \pi_2(\omega) \geq min\{\pi_1(\omega), \pi_2(\omega)\}$.
Moreover: $\pi_2(\omega) \oplus \pi_2(\omega) \geq \pi_1(\omega) \oplus \pi_2(\omega)$ when $\pi_2(\omega) \geq \pi_1(\omega)$
$\Rightarrow \pi_2(\omega) \geq \pi_1(\omega) \oplus \pi_2(\omega)$
$\Rightarrow max\{\pi_1(\omega), \pi_2(\omega)\} \geq \pi_1(\omega) \oplus \pi_2(\omega)$.

Obviously, idempotent operators ignore redundant sources. Namely, if the sources provide the same possibility distribution π, then the result of fusion is simply the same possibility distribution.
Now, let us consider the case of local redundancies between two sources. Namely, the case of two different possibility distributions but they share the same information A. Let $A \subseteq \Omega$, namely:

$$N_{\pi_1}(A) = \alpha > 0 \text{ and } N_{\pi_2}(A) = \alpha > 0.$$

The natural question is to determine conditions under which $N_{\pi_1 \oplus \pi_2}(A) = \alpha$ holds. Proposition 8 considers the case where one possibility distribution is less specific than the other:

Proposition 8 *Let π_1 and π_2 be two possibility distributions such that π_1 is less specific than π_2. Let $\oplus$ be an idempotent operator. Let $A \subseteq \Omega$ such that $N_{\pi_1}(A) = \alpha > 0$ and $N_{\pi_2}(A) = \alpha > 0$. Then,*

$$N_{\pi_1 \oplus \pi_2}(A) = \alpha.$$

Proof.
Let $A \subseteq \Omega$ such that $N_{\pi_1}(A) = \alpha > 0$ and $N_{\pi_2}(A) = \alpha > 0$.
Assume that $N_{\pi_1 \oplus \pi_2}(A) > \alpha$ (resp. $N_{\pi_1 \oplus \pi_2}(A) < \alpha$). Then,
$1 - max\{\pi_1(\omega) \oplus \pi_2(\omega) : \omega \notin A\} > \alpha$ (resp. $1 - max\{\pi_1(\omega) \oplus \pi_2(\omega) : \omega \notin A\} < \alpha$).
Since any idempotent operator is such that:

$$min(\pi_1(\omega), \pi_2(\omega)) \leq \pi_1(\omega) \oplus \pi_2(\omega) \leq max(\pi_1(\omega), \pi_2(\omega)).$$

Then, $1 - max\{min(\pi_1(\omega), \pi_2(\omega)) : \omega \notin A\} > \alpha$ (resp. $1 - max\{max(\pi_1(\omega), \pi_2(\omega))$ $\omega \notin A\} < \alpha$).
Now, π_1 is less specific than π_2 then we get: $1 - max\{\pi_2(\omega) : \omega \notin A\} > \alpha$ (resp. $1 - max\{\pi_1(\omega) : \omega \notin A\} < \alpha$). Hence, $N_{\pi_2}(A) > \alpha$ (resp. $N_{\pi_1}(\omega) < \alpha$), hence a contradiction.

However, the previous proposition cannot be generalized to any pair of possibility distributions due to the fact that there may exist new pieces of information which can be obtained from the two sources altogether, but not from each separately, as it is illustrated by the following example:

Example 11. Assume that we have the two possibility distributions:

Table 5.

ω	$\pi_1(\omega)$	$\pi_2(\omega)$	$\pi_1(\omega) \oplus \pi_2(\omega)$
ab	1	.2	[.2, 1]
$a\neg b$	.4	.2	[.2, .4]
$\neg ab$	1	1	1
$\neg a\neg b$	.2	.4	[.2, .4]

Clearly, $N_{\pi_1}(b) = 1 - max\{\pi_1(\omega) : \omega \not\models b\} = .6$ and $N_{\pi_2}(b) = .6$ but $N_{\pi_1 \oplus \pi_2}(b) = .8 > .6$ if for instance $\oplus = min$.
The reason is due to the fact that we have $N_{\pi_1}(a \vee b) = .8$ and $N_{\pi_2}(\neg a) = .8$ and combining this complementary pieces of information leads to $N_{\pi_1 \oplus \pi_2}(b) = .8$

Reinforcement operators
Reinforcement operators in case of repeated information have a confirmation effect. Namely, if the same piece of information is supported by two different sources, then the priority attached to this piece of information should be strictly greater than the one provided by the sources. A first formal class of reinforcement operators can be defined as follows:

Definition 18 *$\oplus$ is said to be a reinforcement operator if*
$$\forall a, b \neq 1 \text{ and } a, b \neq 0, a \oplus b < min(a, b).$$

Besides, one can require that reinforcement operators recover all the common information with a higher weight. Namely if the same formula is a plausible consequence of each base, then this formula should be accepted in the fused base with a higher priority. The following proposition shows a first case where this result holds:

Proposition 9 *Let π_1 and π_2 be two possibility distributions which are consistent together. Let $\oplus$ be a reinforcement operator. Let $A \subseteq \Omega$ such that $N_{\pi_1}(A) = \alpha > 0$ and $N_{\pi_2}(A) = \beta > 0$. Then,*
$$N_{\pi_1 \oplus \pi_2}(A) > max(\alpha, \beta),$$
and $N_{\pi_1 \oplus \pi_2}(A) = 1$ if $\alpha = 1$ or $\beta = 1$.

Now, in case of conflicts, and more precisely, in case of a strong conflict i.e., the most certain information expressed by π_1 and π_2 conflict. Namely if $h(\pi_{min}) = 0$, where π_{min} is the result of combining π_1 and π_2 with the *min* operator, then the above proposition does not hold, as illustrated below.

Example 12. Let p and q be two propositional symbols.
Let π_1 and π_2 be two possibility distributions defined as follows:

Table 6.

ω	$\pi_1(\omega)$	$\pi_2(\omega)$
ω_0 $\neg p \neg q$	0	.3
ω_1 $\neg p q$	0	.3
ω_2 $p \neg q$	.2	0
ω_3 pq	.2	0

We have $N_{\pi_1}(q) = .8$ and $N_{\pi_2}(q) = .7$.
Let $\oplus$ be the product operator which is a reinforcement operator. Then, $\pi_1(\omega) \oplus \pi_2(\omega) = 0$, $\forall \omega \in \Omega$. Hence, $N_{\pi_1 \oplus \pi_2}(q) = 0$ if q is different of $\top$ [9].

So the first condition is to avoid that π_1 and π_2 together to be fully inconsistent (i.e., $h(\pi_{min}) = 0$). But this is not enough since even if $h(\pi_{min}) > 0$ one can have $h(\pi_1 \oplus \pi_2)= 0$ due to the reinforcement effect which can push the priority of conflicting information to the maximal priority allowed, namely 1. For instance, let us consider the excessively optimistic reinforcement operator defined by:

$$\forall a, \forall b, a \neq 1, b \neq 1, a \oplus b = b \oplus a = 0.$$

Then we can check that as soon as there is a conflict between the possibility distributions to be aggregated, the consistency level of the fused possibility distribution will reach the minimal value.
The following definition focuses on a more interesting class of reinforcement operations:

Definition 19 *The reinforcement operation $\oplus$ is said to be progressive if*
$$\forall a, b \neq 0, a \oplus b \neq 0.$$

At the syntactic level, the reinforcement progressive operation guarantees that if some formula (ϕ, α) with $\alpha > 0$ is inferred by the sources then this formula belongs to $\mathcal{B}_\oplus$ with a weight β such that $\alpha < \beta < 1$. However, this new weight β can be less than the inconsistency degree of $\mathcal{B}_\oplus$ and therefore ϕ will be drowned by the inconsistency of the database. This situation is illustrated by the following example:

Example 13. Let p, q, r be three propositional symbols. Let π_1 and π_2 be two possibility distributions. In the following table, we give π_1, π_2 and $\pi_1 \oplus \pi_2$ the result of combining π_1 and π_2 with the product operator.

Table 7.

ω	$\pi_1(\omega)$	$\pi_2(\omega)$	$\pi_1(\omega) \oplus \pi_2(\omega)$
ω_0 $\neg p \neg q \neg r$	.1	.9	.09
ω_1 $\neg p \neg q r$	.1	1	.1
ω_2 $\neg p q \neg r$	.5	.5	.25
ω_3 $\neg p q r$	.5	.5	.25
ω_4 $p \neg q \neg r$	.5	.5	.25
ω_5 $p \neg q r$	.5	.5	.25
ω_6 $p q \neg r$	.9	.1	.09
ω_7 $p q r$	1	.1	.1

We have $N_{\pi_1}(r) = .1$ and $N_{\pi_2}(r) = .1$ but $N_{\pi_1 \oplus \pi_2}(r) = 0$ since $\mathcal{K}_{\pi_1 \oplus \pi_2} = \{\omega_2, \omega_3, \omega_4, \omega_5\}$ and $Mod(r) = \{\omega_1, \omega_3, \omega_5, \omega_7\}$. Then, $\mathcal{K}_{\pi_1 \oplus \pi_2} \not\subseteq Mod(r)$.
This is due to the fact that the consistency level of π_{min}, the result of combining π_1 and π_2 with the *min* operator, is equal to .5 and $h(\pi_1 \oplus \pi_2) = .25$ which is less than .5.

The following proposition generalizes Proposition 9 and shows that if the consistency level does not decrease, then the common knowledge is entailed.

Proposition 10 *Let π_1 and π_2 be two possibility distributions, and π_{min} be such that $\forall \omega, \pi_{min}(\omega) = min(\pi_1(\omega), \pi_2(\omega))$. Let $h(\pi_{min}) > 0$.
Let $A \subseteq \Omega$ such that $N_{\pi_1}(A) = \alpha > 0$ and $N_{\pi_2}(A) = \beta > 0$. Let $\oplus$ be a reinforcement operator. Then, if $h(\pi_{min}) = h(\pi_1 \oplus \pi_2)$ then,*

$$N_{\pi_1 \oplus \pi_2}(A) > max(\alpha, \beta),$$

and $N_{\pi_1 \oplus \pi_2}(A) = 1$ if $\alpha = 1$ or $\beta = 1$.

The following proposition shed light on the nature of the reinforcement operator:

Proposition 11 *Let π_1 and π_2 be two possibility distributions.
Let $\forall \omega, \pi_{min}(\omega) = min(\pi_1(\omega), \pi_2(\omega))$. Let $h(\pi_{min}) > 0$.
Let $\oplus$ be a progressive operator. Denote by π_2^n the combination of π_2 n times with $\oplus$. Then, there exists n such that*

$$\pi_1 \oplus \pi_2^n \models \pi_2.$$

This proposition means that reinforcement operators are majority operators, in the sense that if the same piece of information is repeated enough time then this piece of information will be believed at a high degree.

Adaptive merging operators

The regular disjunctive operators appear to be appropriate when the sources are completely conflicting. However, in the case of consistency, or of a low level of inconsistency, regular disjunctive operators are very cautious. Besides, reinforcement is not appropriate in the case of complete conflicts.
The aim of adaptive operators is to have a disjunctive behaviour in case of complete contradiction and the progressive reinforcement behaviour in the other case. Let $\oplus_d$ and $\oplus_r$ be respectively a regular disjunctive and a progressive reinforcement operator. Let v be either equal to 1 or to 0. Then we define an adaptive operation, denoted by $\oplus_v$, can be defined as follows:

$$a \oplus_v b = max(min(v, (a \oplus_d b)), min(1 - v, (a \oplus_r b)))$$

Then we have the following result:

Proposition 12 *Let π_1 and π_2 be two possibility distributions. Let π_{min} be a possibility distribution such that $\forall \omega, \pi_{min}(\omega) = min(\pi_1(\omega), \pi_2(\omega))$.*

Let v be equal to 1 if $h(\pi_{min}) = 0$ and equal to 0 otherwise. Let $\oplus_v$ be an adaptive operator. If $h(\pi_1 \oplus_h \pi_2) \geq h(\pi_{min})$ then for each $A \subseteq \Omega$,
if $N_{\pi_1}(A) = \alpha > 0$ and $N_{\pi_2}(A) = \beta > 0$, then $N_{\pi_1 \oplus_v \pi_2}(A) = \gamma > 0$.

Averaging operators
Another class of combination operators of interest are *averaging operators*, well known for aggregating preferences. They are defined as follows:

Definition 20 *$\oplus$ is called an averaging operator if*
$$max(a, b) \geq a \oplus b \geq min(a, b)$$
(and $\oplus \neq max$, $\oplus \neq min$).

One example of averaging operators is the arithmetic mean $a \oplus b = \frac{a+b}{2}$.
In case of consistency we can check that for $A \subseteq \Omega$, if $N_{\pi_1}(A) = \alpha > 0$ and $N_{\pi_2}(A) = \beta > 0$ then $0 < N_{\pi_1 \oplus \pi_2}(A) \leq \frac{\alpha+\beta}{2}$.
In this case, at the syntactic level, the result of combining B_1 and B_2 writes:

$$\{(\phi_i, \frac{\alpha_i}{2})\} \cup \{(\psi_j, \frac{\beta_j}{2})\} \cup \{(\phi_i \vee \psi_j, \frac{\alpha_i+\beta_j}{2})\}.$$

4 Conclusion

This paper has described a logical counterpart of the aggregation of possibility distributions or fuzzy set with membership grades in a finite scale. Then, combination can be directly performed at the syntactic level. The possibilistic logic format, which can model either beliefs with their certainty levels, or goals with their priority levels, appear to be convenient for expressing available knowledge, or preference in a granular and synthetic ways, as local constraints (or necessity measures). This framework provides an approach which unifies numerically oriented views and logical approaches indeed. In [3], it has been shown how an implicit possibility distribution can be naturally extracted from a classical propositional logic base. Thus, recent approaches that have been proposed for merging classical propositional bases, or sets of (non-prioritized) goals, can be embedded in the possibilistic logic framework as well.

References

1. M.A. Abidi, R.C. Gonzalez (Eds.). Data Fusion in Robotics and Machine Intelligence. Academic Press, New York.
2. C. Baral, S. Kraus, J. Minker, Subrahmanian. Combining knowledge bases consisting in first order theories, Computational Intelligence, 8(1), 45-71, 1992.
3. S. Benferhat, D. Dubois, S. Kaci, H. Prade. Encoding information fusion in possibilistic logic: A general framework for rational syntactic merging. In Proceedings of 14th ECAI, 3-7, 2000.

4. S. Benferhat, D. Dubois, H. Prade. How to infer from inconsistent beliefs without revising ? In Proceedings of 14th IJCAI, 20-25, 1449-1455, 1995.
5. S. Benferhat, D. Dubois, H. Prade. From semantic to syntactic approaches to information combination in possibilistic logic, Aggregation and Fusion of Imperfect Information (B. Bouchon-Meunier, Ed.), Physica-Verlag, Heidelberg, Germany, 141-161, 1997.
6. S. Benferhat, D. Dubois, H. Prade, M. Williams, A practical approach to fusing and revising prioritized belief bases. In Proceedings of EPIA 99. LNAI n° 1695, Springer Verlag, 222-236.
7. B. Bouchon-Meunier, Ed. Aggregation and Fusion of Imperfect Information, Physica-Verlag, 1997.
8. L. Cholvy. A logical approach to multi-sources reasoning. In Applied Logic Conference: Logic at Work, Amsterdam.
9. D. Dubois, J. Lang, H. Prade, Possibilistic logic. In Handbook of Logic in Artificial Intelligence and Logic Programming, Vol. 3, 439-513, 1994.
10. D. Dubois, H. Prade. Possibility theory and data fusion in poorly informed environments, Control Engineering Practice, 2(5), 811-823, 1994.
11. D. Dubois, H. Prade, R. Yager. Merging fuzzy information, Fuzzy Sets in Approximative Reasoning and Information Systems, (J.C. Bezdek, D. Dubois, H. Prade Eds.). The Handbboks of Fuzzy Sets Series, Kluwer Academic Publisher, Dordrecht, 335-401, 1999.
12. J. Flamm, T. Luisi (Eds.) Reliability Data and Analysis. Kluwer Academic Publishers.
13. M. Grabisch, S. Orlovski, R. Yager. Fuzzy aggregations of numerical preferences, Fuzzy Sets in Decision Analysis, Operations Research and Statistics, (R. Slowinski, Ed). The Handbooks of Fuzzy Sets Series, Kluwer Academic Publishers, Dordrecht, 31-68, 1998.
14. J. Kacprzyk, H. Nurmi. Group decision making under fuzziness, Fuzzy Sets in Decision Analysis, Operations Research and Statistics, (R. Slowinski Ed.). The Handbooks of Fuzzy Sets Series, Kluwer Academic Publisher, Dordrecht, The Netherlands, 103-136, 1998.
15. S. Konieczny, R.Pino Pérez, On the logic of merging. In Proceedings of the 6th International Conference on Principles of Knowledge Representation and Reasoning (KR'98), 488-498, 1998.
16. S. Konieczny, R. Pino Pérez, Merging with integrity constraints. In Proceedings of ECSQARU'99, LNAI n° 1638, Springer Verlag, 233-244, 1999.
17. J. Lin, Integration of weighted knowledge bases, Artificial Intelligence 83, 363-378, 1996.
18. J. Lin, A.O. Mendelzon, Merging databases under constraints,1998. International Journal of Cooperative Information Systems, 7(1):55-76, 1998.
19. P. Z. Revesz. On the semantics of theory change: arbitration between old and new information. Proceedings of the 12th ACM SIGACT-SIGMOD-SIGART symposium on Principles of Databases, 71-92, 1993.
20. P. Z. Revesz. On the semantics of arbitration. International Journal of Algebra and Computation, 7(2), 133-160, 1997.

Automated Adaptive Situation Assessment

Mark G. Rutten and Dale A. Lambert

Defence Science and Technology Organisation, Salisbury S.A., 5108, Australia.

Abstract. The product of understanding events in any chosen environment is often referred to as Situation Assessment. Situation Assessment involves building a description of events, or sequences of events, in terms of relationships between the objects in the environment. While people perform Situation Assessment as a matter of course in everyday actions, automating the process within a machine is difficult. Situation Assessment adaptation involves modifying the Situation Assessment process to provide an improvement in the assessment of a situation compared with the best approximate assessment. A technique, founded upon Case-Based Reasoning, is presented to facilitate adaptation of Situation Assessment processes.

1 Introduction

Adaptation of Situation Assessment refers to the methods employed to perform Situation Assessment and the improvement of those methods over time. The designer of a Situation Assessment system will write a set of routines that perform assessments for different situations.[1] It is not possible for the designer to write a dedicated routine for every possible condition, so routines are written for the most likely scenarios. A set of routines can then be carefully constructed such that they will provide, at the very least, an approximate assessment of any possible situation. Situation Assessment Adaptation involves modifying the Situation Assessment process to provide an improvement in the assessment of a situation compared with the approximate assessment which may be produced by the original set of routines.

Section 2 introduces Situation Assessment and Refinement of Situation Assessment processes as components of a Data Fusion process. Section 3 introduces Case-Based Reasoning, the methodology under which Situation Assessment processes will be adapted. The description of Case-Based Reasoning highlights the reasons why this theory is considered suitable in a Situation Assessment context. Section 4 describes the way in which Situation Assessment routines can be characterised in order to provide relevant selection criteria, Section 5 outlines the difficulties in selecting the best routine using the routine context and provides an algorithm for selection, and Section 6 is a summary of the routine adaptation process.

[1] *An Exegesis of Data Fusion* (Dale A. Lambert), a companion paper in this volume, discusses the use of routines for Situation Assessment.

2 Situation Assessment

2.1 The JDL Model of Data Fusion

Data fusion is the process of utilising one or more data sources over time to assemble a representation of aspects of interest in an environment. By fusing data from different sources it is possible to exploit unique information gained from each of the sources and refine missing or uncertain information.

Data fusion was first formalised for defence by the Data Fusion Sub-Panel of the United States Joint Directors of Laboratories (JDL) in the late 1980's [1]. The revised JDL Fusion model [2] is illustrated in Figure 1. Level Zero

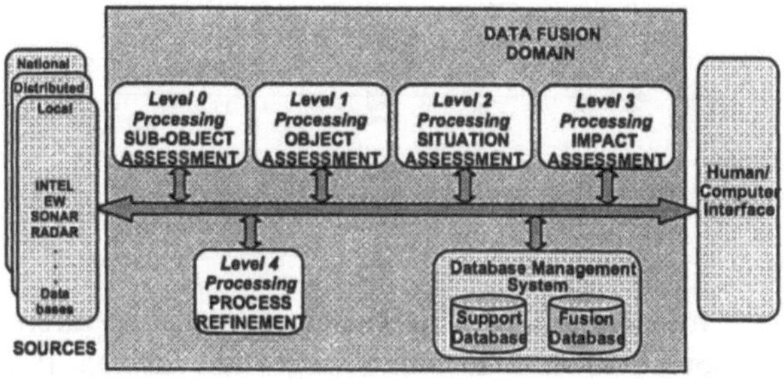

Fig. 1. The Revised JDL Model of Data Fusion

fusion performs signal and image processing. Level One fusion provides estimates of identity and kinematic state of objects within the current situation. Situation Assessment and Impact Assessment both automatically provide interpretations and consequences of the situation based upon data obtained from the Level One process, aided by knowledge and past experience of the domain that is being considered. The identities and properties of each of the objects from Level One would be judged to determine their role in the situation, predict future actions and determine threats. The fourth level of fusion straddles each of the other levels as a general feedback and control mechanism.

This paper describes an automated process of Situation Assessment routine adaptation to provide a level of learning and hence the ability to improve the capabilities of the system. Among the JDL fusion levels, adaptation of routines is a Level Four process. Hence the content of this paper aims to solve problems of Level Four fusion, with relation to routines which perform Level Two fusion.

2.2 An Air-Defence Model of Situation Assessment

The Situation Assessment example introduced here will be used throughout the remainder of this paper to illustrate the techniques developed to provide a Situation Assessment and progressively refine that Situation Assessment process. Figure 2 shows the course mapped out by an aircraft from the posi-

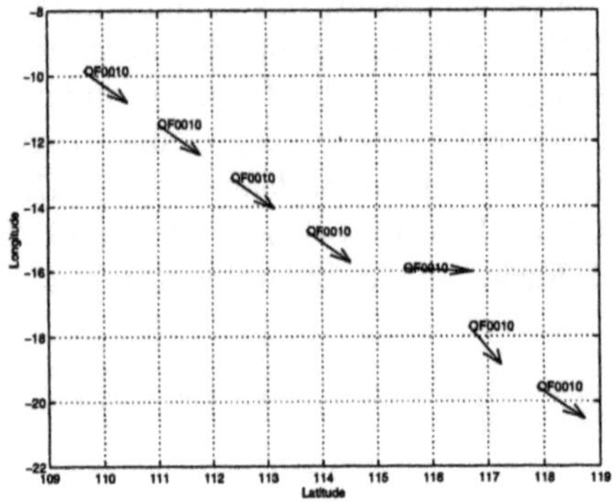

Fig. 2. An Aircraft Diverging Off Course

tion data gained from some Level One fusion process. The aircraft is initially flying from the north-west in a normal, straight line fashion. It then deviates from its normal course, returning shortly after.

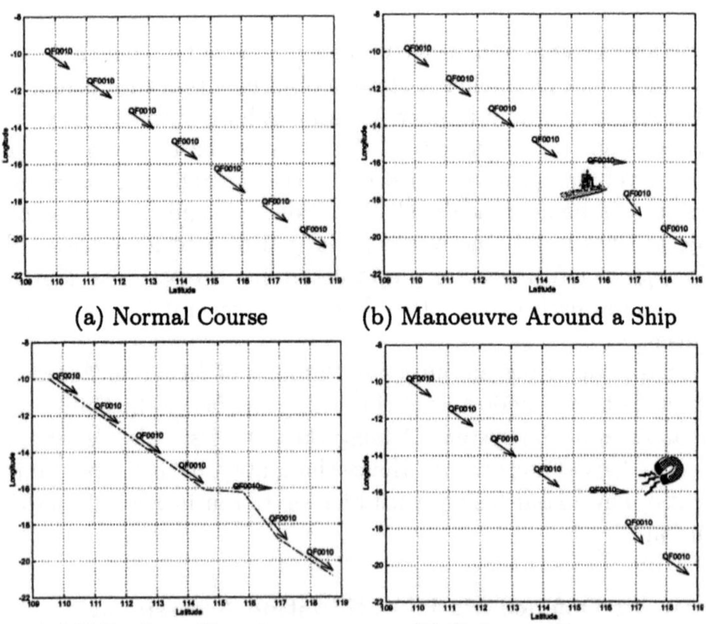

(a) Normal Course (b) Manoeuvre Around a Ship

(c) Following a Boundary (d) Unknown Reason

Fig. 3. Different Possible Manoeuvres

The aim of the Situation Assessment is to explain why the aircraft manoeuvred in such a fashion. The possible situations dealt with by this Situation Assessment process are shown in Figure 3 and summarised by the following:

(a) Shows the aircraft flying on a normal course.
(b) Shows the aircraft manoeuvring from its normal course to avoid an obstacle, in this case a ship.
(c) Shows the aircraft manoeuvring to continue following a boundary.
(d) Shows the aircraft manoeuvring for an unknown reason.

The routines used in this paper are written in ATTITUDE, a real-time reactive BDI (beliefs, desires and intentions) [3] multi-agent system developed at DSTO. An ATTITUDE agent has a self-contained internal representation of the world in its beliefs, and a sense of what it wants to achieve and how to achieve it, through its desires and intentions. ATTITUDE is programmed using propositional attitudes [4] as the basic instructions for the agent. Each propositional attitude can either succeed or fail, giving program flow control depending on the outcome.

```
( ( explain-manoeuvre )
  ( .
    ( ask-if-believe ( close-to ?object ) )
    ( ask-if-believe ( type ?object ship ) )
    ( ask-if-believe ( class ?object ?class ) )
    ( |
      ( .
        ( ask-if-believe ( within-range ?class ) )
        ( believe ( manoeuvre around ?object ?class ) )
      )
      ( believe ( manoeuvre around-radar ?object ?class ) )
    )
  )
)
```

(i) Manoeuvre Around a Ship

```
( ( explain-manoeuvre )
  ( .
    ( ask-if-believe ( on-boundary ?boundary ) )
    ( |
      ( .
        ( ask-if-believe ( following-boundary ?boundary ) )
        ( believe ( manoeuvre follow-boundary ?boundary ) )
      )
      ( believe ( manoeuvre onto-boundary ?boundary ) )
    )
  )
)
```

(ii) Following a Boundary

```
( ( explain-manoeuvre )
  ( |
    ( .
      ( ask-if-believe ( following-boundary ?boundary ) )
      ( believe ( manoeuvre off-boundary ?boundary ) )
    )
    ( believe ( manoeuvre unknown ) )
  )
)
```

(iii) Unknown Reason

Fig. 4. Routines for Situation Assessment

The rudiments of the language necessary for the example are best illustrated through the example routines themselves, which are listed in Figure 4.[2] The first proposition of each routine (`explain-manoeuvre`) is the goal of the routine, which is a statement of what the routine will attempt to

[2] Note that these routines are meant only for illustrative purposes. Routines written in ATTITUDE may be significantly more complex than those listed here, making automatic adaptation similarly more complicated. ATTITUDE routines are general purpose, so are not limited to a Situation Assessment domain.

achieve. Within the routine a concatenation of commands is achieved through (. a b...j), which will attempt to achieve each goal $a, b \dots j$ in succession. A failure to achieve any one of the goals will cause the construct to fail. (| a b...j) provides a selection construct, which will attempt each goal in turn; if one of the goals succeeds then the construct succeeds, otherwise the next goal in the list is attempted. The construct fails if every goal, a through to j, fails. Variables and constants are used within the propositions of these routines. A constant is a unvarying label and a variable is an element who's value may be changed, signified by a `?` in front of the label. The propositional attitude **`believe`** inserts a proposition into the knowledge base, which will always succeed. Conversely **`ask-if-believe`** queries the knowledge base for facts matching the attitude's proposition. The propositional attitude will succeed only if a match exists in the knowledge base. An unbound variable in the proposition of an **`ask-if-believe`** will be assigned the value of a matching fact from the knowledge base, if the attitude succeeds.

The Ship routine of Figure 4(i) firstly asks if the aircraft is close to an object and fails if it is not. Then the routine determines the the type of the object is a ship and the class of the ship If the aircraft is within the radar coverage of that class of ship, then the manoeuvre is described as avoiding the ship, otherwise the manoeuvre is to avoid detection by the ship's radar. The Boundary routine, Figure 4(ii), asks if the aircraft is on a boundary. If the aircraft was already following a boundary, then the explanation is that the aircraft was manoeuvring to continue following a boundary. Otherwise, the reason is described as the aircraft moving onto a boundary. The Unknown routine, Figure 4(iii), tests to see if the aircraft was previously traveling along a boundary, in which case the manoeuvre is labelled as **`off-boundary`**, otherwise the manoeuvre is completely **`unknown`**.

The normal straight-line case has a separate routine not listed here, since it is not an explanation of a manoeuvre, which determines when a unexpected manoeuvre occurs. This routine will attempt to determine a reason for the manoeuvre, by invoking the above explanatory routines. Each of the routines is attempted in the order above, giving a priority to each of the meanings. So if the aircraft is avoiding a ship *and* following a boundary, then the ship explanation is deemed to be the "real" reason. This set of routines will always supply a reason, even in an unusual situation, due to the existence of the Unknown routine. The data that the routines use (**`close-to`**, **`type`**, **`class`**, **`within-range`**, **`on-boundary`** and **`following-boundary`**) are assumed to be readily accessible and provided or calculated by another part of the system, or an external source.

3 Case-Based Reasoning

Case-Based Reasoning [5] has been proposed as a plausible model of human cognition. The theory is founded upon the idea that people follow mental scripts of actions that have been stored away in memory and that seem helpful for what we want to achieve. Intelligence is then derived from how well people can draw analogies [6] between their current problem and the previous experiences that they have stored in memory. Reminding is the way in which our memories prompt us with past experiences. A person's problem solving ability is based upon how well they can create a new plan by adapting that case to the current problem.

People often reason with limited and uncertain information. But people reason robustly, so that they always find an answer, even though it may be sub-optimal. This is why a system which models the way in which people think is sought after and why Case-Based Reasoning's base in human psychology is appealing. The problem solving process uses past experience while generating experience. Each time a case is used in a new context, the success, failure, adaptation techniques, or reasoning required to apply the case in the new situation, can be recorded for the next time this context is encountered.

Plaza and Aamodt [7] describe what they believe to be the four basic components of the Case-Based cycle as being:

1. *Retrieve* the most similar case or cases.
2. *Reuse* the information and knowledge in that case to solve the problem.
3. *Revise* the proposed solution, based upon the extent of the solution's success.
4. *Retain* the parts of this experience likely to be useful for future problem solving.

The "four Re's" give a useful insight into the basic process which is required by a machine which uses Case-Based Reasoning.

It is proposed that Case-Based Reasoning be used in the context of Situation Assessment to improve the way in which it is performed. This requires improvement of the routines which perform Situation Assessment, and so the process of retrieving, adapting and storing cases prescribed by Case-Based Reasoning must be performed on the Situation Assessment routines. An improved routine will be one which either performs its function better or one which performs its function under a broader set of situations. Based on the knowledge of the performance of the routines in different situations and the ability of those routines to perform the desired assessment, adaptations can be made to either extend the routine's strengths or ameliorate the weaknesses.

4 Describing a Routine Context

As a first step in a Case-Based Reasoning process it is necessary to choose one (or more) routine(s) to adapt or apply to the situation from among the available routines. A choice between routines requires criteria which can highlight relevant differences between those routines in order to make a selection from them. The criteria generated for each routine must have the following features:

1. they must capture parts of the initial system state which influence the final state of the system after the routine has been run,
2. they must be relevant on instantiation of the routine, so that time-dependent aspects, such as waiting for new information, become irrelevant to the selection conditions,
3. in order to be both as general as possible and to conserve system resources they should be the minimum subset of relevant factors which influence the routine's execution, and
4. they must encapsulate knowledge about the situations under which the routine will either succeed or fail to achieve a satisfactory result.

4.1 Weakest Preconditions

A routine, R, is denoted as follows:

$$\alpha \rightarrow_R \beta$$

The preconditions of the routine, α, describe the initial state of the system before the routine has been applied. The postconditions, β, describe the system state after the routine has been performed. The routine itself, R, describes the actions by which the system is transformed from a state described by α to a state described by β. It is the preconditions, α, along with the postconditions, β, that form the routine selection context. Selection of a routine becomes a process of determining the desired state and then choosing a routine from among the set of possible routines which can achieve that state from the current state.

Weakest preconditions [8] provide a method for defining the semantics of a language. Given a specific final state, the weakest precondition functions allow the state before the execution of the routine to be described giving the routine preconditions, α. All elements of the system that must have been present on routine instantiation, for the routine to produce a given outcome, are given by the weakest preconditions. Weakest preconditions project the essential parts of the routine back to the start state, thus providing an explanation as to why a particular routine will either succeed or fail for a given system state. They are called weakest preconditions since the weaker the condition, the more states there are that satisfy it.

If a propositional attitude instruction is denoted by s and the desired postcondition is denoted by δ, then the weakest preconditions are given by

$$\mathrm{wp}(s, \delta)$$

The properties of weakest preconditions allow the preconditions for a sequence of statements to be derived. In this way:

$$\mathrm{wp}(ab, \delta) = \mathrm{wp}(a, \mathrm{wp}(b, \delta))$$

The technique used in this work is to find the preconditions dynamically, by using information available during system run-time. By finding the preconditions dynamically the exponential explosion of possibilities evident by statically deriving the weakest preconditions is removed and is replace by a learning process. The method uses an execution trace which provides both a sequence of routine statements and environmental information. This implies that only a specific sequence of preconditions will be collected for each execution and a learning process is required to generate an approximation to the weakest preconditions. As the system tries different routines in different situations, more information about the different parts of the system state which effect the routine can be determined. By using information available from the routine, specific instances of precondition information can be generalised such that just a single successful transition through a path in a routine is required to produce preconditions for that path.

4.2 Selection Trees

Given a sequence of preconditions from a single execution trace, the next step is to combine separate sequences together, in order to gain more information about the routine. The resulting structure, named a Selection Tree, summarises the system's previous experience in attempting to achieve a certain goal. Information from all routines sharing the same goal is collected together under the same structure.

The algorithm for combining a new sequence to a Selection Tree is straight forward. In the case of an empty Selection Tree the precondition sequence which is being merged becomes the new Selection Tree. If the Selection Tree is not empty, then the transitions from the initial state in the tree are compared with the first transition of the precondition sequence, adding a new branch to the tree if the sequence provides new information. Merging proceeds down the tree, with a transition in the tree reflected by a corresponding transition in the sequence.

Although a particular sequence may not contribute new information to a Selection Tree, it aids in gaining knowledge about how often particular execution paths in a routine are pursued. By counting the number of times a particular node in the Selection Tree has been utilised in the past, through either success or failure, the most likely routine to be used and the most likely path through that routine can be found. Figure 5 shows the selection tree resulting from 69 iterations of the three example routines. Each transition is labelled with a selection proposition and each node is labelled with a visitation count. The terminal contains a pair of symbols being either S, or F signifying success or failure and a number representing the routine index.

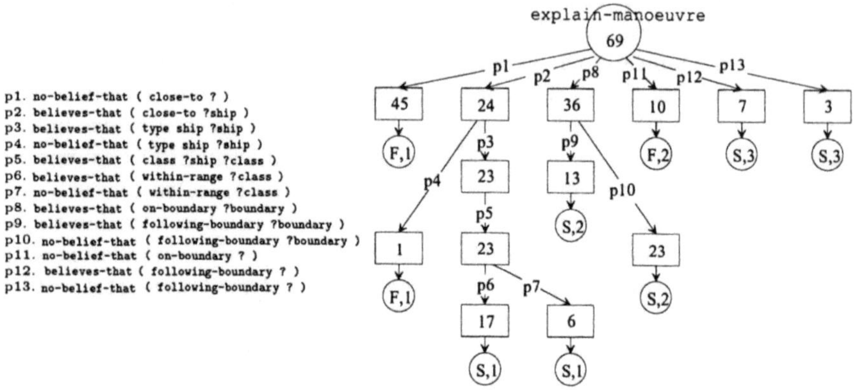

Fig. 5. A Complete Selection Tree

5 Routine Selection

The information about the system states for which routines are applicable is summarised in the Selection Tree, which encapsulates the knowledge about the success and failure of different routines in different situations built up through the system's experience. Having provided a means of characterising routines this information must be used in order to select an appropriate routine.

5.1 Problems of Selection

The first problem to recognise is that it is not necessary for the current system state to meet every precondition for a routine. There may be a routine which is applicable in the current situation, but the relevant path through that routine has not been explored previously or it will be possible to adapt the routine to the situation. The question which this raises is the distinction that should be made between those conditions which match exactly, those conditions which are partially met and those conditions which are not met at all.

Secondly, it is possible for the system to contain more than one piece of information which will satisfy a particular condition. In these cases it will be necessary to explore all possible options to determine which will provide the best match when the entire precondition sequence is evaluated.

The third problem lies in the fact that the Situation Assessment routines must have priorities associated with them. There may be several explanations of a situation, each of which are equally valid, but only one of which would describe the "real" reason. The problem with this is how to apply those priorities within the matching algorithm.

The fourth problem is how to utilise the usage statistics maintained for each of the conditions within a Selection Tree. The statistics are kept in order to indicate which situations occur most often and the routines that have been most successful in explaining those situations.

5.2 The Viterbi Algorithm

A Markov Model is a simple discrete-time Finite State Machine with the special quality that the transition probabilities depend only on a single previous state. An analogy can be drawn between the more redundant trellis form of

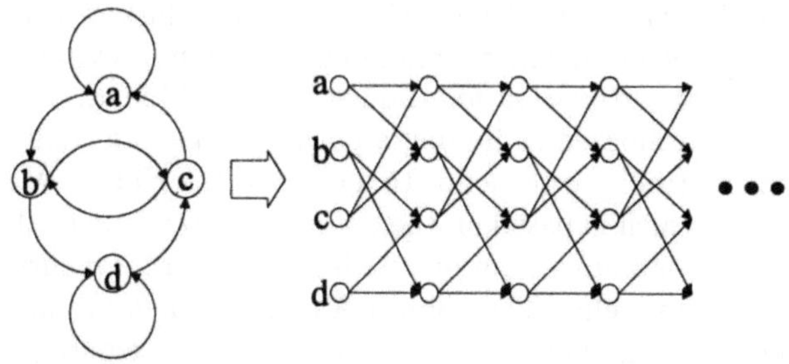

Fig. 6. A Finite State Machine and its Corresponding Trellis.

a FSM, shown in Figure 6 and a Selection Tree. The probability of transition from node i to node j at level k in the tree, a_{ij}^k, is given by:

$$a_{ij}^k = P(x_i^k | x_j^{k-1})$$

where $A^k = \{a_{ij}^k\}$ is the set of transition probabilities, here the success statistics of the Selection Tree, and $x_i^k \in X^k = \{x_1^k, ..., x_{n_k}^k\}$ is the finite set of nodes at the kth level in the Selection Tree. A Hidden Markov Model is a Markov Model which has been sampled through some memoryless noise. $Y^k = \{y_1^k, ..., y_{m_k}^k\}$ is a set of possible measurement symbols, which are the samples from the output of the Hidden Markov Model. The probability of measuring the output symbol j when the system is in state i at the kth level is

$$b_i^k(j) = P(y_j^k | x_i^k)$$

where $B^k = \{b_i^k(j)\}$ is the set of measurement probabilities.

The noise characteristics of a Selection Tree are derived from trying to find the closest match, rather than an exact match. Hence the B^k's will not be calculated as measurement probabilities, they will be a more general weighting function. This may be written as

$$b_j^k(i) = \prod_{a=1}^{A^k} \theta_a(i,j)$$

where $\Theta^k = \{\theta_1, ..., \theta_{A^k}\}$ is the set of weighting functions, defined such that $0 \leq \theta_a(i,j) \leq 1$. There are three such component weighting functions

1. $\theta_{match}(i,j)$ represents the weighting given to the mismatch between the condition and the current system state.
2. $\theta_{pri}(i,j)$ will take into account the priority assigned to each of the routines.
3. $\theta_f(i,j)$ will incorporate the failure statistics of the transition. Note that the success statistics are used as the transition probabilities, a_{ij}^k in the Viterbi Algorithm.

The Viterbi Algorithm [9] can be used to find an optimal series of states in a system which can be modelled using a Hidden Markov Model, such as a Selection Tree. It is a dynamic programming solution, providing a maximum a *priori* state series estimation for Hidden Markov Models. The Viterbi Algorithm can be considered as finding a path through the trellis of Figure 6. Each node in the trellis corresponds to a state of the Hidden Markov Model. A cost function, which is based on the noise characteristics of the model and the transition probabilities from one node to another, is recursively applied to each node in the trellis. The Viterbi Algorithm then minimises the total cost for a certain path. The algorithm consists of four separate steps:

1. Initialisation: assigns a metric to each of the initial states of the trellis.
2. Recursion: assigns a metric to each of the states in the trellis, based on a combination of the metrics of previous states and transition metrics.
3. Termination: selects the node with the minimum path weight from the final node set in the trellis.
4. Backtracking: retrieves the optimal path by backtracking along the trellis, starting at the node selected by the termination step.

6 Routine Adaptation

Adaptation is necessary after a selected routine has been attempted, but has failed. There are four questions which need to be answered in order to adapt around that failure:

1. What information needs to be stored for adaptation?
2. Where should adaptation be applied (credit assignment problem)?
3. How is the routine repaired?
4. How should the new adapted routine be stored?

The basic intention of this adaptation mechanism is to use information from previous experience in similar problem solving exercises to adapt around the current failure.[3] All possible sub-sequences are generated from execution

[3] The technique for adaptation used here follows the derivational analogy approach of Carbonell [10] and Veloso [6]. A system using derivational analogy replays a previous solution to the problem in the current context using previous information stored about the problem solving process to guide the current solution.

traces derived from previously instantiated routines. Preconditions and postconditions are produced for each sub-sequence, storing the information for the repair phase of adaptation. When a location for repair is identified, a goal condition is also identified. By using the stored sub-sequences, along with their pre- and post-conditions, a classical planning technique is used to join several sub-sequences together, attempting to direct the system from its current state to the identified goal state.

The information required for adaptation can then be summarised as follows:

1. Postcondition trees, giving the goal state β (c.f. $\alpha \rightarrow_R \beta$), are required as descriptions of the termination points for each routine.
2. Each possible sub-sequence is generated from the execution trace, along with the generalised preconditions and postconditions of that sequence. These sub-trace sequences are stored for use in the repair phase.
3. The preconditions are required for each step in a routine, in addition to a selection tree being calculated for a routine. These are stored along with each routine, resulting in a precondition tree built from system experience associated with each instruction in the routine.

Postconditions are generated in a similar fashion to the preconditions, but with a forward pass through the execution trace, rather than a reverse pass. The postconditions provide complete information about the routine's desired final system state. Preconditions identify the requirements of the system at the beginning of the sub-trace in order for that sequence of propositional attitudes to succeed. Postconditions summarise the parts of the system state modified by the sequence. For example, beliefs added or removed from the knowledge-base would appear as part of the postconditions.

Assigning blame for the failure to a portion of the routine which has run (the credit assignment problem) provides a starting point for the adaptation. An end-point for repair of a routine section must also be chosen. Since the preconditions describe the required information for success, a mismatch between the preconditions of previously successful routines and the current execution context gives indication of a reason for failure. The preconditions previously acquired for that routine are compared to the system state at each instruction in the routine during run-time, through a process identical to that used for routine selection. Any mismatches which produce a cost above a certain threshold are marked as possible reasons for the routine failure. The end-points of the reasons for failure are the possible terminations of the routine from the starting points for adaptation. Postcondition information is required for routine repair, so only those paths through the routine for which the system has prior experience are selected as possible paths for repair.

Credit assignment will generate several possibilities for repair, due to the multiple assignments of blame and the multiple routine terminations for each of the reasons for failure. Routine repair uses a classical planning method, aided by heuristic search. The store of routine sub-traces is used to select a sequence of sub-traces which will provide a path from the precondition information required by the starting point for repair to the postcondition information required by the end-point. Each possible path assigned blame is considered, and the best solution from among them is kept. The best solution is measured by the planning cost, based on the length of the final routine and the precondition/postcondition mismatch at the start and end-points of each of the joined sub-traces.

The repaired routine is separately stored exactly as a routine written by the designer of the system. Storing the routine separately is necessitated by the fact that the new routine may do no better than the original routine and the change made to the routine may actually be detrimental. Prior experience of a particular path through a routine is required for routine repair, hence an attempt at repairing a previously repaired routine will not be made until it has been successful at least once. Feedback to the routine selection algorithm is supplied through usage statistics, the possible increased information about the precondition and postcondition states and the addition of new routines.

To complete the example, a situation summarised by the following relevant beliefs needs to be explained:

```
( following-boundary J87 )
( close-to weather )
( class weather storm )
```

Using the routine selection algorithm of Section 5, the first routine, 4(i), has been chosen as the most appropriate for an explanation in the current context. The postcondition of the routine is one of two possibilities:

```
believes-that ( manoeuvre around ?object ?class )
bound-variable ?object
bound-variable ?class
```

or

```
believes-that ( manoeuvre around-radar ?object ?class )
bound-variable ?object
bound-variable ?class
```

The first and second instructions of the routine are attempted, but a failure occurs:

```
( ask-if-believe ( close-to ?object ) )  → ?object = weather
( ask-if-believe ( type ?object ship ) )   fails
```

Since **believes-that (type ?object ship)** is a precondition of the routine, the first propositional attitude of the routine is blamed for the failure. The two postcondition sequences outlined above then form the goals of the planning process invoked to attempt to solve the failure. One result, among the many possibilities using the complete set of generated sub-sequences, produces the following routine:

```
( ( explain-manoeuvre )
  ( .
    ( ask-if-believe ( close-to ?object ) )
    ( ask-if-believe ( class ?object ?class ) )
    ( believe ( manoeuvre around ?object ?class ) )
  )
)
```

which achieves the goal described by the post-condition from the current state.

This is a very simple, although plausible, example since the final result has removed propositional attitudes from the original routine. The beauty of this technique is that, given a more complex set of routines, the resulting routine may in fact be more intricate that the original and reuse components of relevant experience within any other routine. If a successful adapted routine does not exist, then ATTITUDE can resort to its basic reactive planning process.

7 Summary and Further Work

The question remains: having automatically adapted a particular routine to provide an explanation of a situation, is that explanation accurate? This depends on several factors:

1. the quality of the process used to adapt the routine, and
2. the robustness and reuseability of the original routines.

The ultimate aim is to let the system run autonomously regardless of the quality of the adaptation and the quality of the original routines, and rely on the generated results. However, in a practical situation there needs to be control over the results. In an extreme case, it would be highly undesirable to take the same action over `( close-to flock-of-birds )` and `( close-to unfriendly-aircraft)`, launching full-scale air defence. During production of the system, the basic routines are written, followed by thorough testing, during which the automatic adaptation is active. An expert can then check the generated routines, determining their fitness for use in a fielded system.

Significant advantages can be gained from having a system which improves its adaptation ability [11]. Implementing the adaptation algorithm within ATTITUDE will allow the adaptation routines to improve automatically, in exactly the same way as the Situation Assessment routines can be adapted. The effect of this factor on overall system capability and stability is yet to be determined.

This paper has described a system performing situation assessment, which can automatically adapt to new situations and new contexts. Following a Case-Based Reasoning methodology, techniques have been proposed for characterising a Situation Assessment routine, selecting a routine for a particular context and then repairing that routine if it does not succeed. In the example presented throughout this paper, a new situation, manoeuvring around bad weather, was not anticipated during system design. Through adaptation of the original routines, the system maintained an adequate Situation Assessment.

References

1. Edward Waltz and James Llinas. *Multisensor Data Fusion.* Artech House, Norwood, MA, 1990.
2. Alan N. Steinberg, Christopher L. Bowman, and Franklin E. White. Revisions to the JDL Data Fusion Model. Presented at the Joint NATO/IRIS Conference, Quebec City, Quebec, 19-23 October 1998.
3. M. E. Bratman, D.J. Israel, and M.E. Pollack. Plans and resource-bounded practical reasoning. *Computational Intelligence*, 4:349–355, 1988.
4. J.A. Fodor. *Psychosemantics: The Problem of Meaning in the Philosophy of Mind.* The MIT Press, Cambridge, Massachusetts, 1987.
5. Roger C. Schank and Robert P. Abelson. *Scripts, Plans, Goals and Understanding: An Inquiry into Human Knowledge Structures.* Lawrence Erlbaum, Hillsdale, N.J., 1977.
6. Manuela M. Veloso. *Planning and Learning by Analogical Reasoning.* Number 886 in Lecture Notes in Artificial Intelligence. Springer-Verlag, 1994.
7. Agnar Aamodt and Enric Plaza. Case-based reasoning: Foundational issues, methodological variations and system approaches. *AI Communications*, 7(1):39–59, 1994.
8. Edsger W Dijkstra. *A Discipline of Programming.* Prentice-Hall, 1976.
9. M. S. Ryan and G. R. Nudd. The Viterbi algorithm. Technical report, Department of Computer Science, University of Warwick, Coventry, CV4 7AL, England, 1993. Warwick Research Report RR238.
10. Jaime G Carbonell. Derivational analogy: A theory of reconstructive problem solving and expertise acquisition. In J. G. Carbonell R. S. Michalski and T. M. Mitchell, editors, *Machine Learning: An AI Approach, Volume II*, pages 137–161. Morgan Kaufmann, 1986.
11. David B. Leake, A. Kinley, and D. Wilson. A case-study of case-based CBR. In David B. Leake and Enric Plaza, editors, *Case based reasoning research and development: second international conference on Case-Based Reasoning, ICCBR-97*, volume 1266 of *Lecture Notes in Artifical Intelligence*, pages 371–384. Springer, 1997.

Soft Computing, Real-Time Measurement and Information Processing in a Modern Brewery

Duncan Campbell[1] and Michael Lees[2]

[1] Griffith University, Nathan QLD 4111, Australia
[2] Carlton & United Breweries, Abbotsford VIC 3067, Australia

Abstract. Computational intelligence methods provide mechanisms by which human expertise and learning can be embedded and implemented to solve problems, provide assessment of process performance from input data, and provide intelligent control. The use of sophisticated analytical techniques to monitor quality and processes in many manufacturing environments is becoming well established. In particular, soft computing concepts coupled with developments in real-time measurement of biological parameters are now allowing significant progress to be made in the historically challenging food and beverage industries. This chapter discusses specific developments which engage these technologies within a brewery.

1 Introduction

In today's competitive marketplace, the push for sustaining or even increasing profitability places extra pressure on factories and production plants. Many industries are now facing common issues such as minimising production costs, maximising plant utilisation, managing knowledge and technical skills, and providing increasingly safe working environments. In complex production and manufacturing environments, these objectives can only be met with the assistance of highly integrated and intelligent information exchange and processing systems.

The brewing industry is not alone in experiencing an increasing focus on issues such as resource management and wastage, right-first-time production, minimisation of plant down-time, and automation. The changing work-force profile has also contributed to the list of challenges that are faced by the modern production facility. Many plants have historically relied on work-force skilling and knowledge of the processes involved. As the work-force profile changes, often due to so-called rationalisations, many plants are having to address the issue of knowledge management. There is a need to capture and retain the often plant-critical tacit knowledge of its employees. Soft computing techniques provide a means by which these issues can be addressed.

There are many areas within a high-volume brewery to which soft computing can contribute. Intelligent control can be applied to optimise particular processes, intelligent sensing can be employed to monitor parameters on-line, knowledge management techniques used to capture and share knowledge, and the global integration of corporate enterprise and production systems to achieve global plant-wide optimisations. Many of these systems are required to operate in real-time, in the production sense, and so attention must be paid to data connectivity and the real-time performance of these systems.

The following sections provide topical discussions on these concepts and advances made in a brewery.

2 The Intelligent Factory

The intelligent factory capitalises on intelligent systems, global integration of all systems and the ability to transparently access and share information. This section visits these three topics in the context of the intelligent factory.

2.1 Intelligent Systems

There are many interpretations of what constitutes intelligence in a product or system. Products are sometimes marketed as being intelligent due to advanced feature levels, even though there may be a total absence of soft computing involvement. Alternatively, a high degree of artificial intelligence is incorporated in order to mimic biological intelligence. Between these two extremes lies a compromise which may be regarded as being the practical application of computationally intelligent methods borne of the efforts toward artificial intelligence.

Different industry and academic communities will have their own interpretations. In the context of factory automation, the authors adopt the notion that an intelligent system is one that employs one or more computationally intelligent techniques. The four most pertinent of these are expert systems, fuzzy logic, artificial neural networks and evolutionary computing.

2.2 Factory Global Integration

Factory global integration is intended to represent the seamless integration of data from the plant process controllers (PLCs, SCADA systems, embedded controllers etc.) and enterprise management systems. To achieve global optimisation of the factory and to support e-commerce activity, corporate enterprise systems must have access to data and information from the plant control systems. Information regarding performance indicators such as: resource usage, plant running costs, production output, production performance and production capacity can assist the company to respond quicker to market demands and opportunities.

The path towards global integration often encounters conflicting requirements. Whilst it is desirable to have seamless data flow between the factory floor and enterprise systems, access rights and security issues must be resolved. For instance, it would not be desirable for corporate managers to have over-riding control of process controls, nor would it be desirable for plant staff to advertently or inadvertently have access to corporate accounts and other management applications. This situation is often resolved by using an intermediate SQL database (or the like) as neutral or common ground for the systems.

The *SP95 - Enterprise/Control Integration Committee* has been defining interface standards between factory automation systems and corporate enterprise systems. The first of the standards has been published (ISA-S95.00.01-2000, Enterprise-Control System Integration) and addresses terminology and models for integrating control and enterprise systems. [1]

2.3 Protocols & Networks

In complex production and manufacturing environments, such as a high-volume production brewery, highly integrated information flow between the factory floor and enterprise systems can only be met with the assistance of highly integrated and intelligent information exchange and processing systems. Publicly accessible protocol and network standards have provided the open architectures necessary to allow transparent information connectivity from the global internet to items of plant on the factory floor.

The open fieldbusses and communication protocols have led to global support of these technologies rather than the inaccessibility of the many vendor proprietary systems previously available. This has allowed a virtual flattening of the control, device and bit layers to occur with corresponding transparency in data flow. These layers are characterised by their time-critical performance and relatively low data communication volumes. Conversely, the enterprise networks have traditionally not been time-critical and have higher data communication volumes.

The seamless integration of systems and data exchange has taken advantage of recent developments such as OPC (OLE for Process Control) and Java. Recent developments in hardware-based Java virtual machines will soon allow cost-effective and time-critical Java Beans and Java Applets to be implemented at the automation level in real-time.

These developments have enabled e-business and e-commerce technologies to be available to production and manufacturing industries to gain access to suppliers, distributors, remote plants and to conduct sales directly across the World-Wide-Web.

3 Knowledge Management

With increasing pressures on manufacturing based organisations to reduce the *cost of goods sold*, and through the natural attrition of retiring experienced personnel, the factory workforce profile often changes to one of less experience or of different a skillbase. The consequential erosion of tacit knowledge has highlighted the significance of knowledge management.

Knowledge management and knowledge sharing can also assist with the attainability of consistency, both in terms of process control and in general organisational responses and actions. Knowledge management can also be harnessed to provide training for staff from a collective expert knowledge-base.

The boundaries between knowledge and information are often nebulous. Bellinger [2] provides an attempt to categorise the roles and differences between *knowledge*, *information* and *data*. A fourth category, *Wisdom*, is also categorised.

- **Data:** Any values, numbers or words that are presented out of context can all be classified as data.
- **Information:** The interpretation, description, definition, or the understanding of the relationships between elements of data can be described as *information*. To be classified as information something would typically address such questions as *what, who, when* or *where*.
- **Knowledge:** To be classed as knowledge, issues such as strategy, approach, practice or method would be addressed. In essence, it would have to deal with the question of *how*.
- **Wisdom:** Related to the question of *why*, and covers the issues of reason, principle, morality or insight.

An interpretation of this relationship between data, information and knowledge is given in Fig. 1.

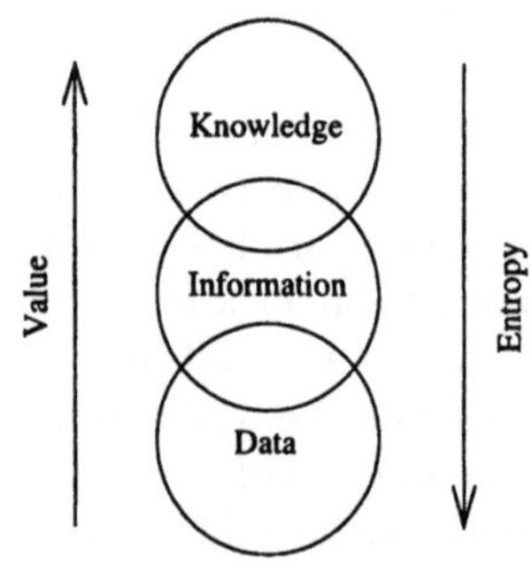

Fig. 1. Relationships between data, information and knowledge

The following example demonstrates the delineation of data, information, knowledge and wisdom for the intelligent control of a primary beer filter in a brewery [3][4].

Data: Current values of key filter parameters (ie. filter differential pressure,

rate of change of differential pressure (dp/dt), filter inlet turbidity and current dose rate of filter-aid).
Information: An example of information that could be based on such data would be "the filter dp/dt is currently below the optimum level".
Knowledge: A corresponding example of knowledge may be "decrease the dosage of filter aid by a certain percentage".
Wisdom: The optimal dp/dt for the filter is also the point of lowest operational cost, hence wherever possible the optimal dp/dt should be sought.

3.1 Knowledge Discovery & Creation

Acquired knowledge can be categorised into two broad groups, *explicit knowledge* and *implicit knowledge* (depicted in Table. 1). These categories can be further decomposed into *procedural knowledge*, *declarative knowledge*, *meta-knowledge*, *heuristic knowledge* and *structural knowledge* [5]. In the broad sense, potential knowledge, or yet-to-be-discovered knowledge, falls into the category of *discovery* as presented in Table. 1. This category in particular benefits from computationally intelligent methods to assist in extracting knowledge due to the uncertainties involved.

Table 1. Categories of knowledge

Category	Explicit	Implicit	Discovery
Description	Codified/Documented	Tacit	Yet to be discovered
Method of Acquisition	Already acquired	Externalisation	Knowledge creation or discovery

Human experience, investigation and analysis are the more traditional sources of human knowledge and account for a significant portion of knowledge creation. Psychological techniques are often employed to assist in acquiring tacit knowledge from the expert, end-user, and multiple experts. [5]

Knowledge can also be *discovered* from process data using computationally intelligent techniques such as:

1. **Rule induction:** Rule induction describes the automated generation of decision trees from large volumes of collected data. The resulting decision tree is a somewhat minimalised representation of relationships and correlations that were often previously obscure within the data. Rule induction is often used in data mining applications. [6]
2. **Genetic algorithms:** Genetic algorithms provide a means of identifying a set of optimal values for a group of related variables based on reproduction and evolutionary techniques (for a given *fitness* function).

3. **Artificial neural networks:** Artificial neural networks (ANNs) provide means of *learning* non-linear relationships between variables. However, ANNs represent the discovered relationships using neuronal weights (and biases) which can not be easily interpreted by humans.
4. **Neuro-fuzzy systems:** Neuro-fuzzy systems provide a means by which rule-based fuzzy systems can be automatically configured and optimised based on a given data set. In terms of knowledge creation, the advantage of these systems is that they can learn in a similar manner to that of ANNs, however the internal representation of the discovered relationships lies primarily in human readable rules. [7]

3.2 Knowledge Representation and Application

The effectiveness of knowledge capture and acquisition is pivotal to the effectiveness of the transfer, sharing and dissemination of that knowledge.

Knowledge can be stored either in a *passive* form or an *active* form. Passive knowledge storage includes the traditional means of documentation, such as books, papers, procedures, manuals etc. Conversely, *live* or operational knowledge-based systems are active forms of knowledge storage. An expert system includes a knowledge-base and some mechanism of interpretation or reasoning. Ideally, the knowledge-base can be accessed and interrogated to provide diagnostic and decision support for processes. It can give the factory process operator access to site-specific, expert knowledge - a guardian angel concept.

Active knowledge storage mechanisms include decision trees, deductive logic and rules of inference. *Common sense reasoning* can be implemented using fuzzy logic and mimics human reasoning. This approach is particularly useful in encapsulating tacit knowledge of factory processes. [8]

4 Real-Time Solutions

Many soft computing applications in the factory are on-line systems and require access to plant data during operations. Typical applications include regulatory control, diagnostics, predictive, analytical and decision support systems. For most of these systems to be effective, they must be able to respond in a time frame that is considered to be *real-time* for the given application. [9]

In practice, many real-time systems achieve real-time operation due to the computing hardware having sufficient computational ability to process the procedures in real-time. Imposed cost limitations or the bounds of current technology may limit program size and complexity in order to maintain real-time performance on the given hardware platform. If data networking is involved, communication speed and latency must also be considered.

Real-time performance also depends on the adequacy of the operating system and its ability to respond rapidly to events. QNX is an example of a *real-time operating system* (RTOS) often used in the factory. Linux is increasingly being used in the factory, primarily for the stability of the kernel. Although Linux systems have not been strictly real-time, the performance is more than adequate in many situations. Real-time Linux systems are now available. Embedded RTOSs are often used in embedded process controllers. Computationally intelligent algorithms must be efficient and time-constrained implementations, such as time-constrained reasoning, used when possible.

4.1 Time Constrained Reasoning

Time constrained reasoning guarantees a solution within a given processing time (typically defined by the real-time requirements of the system). Solutions may be required at a specific time or they may be required at any time.[1]

Three of the more common methods of time constrained reasoning are depicted in Fig. 2. *Traditional reasoning* will provide a solution to the problem after time *t=1* and no sooner. The algorithm would only be considered to be a real-time algorithm for applications where a solution would not be required prior to time *t=1*. More advanced algorithms enable rule-based reasoning to be performed within a deadline. *Default reasoning* is often used with these algorithms, to ensure that a default solution is available from time *t=0*. At time *t=1*, a better solution has been calculated based on limited input data and at time *t=n*, the algorithm's highest quality solution has been calculated based on all available input data. Most significantly, a solution is available at any time from *t=0*. *Genetic algorithm (GA) based reasoning* differs from rule-based reasoning within a deadline in that all input data is considered at all times with regular incremental improvements made to the solution.

4.2 Soft Sensing

Soft sensing provides a means of inferring parameters that cannot be directly measured due to technological or economic constraints [10]. Known and measurable process parameters are obtained and an unmeasurable parameter inferred primarily using computational intelligence techniques. An example of soft sensing is the determination of diacetyl concentration in beer during fermentation. Diacetyl is an undesirable by-product of brewing fermentation. During fermentation the concentration of diacetyl increases to a peak, and then decreases (it is removed) as fermentation continues. Diacetyl is one of the endpoint indicators for the fermentation process. Fermentation is continued until the diacetyl concentration is below a pre-determined threshold.

[1] Time constrained reasoning can be implemented using real-time intelligent development platforms such as *G2* by the Gensym Corporation.

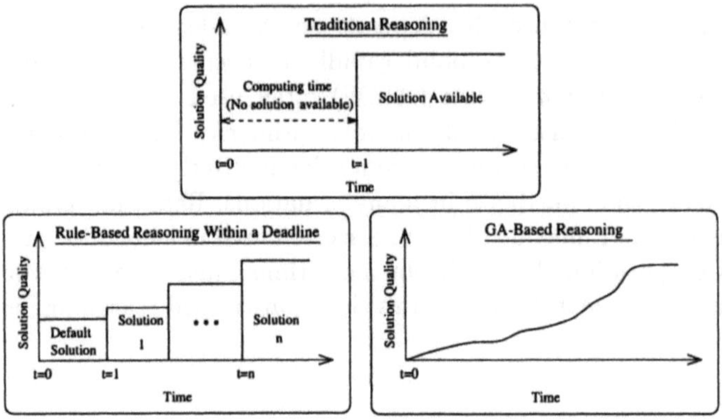

Fig. 2. Time-constrained reasoning

The direct measurement of diacetyl requires off-line laboratory analysis with costly instrumentation. It is not currently feasible to perform direct measurements in real-time using on-line sensors.

One alternative approach is to use a number of easily measured parameters, such as temperature, pressure, gravity, turbidity and pH, to train an ANN-based predictor which is able to predict the instantaneous levels of diacetyl [11]. A similar solution uses a multi-level fuzzy system with a rule-base of 10 rules and only requires real-time input measurements of pH and specific gravity [12].

Soft sensing can also be used to avoid installation and running costs of additional sensors in situations where inferential values are sufficient. [10]

4.3 Sensor Validation

Sensor failures can lead to significant cost burdens due to complete and unplanned process shutdowns or by producing product out of specification. Common modes of sensor failure include [9]:

- **No value** - the sensor is not transmitting a signal
- **Dead value** - the sensor is incorrectly frozen within a narrow margin
- **Abnormal change** - the sensor indicates abnormal fluctuations not consistent with changes in the measured parameter
- **Incongruous readings** - the sensor gives readings which occur within the admissible range but do not reflect the parameter being measured

A number of techniques can be used to validate a sensor reading. Statistical process control (SPC) techniques can be used to detect statistical deviations between actual and expected values from sensors. SPC techniques are useful for detecting problems which present as changes in statistical properties. However, this is very process specific and requires a knowledge of the statistically expected (and permissible) values.

As an alternative, computationally intelligent techniques can be implemented which are able to determine expected values on a non-statistical basis when appropriate. This would be an ideal situation in which to embed knowledge of the process and to determine the regime in which the sensor should be operating. It could also be used to diagnose a sensor fault and possibly advise on appropriate corrective action [13].

There may be an opportunity to infer values from other sensors using some form of knowledge base and to therefore validate a sensor reading. An extended form of this solution could involve a number of intelligent sensor validation nodes. These nodes would interact and validate the corresponding group of sensors thereby providing a degree of redundancy, in a virtual sense, in the system.

4.4 Biotechnology

One challenge that faces biological based manufacturing processes is to measure biological parameters in real-time. Process parameters such as pH, temperature, gravity, oxygen, carbon-dioxide, turbidity etc. can currently be measured using off-the-shelf sensors. Analytical technology such as near infrared (NIR), ultra-violet and visible spectral analysers are currently available and provide a means of profiling particular materials. However, these technologies can be costly, and do not directly provide a breakdown of targeted biological components. They would have to be inferred from the spectral profile and the analysers often require frequent calibration.

Access to real-time measurements of biological components such as protein, carbohydrate, polyphenols, enzymes, and flavour and taste are only just beginning to emerge with developments in biosensor technology. Ideally these sensors should operate on-line, but even if they operate ***at-line***, they will still allow the development of intelligent systems to manage quality and control. Many of these developments are originating in the medical and pharmaceutical sectors, however it is expected that they will flow on into other bio-industries such as food and beverage.

Precise analytic real-time information on particular biological components allows much greater scope for process optimisations and right-first-time practice. Deficiencies in a critical biological parameter which may relate directly to product quality could be detected immediately and acted upon by an intelligent diagnostic and decision support system.

Some of the emerging biosensor technologies are based on amphometric biosensors, molecular recognition, fluorescence, ion channels, plasmon resonance and olfactory models. Combinational chemistry technology will provide novel ligand technologies. Micro-array developments (for molecular biology applications) will continue to expand the market and decrease in cost. [4]

5 Toward an Intelligent Brewery

Brewing is a food and beverage industry. Breweries are typically high volume batch process operations with time-division product and brand processing. They are based on highly traditional practices with many processes still run intuitively. The product involves many biological processes which can not be measured directly in real-time. Soft computing methods and emerging biotechnology developments have now positioned breweries in an ideal position to capitalise on these technologies and to reduce the cost of goods sold and to maintain consistent product quality.

This section presents a brief review of some of the soft computing applications in breweries world-wide followed by a progress report on some developments made at *Carlton & United Breweries, Australia.*

5.1 Soft Computing in Breweries

Maintaining consistent product quality poses many challenges to the brewer. For instance, the seasonal variation of malt represents an often significant biochemical variation. Hence various components of the brewing process have to be adjusted to compensate for these variations. These required adjustments offer many opportunities for soft computing based solutions.

Brewhouse

The brewhouse is one of the major components of the brewery, and is responsible for preparation of the malt into a fermentable extract solution known as wort. An Expert System for Malt Allocation (ESMA) has been developed by Bull *et. al.* [14]. It is a rule-based expert system which is based on the knowledge of experts in the field of malt categorisation and blending control of malt silos.

Fermentation

Fermentation is the process through which fermentable sugars are converted into alcohol. The fermentation process is also responsible for the development of many of the flavour compounds and other important characteristics of the beer. The process is influenced by many variables such as malt quality, yeast characteristics and quantity, temperature and pH. Many computationally intelligent systems have been developed for fermentation based on: neural networks [11][15][16][17][18], fuzzy systems [12][19][20][21] and hybrid neuro-fuzzy systems [22][23]. The majority of these systems are in some way related to state prediction of the fermentation process.

Whitnell *et. al.* devised a multi-stage fuzzy system [12]. This was a two layer system that was comprised of three individual fuzzy predictors. The third fuzzy system accepts the outputs of the other two, to produce a refined prediction of fermentation time.

The Gensym corporation has developed a commercially available package called *G2 Fermentation Expert*. It is an advanced real-time diagnostic tool based on a combination of neural network, fuzzy logic and rule inference technologies to reduce process variability and increase the yield of a fermentation process. [24]

Scheduling

To produce multiple product batches with shared resources, manufacturing schedules can be arranged and sequenced so as to optimise a number of constraints such as minimising changeover costs, maximising the use of available plant capacity and minimising inventory costs. This can be achieved through the development of dynamic scheduling systems using soft computing techniques such as *constraint-directed reasoning*. [25][26]

5.2 Progress Towards an Intelligent Brewery

A factory-wide *virtual brewer* concept has been proposed at *Carlton & United Breweries, Australia* [4]. It is a soft computing based distributed intelligent system that monitors and oversees the process, plant and product quality throughout the factory. It is a hierarchical structure which comprises a series of different types of *expert nodes* (enodes). The system architecture is illustrated in Fig. 3.

Physical expert nodes (p-enodes) are connected to the individual sensors and process control systems. They deal with operational issues of their respective specific process plant. The *intermediate expert nodes* (i-enodes) manage entire functions or areas of the process such as fermentation or filtration. The i-enodes collate, co-ordinate and control the information from the p-enodes and are the interface between the p-enodes and the factory wide, high-level *super expert-node* (s-enode).

Final product quality is one of the more significant process performance measures. An array of *quality expert-nodes* (q-enodes) is responsible for the key quality parameters of the final product. Each q-enode can interact with any other enode to provide a real-time measurement or prediction of the respective quality parameter wherever possible throughout the process. The structure of a typical q-enode is depicted in Fig. 4.

Expert Node Examples

The design and implementation of a system the size of the distributed virtual brewer is a significant undertaking. At *Carlton & United Breweries, Australia*, progress has been made toward the development of a number of

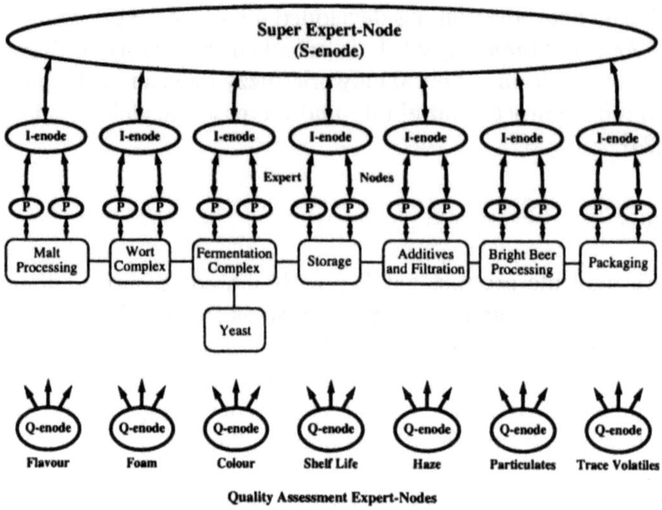

Fig. 3. Topology of the distributed virtual brewer [27]

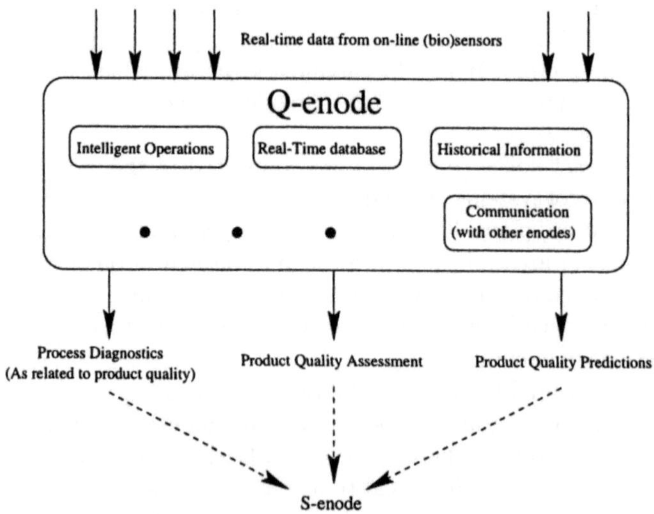

Fig. 4. Quality expert' node [4]

enodes. A three-input, single output, real-time fuzzy controller has been developed for dosage control of the filtration agent on the brewery's primary beer filter [3][28]. This is essentially one of the p-enodes of the filtration i-enode. Knowledge of experts in the field of beer filtration is embedded within the fuzzy controller rule-base. Not only is the controller designed to provide real-time optimisation of the filter agent dosage, but it has also captured and maintained some of the company's intellectual capital in the area of beer filtration.

One of the key components of the brewhouse is the mashing process. This is the process where the complex sugars and starches within the malt are broken down into fermentable sugars. The quantity of sugar that is converted is proportional to the time spent in the mashing process. Ideally, a certain ratio of fermentable and non-fermentable sugars is desired, as a good quality final beer requires some characteristics of both sugar types. The required mashing time is very difficult to determine, and is sensitive to some regular variations in malt characteristics (among other parameters). A neuro-fuzzy expert system was developed which could 'learn' what the required mash stand-time would be [4][27]. Progress has also been made on the infrastructure and operational platform of the *Virtual Brewer*. Experimental evaluation of the *G2 Intelligent Operations Management System* has been carried out in terms of its suitability for this purpose. It is a robust, real-time, intelligent system development and deployment platform. It has a wide range of connectivity capabilities to enable interfacing with existing plant PLC, plant control systems, databases and corporate applications. [4]

6 Conclusion

The modern brewery in today's highly competitive and global environment must be able to produce with minimal cost and consistent quality. Not only do they have to maintain market share but, where applicable, shareholders expect good returns for their investment in the organisation. More stringent workplace safety and environmental compatibility requirements must also be adhered to. It is being recognised by more and more industries that these issues can only be addressed through plant-wide information sharing and optimisations. Developments in data communication technologies have provided the necessary physical layers in data transport. Increasingly, soft computing techniques are being used to provide intelligent control, diagnostic and decision support systems. Computationally intelligent methods are being used with a major emphasis being placed on knowledge management.

Many industries have already made substantial progress in these areas. Real-time performance of these systems is essential. One of the difficulties encountered by many industries is the ability to measure product parameters on-line (or at-line) and in real-time. For instance, the food and beverage industries would benefit greatly from the ability to measure many different biological parameters on-line. To date, this has largely not been possible. Recent biotechnological developments emerging from the medical and pharmaceutical industries have produced biosensors which have direct application in the brewery. Also, developments in immersive sensor technology in the form of micromachines would allow real-time sensing in-process and could measure beer process parameters in situ.

Brewing is one beverage industry which is now beginning to seriously adopt soft computing, real-time sensing and measurement, and information processing techniques in the factory and is now well poised to gain from all of the inherent benefits.

7 Acknowledgements

The authors wish to thank Dr Peter Rogers (team leader - BrewTech, Carlton and United Breweries, Australia), Dr Kerry Gutowski (team leader) and Mr Adrian Gardner (Vice President) for their vision and commitment to the intelligent brewery. The authors acknowledge the technical contributions made to various intelligent processes within the breweries by Mr Michael Pecar and Dr David Sudarmana (senior process engineers).

References

1. International Society for Measurement and Control (ISA), "SP95, Enterprise/Control Integration Committee." http://www.isa.org/sc/committee/1,1512,145,00.html, June 2000.
2. G. Bellinger, "Knowledge Management - Emerging Perspectives." http://www.outsights.com/systems/kmgmt/kmgmt.htm.
3. D. Campbell, M. Pecar, and M. Lees, "Intelligently Controlled Beer Filtration," *Proc. Second International Workshop on Intelligent Control*, vol. 1, pp. 313–316, Durham, USA 1998.
4. P. Rogers, M. Lees, D. Campbell, D. Sudarmana, and M. Pecar, "The Development of Assessment and Control Systems for the Brewery based on Real-Time Measurement of Biological Parameters and Expert System Technology," *Master Brewers' Association of the Americas Technical Quarterly*, vol. 37, no. 2, pp. 183–198, 2000.
5. J. Durkin, *Expert Systems - Design and Development.* Prentice Hall, New Jersey, 1994.
6. K. Crockett, Z. Bandar, and A. Al-Attar, "Fuzzy Rule Induction from Data Sets," *Proceedings of the 10th Annual Florida Artificial Intelligence International Conference (FLAIRS 97)*, pp. 332–336, May 1997.
7. J.-S. R. Jang, C.-T. Sun, and E. Mizutani, *Neuro-Fuzzy and Soft Computing.* Prentice Hall, New Jersey, 1997.
8. J. Giarratano and G. Riley, *Expert Systems - Principles and Programming.* PWS Publishing Company, Boston, 1998.
9. J. Soto, A. Ayerbe, and M. Alejo, "Real Time Intelligent Systems." Presented at Expert Systems '91 Avignon France, May 1991.
10. K. Fouhy, "Optimization goes Enterprise-wide," *Chemical Engineering*, pp. 153–156, April 2000.
11. V. Denk, "New Method of Online-Determination of Diacetyl in Real Time by means of a "Software-Sensor"," *Presentation at the J. De Clerk Chair VII in Leuven/Belgium*, pp. 30–35, September 1996.

12. G. Whitnell, V. Davidson, R. Brown, and G. Hayward, "Fuzzy Predictor for Fermentation Time in a Commercial Brewery," *Computers chem. Engng*, vol. 17, no. 10, pp. 1025–1029, 1993.
13. G. Stanley, "Experiences Using Knowledge-Based Reasoning in Online Control Systems," *Proceedings of International Federation of Automatic Control (IFAC) Symposium on Computer Aided Design in Control Systems*, pp. 11–14, July 1991.
14. H. Bull, M. Lorrimer-Roberts, C. Pulford, N. Shadbolt, W. Smith, and P. Sunderland, "Knowledge Engineering in the Brewing Industry," *Ferment*, vol. 8, pp. 49–54, February 1995.
15. V. Breusegem, J. Thibault, and A. Cheruy, "Adaptive Neural Models for Online Prediction in Fermentation," *The Canadian Journal of Chemical Engineering*, vol. 69, pp. 481–487, April 1991.
16. T. D'Amore, G. Celotto, G. Austin, and G. Stewart, "Neural Network Modeling: Applications to Brewing Fermentations," *EBC Congress*, pp. 221–230, 1993.
17. L. Garcia, F. Argueso, A. Garcia, and M. Diaz, "Application of Neural Networks for Controlling and Predicting Quality Parameters in Beer Fermentation," *Journal of Industrial Microbiology*, vol. 15, no. 5, pp. 401–406, 1995.
18. G. Gvazdaitis, S. Beil, U. Kreibaum, R. Simutis, I. Havlik, M. Dors, F. Schneider, and A. Lubbert, "Temperature Control in Fermenters: Application of Neural Nets and Feedback Control in Breweries," *J. Inst. Brew.*, vol. 100, pp. 99–104, March-April 1994.
19. B. Postlethwaite, "A Fuzzy State Estimator for Fed-Batch Fermentation," *Chem. Eng. Res. Des.*, vol. 67, pp. 267–272, 1989.
20. S. Vassileva, V. Huong, and J. Votruba, "An Expert System Applied to the Physiological Analysis of Early Stage of Beer Fermentation," *Folia. Microbiol.*, vol. 39, no. 6, pp. 489–492, 1994.
21. C. Venkateswarlu and K. Gangiah, "Fuzzy Modeling and Control of Batch Beer Fermentation," *Chem. Eng. Comm.*, vol. 138, pp. 89–111, 1995.
22. R. Simutis, I. Havlik, and A. Lubbert, "Process State Estimation and Prediction in a Production-Scale Beer Fermentation using Fuzzy Aided Extended Kalman Filter and Neural Networks," *IFAC Modelling and Control of Technical Processes*, pp. 95–100, 1992.
23. R. Simutis, I. Havlik, and A. Lubbert, "Fuzzy-Aided Neural Network for Real-Time State Estimation and Process Prediction in the Alcohol Formation Step of Production-Scale Beer Brewing," *Journal of Biotechnology*, vol. 27, no. 2, pp. 203–215, 1993.
24. Gensym Corporation, "G2 Fermentation Expert." http://www.gensym.com/expert_operations/products/FermentationExpert.htm.
25. H. Rosenof, "Dynamic Scheduling for a Brewery," *World Batch Forum*, May 1995.
26. H. Rosenof, "How to Organise a Schedule in a Brewery," *Expert Systems Applications*, vol. 11, no. 11, pp. 10–12.
27. M. Lees, P. Rogers, D. Campbell, M. Pecar, and D. Sudarmana, "Intelligent Systems for the Brewery based on Real-Time Measurement of Biological Parameters," *Proceedings of the 9th Australian Barley Technical Symposium*, pp. 2.8.1–2.8.4, September 1999.

28. M. Pecar, M. Lees, and D. Campbell, "An Alternative Control Strategy for D.E. Dosing Rates of Primary Beer Filtration," *27th Australian and New Zealand Chemical Engineering Conference CHEMECA'99*, pp. 546–551, September 1999.

The Aggregation of Industrial Performance Information by the Choquet Fuzzy Integral

L. Berrah[(1)] , G. Mauris[(2)] and L. Foulloy[(2)]
[(1)]Laboratoire de Logiciels pour la Productique LLP
[(2)]Laboratoire d'Automatique et de Micro-Informatique Industrielle LAMII
CESALP – Université de Savoie – 41, Avenue de la Plaine, 74016 Annecy, France
E-mail:mauris@esia.univ-savoie.fr

Abstract: This study deals with the aggregation of industrial performance information, i.e. with the mechanism which allows the computation of a global performance knowing the partial ones. The performance information is aimed at controlling the production activity, by indicating how the real objective is reached. The characteristics of this kind of aggregation is that the partial performances to be aggregated are, on the one hand, often information of heterogeneous nature (dimension, format). On the other hand, they are associated to diversified and numerous objectives which interacts in different ways (redundant, complementary, ...). In this sense, the fuzzy subset theory provides tools to deal with: the heterogeneity of the entities involved, the commensurability of the partial performances expressed in the interval [0,1], and the different behaviors of the aggregation operation (compromise effect, optimistic or pessimistic effect, ...). Among all the fuzzy aggregation operators, we consider here, as an illustration, one use of the Choquet fuzzy integral family for modeling the different interactions between the objectives and aggregating their associated performances.

1 Introduction - Context

In the post-taylorian manufacturing enterprises, the traditional financial performance expression is out of step. Indeed, these enterprises not only have to reduce costs, but they also have to continuously improve the quality, the delivery and now the services around their products. To remedy the inadequacies of the information cost system, they turn towards a balanced representation of both financial and technical performance information [Kaplan 96] (fig.1), gathered in the so-called scorecards. The latter complement the former by taking all performance criteria into account, such as customer satisfaction, productivity, ecology, safety, ... Performance indicators are a means to this purpose: they provide the necessary information to compare measures of the enactment of the considered activity with a pre-set objective and to measure the extent of any drift [Fortuin 88] [Kaplan 92].

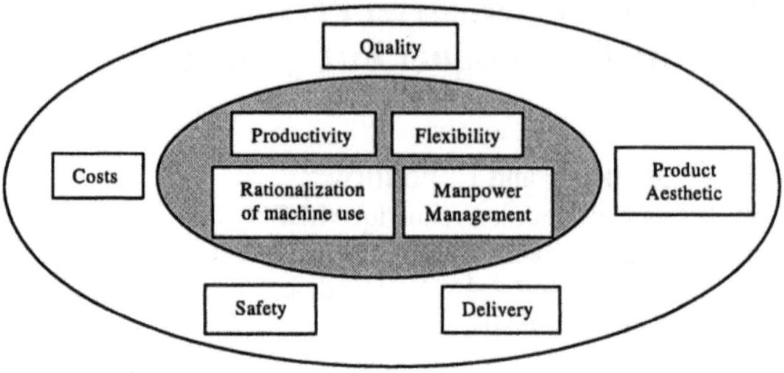

Figure 1. Example of post-Taylorian industrial performance criteria.

Moreover, in addition to this **multi-criteria** aspect, the performance structure is today needed to be essentially a support for a new vision of the control; while in the conventional systems, performance measures were only used as a way to verify the productivity of the resources and to assess the workmanship. In this sense, the new structure of performance indicators must give information which is able to indicate quickly how sound the considered activities or processes are; hence allowing a more reactive analysis than the one provided by financial reports. Indeed, whenever one drift is detected by one performance indicator, supervisors can immediately react, rather than passively wait for the subsequent financial information.

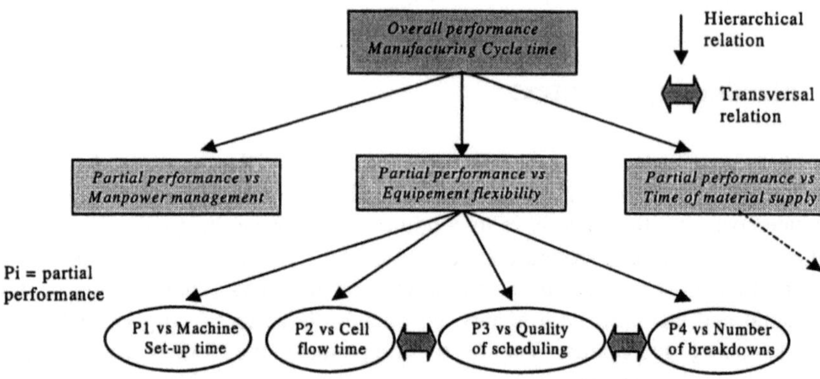

Figure 2. One example of relationship between performance indicators.

To be more precise, in order to be a support for a reactive control, the whole set of performance indicators has to be defined into a **system**, according to the set of enterprise objectives. Generally, a performance indicator system is structured in hierarchical and transversal relations, for example as seen in fig. 2. To benefit from both types of relation, **aggregation** procedures are required.

This paper focuses on this key concept of performance information aggregation. From a general point of view, this kind of aggregation can be seen as a function, applied to a set of partial performance information, and which provides a single overall performance information [Grabisch 98].

In the industrial case, the partial performances are not always expressed **homogeneously** due to the different characteristics of the different entities involved in the performance computation. In fact, they can be expressed in different units, e.g. in *second* for *cell flow time* , in *percent of default products* for *quality*. Moreover, measures and objectives can be defined in an uncertain or/and imprecise way (e.g. with regard to the *machine set-up time* which can be represented by a probability distribution, or with regard to the *assembly duration* which can be expressed as being *short*), and even in a linguistic way when human operators are involved (e.g. with regard to the *product aesthetic*). Therefore, new tools are required to homogeneously deal with these characteristics of information. In this work, according to the nature of the considered problems, the fuzzy formalism is used to deal with numeric, linguistic, imprecise and uncertain aspects of information [Zadeh 65].

Furthermore, the objectives associated to performance indicators are often not **independent**, but **interacting** (which encompasses redundancy and complementarity behaviors) and even **conflicting** (concerning e.g. *delivery* and *cost*). Moreover, trade-offs between their satisfaction, generally made by using the weighted arithmetic mean [Globerson 85], are no longer sufficient due in particular to **veto** effect of a particular objective (e.g. the *safety* objective must be satisfied at a minimal level under which the overall performance cannot be increased by other partial performances). For this purpose, the Choquet fuzzy integral family, by its capability of modeling various aggregation behaviors, provides interesting alternatives to the weighted mean for aggregating objective satisfactions [Marichal 99], as we will see later.

In section II, characteristics of the aggregation of industrial performance information are considered. A fuzzy representation of information involved in performance indicators is developed in the view of their combination in a single overall value. The definition and the main properties of the Choquet fuzzy integral are presented in section III. The focus is both on the modeling of objective importance and interaction. Applications to the industrial performance aggregation are illustrated by examples.

2 The Aggregation of Industrial Performance Information

2.1. The partial performance expression

The performance information is aimed at helping to determine the actions to apply in order to control the production activities in an efficient and reactive way. From a global point of view, its expression means the expression of the realization of the assigned objectives. This role is devoted in industrial companies to so called *performance indicators* which generate the performance information from the measures acquired on the processes [Fortuin 88]. The measure m can be a simple physical entity acquired under the form of a single value (e.g. *number of default products*, *cell flow time*) or obtained by the processing of several measure acquisitions (e.g. *mean number of default products per week*, *mean time of cell flow per day*).

More precisely, the performance expression returned by the indicators results from the comparison between the objective and the measure describing the effective process or activity enactment. Hence, the performance expression can be formalized by the following mapping [Berrah 00]:

$$P: \quad OxM \rightarrow E$$

O, M, E being respectively the sets of the objectives o, of the measures m and of the performance P ($P(o,m)$ denotes the value of the performance evaluated from o and m). In this work, o and m being represented in the fuzzy formalism, the building of the mapping P is also issued from this formalism. The mapping P identifies a fuzzy comparison operator such as the distance ones (e.g. the min-distance, the Haussdorff distance), the similarity ones (e.g. equality operator, intersection operator), ... [Dubois 82]. Naturally, the semantic of the performance expression depends on the semantic of the chosen operator.

Example: Let us consider one objective related to the *delivery*, particularly the duration of one assembly task. The figure 3 illustrates one example of performance expression, concerning the vague objective o:: *duration of the assembly task* = *short*, being represented by the fuzzy subset - called the meaning of the term *short* [Zadeh 71] - $M(short) = [8;10;12;15]$. Let m be the fuzzy measure *around* 14 min. One manner to define the mapping P is to consider an inclusion matching between o and m, by computing the necessity degree [Dubois 88]: $N(o;\ m) = \inf_u \max(1-\mu_m(u), \mu_o(u))$, which is equal to 1 only if $supp(m) \subset ker(o)$. Otherwise, the performance is less than 1, e.g. $P(o,m) = N(o;m) = 0.25$ in the situation considered in fig.3 (i.e. the performance with regard to *duration of the assembly task* , and in a necessity semantic vue, is 0.25).

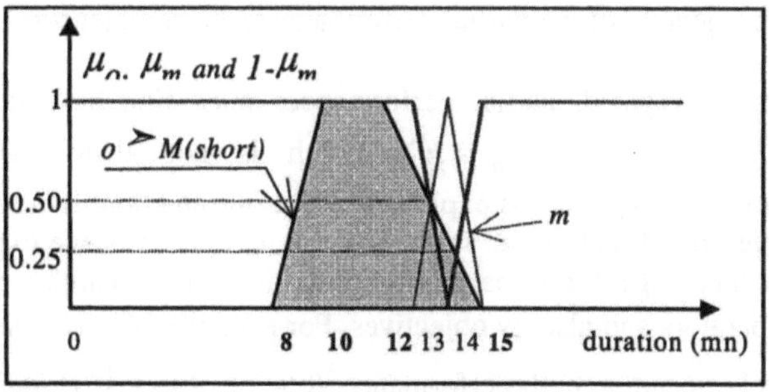

Figure 3. One example of partial performance expression.

- Note: Other situations than the one described in example 1 can arise. Objectives can be flexible, i.e. with gradual satisfaction. This satisfaction or performance function can be defined by a fuzzy subset [Bellmann 70]; the higher the objective membership function value of a measure is, the higher the performance for this measure is. Moreover, in other respects, in particular for objectives being qualitative in nature (e.g. aesthetic ones), a linguistic expression can be useful, e.g. by the terms *low*, *medium*, *high*. This kind of linguistic information can also be treated by fuzzy tools in the same spirit as the numeric case illustrated in example 1 [Berrah 00]. Here, for the sake of conciseness, these situations will not be considered.

Before considering mathematical tools able to deal with the problem of industrial performance information aggregation, it is useful to emphasize first the corresponding requirements.

2.2. The requirements of the industrial performance aggregation

The aggregation of information can be formalized by the following mapping:

$$Ag: \quad E_1 \times E_2 \times ... E_n \rightarrow E$$

E_i being respectively the sets of the elementary information I_i and E of the aggregated information denoted I_{ag}. Regarding to the information involved in the performance indicators, I_i can be the measure or the partial performance, and I_{ag} can be an aggregated measure or an aggregated performance. Here, we will only consider the case where the elementary information is identified to the partial performance, and we will detail the industrial requirements associated to such an aggregation approach, in particular the problem of the heterogeneity of the partial performances, and of the interactions between them.

2.2.1. The heterogeneity of the partial performances: the commensurability notion

In order to be aggregated, partial performances must first have homogeneous expressions ($E_1 = E_2 = ... = E_n = E$), which is not always the case in performance indicator systems as explained in the introduction. In addition, as we said before, even if defined on the same universe, the performance expression can have different meanings due in particular to the different semantics conveyed by comparison operators and also by objectives. For example, P :: *cell flow time* = 0 can respectively mean a null performance (e.g. if $P(o,m)$ is defined by an equality operator), or a total one (e.g. if $P(o,m)$ is defined by a distance operator). These aspects of performance dimension and meaningfulness became identified in the literature with the so-called **commensurability** concept [Grabisch 00]. More formally, this concept can be perceived by the following mapping R :

$$R: \quad \Theta x \Theta \rightarrow E$$

Θ and E being respectively the sets of objectives and of the performance. The mapping R is such that $R(o,o') = 1$ if and only if satisfaction of o and o' are expressed on the same scale type (ordinal, interval, ...). Here, this scale will be an ordinal one defined on $E = [0,1]$, and we will merely make the assumption that the *partial performances have been well acquired in a commensurate way*. The validation of this assumption is related in particular to the choice of the comparison operator for each performance indicator, which is beyond the scope of this paper.

2.2.2. The partial performance interaction

In order to choose the adequate aggregation operator, the problem of performance interaction can be summarized in the following question: *at a given level, what kind of interaction is there between the partial performances?*

For example, considering again the case described in fig. 2, this question becomes: how do *machine set-up time*, *cell flow time* and *quality of scheduling* interact?

In fact, performance indicators are constituted from the three elements [Berrah 00]: objectives, measures, action variables, according to the *control loop* principle. Interactions can be considered for each element. Nevertheless, in this paper, the focus is only on objective interactions through which all the other can be considered. In this sense, we consider that:

$$interaction(partial\ performances) = interaction(objectives).$$

From this point of view, many situations can be distinguished.

- The objectives are completely non-interactive or **independent**, such as e.g. $o{::}\ absenteeism = less\ than\ 5\%$ and $o'{::}\ work - in - process\ level = low$.
- The objectives are **interactive**. One kind of interaction concerns the ***conflict*** notion, such as e.g. $o{::}\ quality\ level\ of\ the\ products = high$ and $o'{::}\ manufacturing\ costs = low$. From another point of view, these objectives are partially *complementary*, i.e. simultaneous satisfactions affect the aggregated performance more significantly than separate satisfactions.
- Another kind of objective interaction is ***redundancy***, i.e. objectives are in some way more or less partially interchangeable because respective performances evolve in the same way, such as e.g. the objectives related to *flexibility* and *reactivity*. Moreover, the redundancy can be due to the use of the same measure in the expression of the performance, e.g. the *machine time avaibility* is involved both for *flexibility* and *reactivity*.
- One objective has a **veto** effect, i.e. the aggregated performance cannot be higher than the partial one associated to this objective, the *safety* objective is often such a one.

An important issue is to have tools able to model and methods to acquire in an quantitative way such interaction information between objectives. Before making propositions for handling such cases, let us consider how performance aggregation is really treated in industrial situations.

2.2.3. The real aggregation practice

In manufacturing enterprises where indicators remain essentially financial ones, the aggregation is simple because it generally only consists in arithmetic operations. Since the recent development of new indicator structures, the Analytic Hierarchy Process method AHP [Saaty 77], widely used in enterprise in decision problems, has been applied to performance context, in particular for aggregation purposes [Bititici 95][Rangone 96]. This method includes the following steps:

- develop a hierarchical structure in terms of overall objectives, sub-objectives, ..., and measures of their satisfaction,
- determine on a pair-wise comparison basis, the relative priorities of objectives, that express their importance in relation to the higher level objective,
- compute the aggregated performance, weighting partial performances with their relative importance, i.e. computing a weighted arithmetic mean.

AHP allows to consider financial, non financial, quantitative and qualitative measures and trade-offs among them taking importance weights into account. But, for an effective application of the standard[1] AHP, it is important for the hierarchical structure to include only objectives that are independent and additive. As explained before, these two points are not always satisfied in practice, because independence implies no redundancy, no conflict, no complementarity, and the weighted mean is not able to identify the veto effect. For overcoming these AHP limitations, rather than trying to build independent and additive hierarchical structure, the Choquet fuzzy integral family, by its capability to represent objective interaction, is an interesting tool presented below.

3 Application: The Choquet Fuzzy Integral

3.1. Introduction

The aggregation operators based on the Choquet fuzzy integral family encompass a lot of operators (weighted mean, min, max, median, OWA) [Grabisch 95]. One advantage of this family is that it provides a unified mathematical framework for all the operators mentioned before. Thus, the identification of the right operators for the aggregation in a specific context can be made in a more methodological way than trying successively all these different operators. In particular, the Choquet fuzzy integral allows to integrate knowledge about the importance of partial information and of their interactions, and also knowledge about the information combination, i.e. different behaviors of the aggregation operators in relation to the decision maker (severe, compromise, tolerant) [Grabisch 95]. For the sake of conciseness, mathematical developments will be restricted to the minimum (further developments can be found in [Grabisch 00]), and the Choquet fuzzy integral is only presented here through our practical case.

3.2. Description of an industrial example

Let us consider again the example given in fig. 2. To obtain the aggregated performance P_{ag} related to the *equipment flexibility* objective (o_{ag}), partial performances of: *machine set-up time* P_1 (related to o_1), *cell flow time* P_2 (o_2), *quality of sheduling* P_3 (o_3), *number of breakdowns* P_4 (o_4), must be aggregated. A first simple idea would be to compute the mean of these partial performances. But this approach is limited, because it does not take into account the relations between the objectives as encountered in the considered example.

[1] Modified AHP seems to partly take into account redundancy and non-additivity [Rangone 96].

Two situations are described hereafter to put in light the interactions to consider (case a), and also an eventual veto behavior for the aggregation operator (case b).

Case a

The different interactions to be considered are the following:

- redundancy between o_2 and o_4, knowing that the *number of breakdowns* and the *cell flow time* partly express the same thing, the *cell flow time* is low when the *number of breakdowns* is low and vice-versa,
- partial complementarity between o_2, o_4 and o_3, knowing that simultaneous high performances for them provide a higher improvement of the aggregated performance than simultaneous high performances for o_1 and o_3,
- independence of o_1, since the *machine set-up time* is inherent only to the characteristics of machines.

Case b

The machines have been changed. Thus *machine set-up time* is negligible. Moreover, o_2 is no more redundant with o_4, the *number of breakdowns* being no more related to the speed of the machines and then to the *cell flow time*. In fact, the *quality of sheduling* o_3 must be satisfied in order to satisfy the global objective, i.e. o_3 is now a veto objective. This veto effect cannot be modeled by the weighted mean.

3.3. The conventional Choquet fuzzy integral representation

In order to have a representation of interaction between the objectives, the Choquet fuzzy integral concept is:

- first, to affect a weight, not only to each objective, but also on each subset of partial objectives thus defining a fuzzy measure g (or Choquet capacity), defined on the power set $\mathcal{P}(O)$ of the set of objectives $O = \{o_1, o_2, \ldots, o_n\}$ by:

$$g: \quad P(O) \to [0,1]$$

with $g(\varnothing)=0$ and $g(O)=1$ and $g(S)\leq g(T)$ whenever $S\subseteq T$.

- second, to aggregate the partial performances P_i's using g by the Choquet formula:

  $CI_g(P_1,P_2,...,P_n)=\sum_{i=1}^{n}P_{(i)}[g(\{O_i\})-g(\{O_{i+1}\})]$, where the notation $P_{(i)}$ indicates a permutation on the partial P_i's such that $P_{(1)}\leq....\leq P_{(i)}\leq...\leq P_{(n)}$ and $O_i=\{o_{(i)},o_{(i+1)},...,o_{(n)}\}$ with $O_{(n+1)}=\varnothing$ (the objectives being associated to their corresponding performance in a one-to-one way, they are permuted and ranked in the same way as the performances). When the fuzzy measure g is additive, i.e. $g(S\cup T)=g(S)+g(T)$ whenever $S\cap T=\varnothing$, thus the associated Choquet integral becomes the classical weighted arithmetic mean $CI_g(P_1,P_2,...,P_n)=\sum_{i=1}^{n}P_i g(\{o_i\})$.

Nevertheless, the application of the Choquet formula requires the determination of the values of g, generally from an expert. This aspect creates difficulties. Indeed, it seems to be difficult to get an intuitive understanding of the meaning of $g(S)$, because $g(\{o_i\})$ defines the importance of the objectives o_i considered **on their own**, and not the global importance of this particular objective o_i in the aggregated performance. For example, $g(\{o_2\})=0.1$ and $g(\{o_2,o_3\})=0.5$ express the fact that, though $g(\{o_2\})=0.1$ is low, o_2 is nevertheless important when associated with o_3. To overcome the problems of the conventional representation, the consideration of a specific Choquet integral based on the 2-additive measure notion has been introduced by [Grabisch 97].

3.4. The 2-additive measure

The notion of 2-additive is based on two types of indice [Grabisch 97]:

- the global importances of each objective in relation to all the others by the so-called *Shapley indices* v_i's, which satisfy $\sum_{i=1}^{n}v_i=1$, which is a natural condition for experts,

- the *interaction indices* $I(S)'s$ of any subset of objectives, which range in $[-1,1]$; a value of 1 means positive synergy, a value of -1 means negative synergy and a value of 0 means no interaction.

The relations of these coefficients with the fuzzy measure g are not of interest here, but can be found in [Grabisch 97]. Moreover, this representation gives at the practical level a way to reduce the exponential complexity of the fuzzy measure. One has to define the real numbers $g(S)$ for each subset S of the set of objectives, i.e. 2^n coefficients, by considering only interactions between pairs of objectives (o_i, o_j) denoted I_{ij}. For this kind of measure, the associated Choquet integral is simplified in:

$$CI_g(P_1, P_2, ..., P_n) = \sum_{I_{ij} \succ 0} \min(P_i, P_j) I_{ij} + \sum_{I_{ij} \prec 0} \max(P_i, P_j) \left| I_{ij} \right| + \sum_{i=1}^{n} P_i (v_i - \frac{1}{2} \sum_{j \neq i} \left| I_{ij} \right|)$$

with the property that $(v_i - \frac{1}{2} \sum_{i \neq j} \left| I_{ij} \right|) \geq 0$.

Thus, this Choquet fuzzy integral is decomposed in a conjunctive, a disjunctive, and an additive parts:

- a positive I_{ij} implies that the simultaneous satisfaction of objectives o_i and o_j is significant for the aggregated performance, but a unilateral satisfaction has no effect,
- a negative I_{ij} implies that the satisfaction of either o_i or o_j is sufficient to have a significant effect on the aggregated performance,
- a null I_{ij} implies that no interaction exists; thus v_i acts as the weights in a classical weighted mean.

This makes the meaning of v_i and I_{ij} clearer, by explaining to experts how these coefficients influence the aggregated performance. Indeed, the v_i values are the weights of an aggregation by the weighted mean when no interaction exists, I_{ij} values performing second order corrections on this weighted mean.

In other respects, this interaction representation of the Choquet fuzzy integral is able to identify the veto effect of an objective o_i, i.e. if the partial performance on objective o_i is high, it has no effect on the aggregated performance, but if it becomes low, the aggregated performance will become low too, whatever the other partial performances are. This is translated by the property: $CI_{g_{veto}}(P_1,\ldots,P_{i-1},P_i,P_{i+1},\ldots,P_n) = \min(P_i, CI_g(P_1,\ldots,P_{i-1},P_{i+1},\ldots,P_n))$.

This property is satisfied for a veto objective o_i if and only if [Grabisch 97]:

$I_{ik} \geq 0 \quad \forall k \neq i$, $I_{kl} = 0 \quad \forall k,l \neq i$, and $v_i = \frac{1}{2} I_{ik} \quad \forall k \neq i$.

3.5. Application to the industrial example

Let us now consider the use of the 2-additive measure for the two cases considered before.

Case a

Let us continue with the industrial example considered, according to the interactions mentioned in the paragraph 3.2., coefficients can be defined by:

$$v_1 = 0.1;\ v_2 = 0.2;\ v_3 = 0.5;\ v_4 = 0.2,$$
$$I_{24} = -0.1;\ I_{23} = 0.2;\ I_{34} = 0.2;\ I_{1k} = 0.$$

They express that o_2 and o_4 are redundant, o_2 and o_3 are complementary, and that o_1 is independent. Obviously, o_3 and o_4 are complementary since o_2 and o_4 are redundant, which explains that $I_{34} = I_{23}$. Then, the Choquet fuzzy integral becomes:

$$CI_g(P_1,P_2,P_3,P_4) = 0.2\min(P_2,P_3) + 0.2\min(P_3,P_4) + 0.1\max(P_2,P_4)$$
$$+ 0.1P_1 + 0.05P_2 + 0.3P_3 + 0.05P_4$$

If the following values of partial performances are considered:

$$P_1 = 0.30,\ P_2 = 0.60,\ P_3 = 0.70,\ P_4 = 0.50, \text{ then } P_{ag} = 0.57,$$

which is coherent to the intuitive judgment of the expert. The objective has a little influence on the global performance, which is therefore close to a weighted mean of the other performances (the redundancy effect being partially compensated by the complementarity effect).

Case b

Now, *machine set-up time* is negligible. Moreover, o_2 is no more redundant with o_4, *number of breakdowns* being no more related to the speed of the machines and then to *cell flow time*. So,

$$v_2 = 0.25,\ v_3 = 0.5,\ v_4 = 0.25,$$
$$I_{24} = 0,\ I_{23} = 0.5,\ I_{34} = 0.5.$$

These values satisfy the preceding requirements and therefore o_3 is now a veto objective. The associated Choquet fuzzy integral becomes:

$$CI_g(P_2, P_3, P_4) = 0.5\min(P_2, P_3) + 0.5\min(P_3, P_4).$$

For the following values of partial performances:

$$P_2 = 0.8,\ P_3 = 0.2;\ P_4 = 0.7,\ \text{then}\ P_{ag} = 0.2,$$

$$\text{for}\ P_2 = 0.8;\ P_3 = 0.9;\ P_4 = 0.7,\ \text{then}\ P_{ag} = 0.75,$$

which is coherent with the veto effect.

4 Conclusion

This paper has presented a brief analysis of the evolution of the industrial performance information, in which are involved not only the traditional financial criteria but also technical ones. More particularly, this new expression has led to new requirements in terms of aggregation. Indeed, the performance criteria and thus the associated objectives being numerous and diversified, it has been noted that the real conventional aggregation practice, based on the weighted arithmetic mean, is not capable to deal with all the new encountered situations. Besides, the weighted arithmetic mean, on the one hand, assumes the independence of the criteria and, on the other hand, identifies a compromise effect. Today, in up-to-date performance systems, we have to deal with other effects such as redundancy, conflict, veto, ...

In this view, the Choquet fuzzy integral provides interesting issues for modeling redundancy, complementarity or veto effect between objectives, as illustrated on some industrial examples. In this sense, two concepts found the basis of this tool: the relative importance of the different objectives and the interactions between them. As an application, we have considered here only the interactions by pairs of objectives, leading thus to a more simple expression of the Choquet formula, i.e. the 2-additive measure.

But, from a global point of view, this approach requires the determination of global importance and interaction coefficients for each objective, which is not always easy for an expert. A further study of the interpretation of these coefficients in the industrial context is now being led. Moreover, the Choquet fuzzy integral approach is based on the assumption of commensurability of partial performance information, and leads to a unique global performance expression. Further work will now be carried out in order to determine the conditions to fulfill for the validation of these assumptions at the practical level. This leads to a more general thinking concerning the definition of the aggregation. Today, the aggregation is built as a combination mapping of partial performances. Why, do not we build it as the converse mapping of the objective decomposition?

References

[Bellman 70] Bellman, R. and L.A. Zadeh, *Decision-making in a fuzzy environment*, Management Sciences 17, 1970, pp. B141-B164.

[Berrah 00] Berrah, L., G. Mauris, L. Foulloy and A. Haurat, *Global vision and performance indicators for an industrial improvement approach*, Computers in Industry, to be published in 2000.

[Bititci 95] Bititci, S., *Modelling of performance measurement systems in manufacturing entreprises*, Int. Journal of Production Economics, Vol. 42, 1995, pp. 137-147.

[Dubois 82] D. Dubois and H. Prade, *A unifying view of comparison indices*, in Fuzzy sets and possibility theory recent advances, eds. Pergamon Press, 1982, pp. 3-13.

[Dubois 88] Dubois, D., H. Prade and C. Testemale, *Weighted fuzzy pattern matching*, Fuzzy Sets and Systems 28, 1988, pp. 313-331.

[Fortuin 88] Fortuin, L., *Performance indicators - why, where and how?*, European Journal of Operational Research, Vol. 34, 1988, pp. 1-9.

[Globerson 85] Globerson, S., *Issues in developing a performance criteria system for an organization*, Int. Journal of Production Research, Vol. 23, No 4, 1985, pp. 639-646.

[Grabisch 95] Grabisch, M., *On equivalence classes of fuzzy connectives – The case of fuzzy integrals*, IEEE Trans. On Fuzzy Systems, Vol. 3, No 1, 1995, pp. 96-109.

[Grabisch 97] Grabisch, M., *Alternative representations of discrete fuzzy measures for decision making*, Int. Journal of Uncertainty, Fuzziness and Knowledge-based Systems, No 5, 1997, pp. 587-607.

[Grabisch 00] Grabisch, M., Murofushi, T. and M. Sugeno, Fuzzy measures and integrals: theory and applications, Physica-Verlag, 2000.

[Kaplan 92] Kaplan, R.S. and D.P. Norton, *The balanced scorecard - measures that drive performance*, Harvard Business Review, Jan/Feb 1992, pp. 71-79.

[Kaplan 96] Kaplan, R.S. and D.P. Norton, *Using the balanced scorecard as a strategic management*, Harvard Business Review, Jan/Feb 1996, pp. 77-85.

[Marichal 99] Marichal, J.L., *Agrégation de critères interactifs au moyen de l'intégrale de Choquet discrète*, Rencontres francoph. sur la logique floue et ses applications LFA'99, Valenciennes, France, 1999, pp. 130-140.

[Rangone 96] Rangone, A., *An analytical hierarchy process framework for comparing the overall performance of manufacturing departments*, Int. Journal of Operations and Production Management, Vol. 16, No 8, 1996, pp. 104-119.

[Saaty 77] Saaty, T., *A scaling method for priorities in hierarchical strutures*, Journal of Mathematical Psychology, Vol. 15, 1977, pp. 234-281.

[Zadeh 65] Zadeh, L.A., *Fuzzy sets*, Information and Control, Vol. 8, 1965, pp. 338-353.

[Zadeh 71] Zadeh L.A., *Quantitative fuzzy semantics*, Information Sciences, 3, 1971, pp. 159-171.

Computing Image with an Analog Circuit Inspired by the Outer Retinal Network

Seiji Kameda and Tetsuya Yagi

Kyushu Institute of Technology, Iizuka, Fukuoka 820-8502, Japan

Keywords: early vision, standard regularization theory, analog circuit, image processing, parallel computation, silicon retina

1 Introduction

The response of neurons of the visual system is much slower, by several orders of magnitude, than that of semiconductor devices. Frogs can, however, visually follow insects in real time and catch them. Such superior function is realized by a unique algorithm/architecture of visual systems of creatures, which provides engineers an important insight to develop a novel image processing systems.

The retina is a part of the central nervous system in the vertebrate and computes the image with its electrical properties of neuronal circuit. Therefore electric circuits models are useful to understand the function of the retina in an early stage of visual information processing. On the other hand, it has been demonstrated that some of early vision problems can be formulated in terms of the standard regularization theory [18,19]. This computational framework is a useful top-down approach to understand the function of the retina. These bottom-up and top-down approaches are closely related since the electric circuit models could provide a natural way of solving the minimization problems dictated by the standard regularization theory [18,10,23]. It became more important and interesting to study the bio-inspired electric circuit models, since recent studies demonstrated that the analog electric circuits could be fabricated in the complementary metal-oxide-semiconductor (CMOS) Very Large Scale Integrated (VLSI) circuit [11,12,14,24].

In the present paper, we first describe the architecture of the outer retinal circuit consisting of photoreceptors, horizontal cells and bipolar cells, with an analog circuit model. The model was developed based on recent physiological observations. Secondly, the function of the analog circuit model is characterized in terms of the standard regularization theory [9,18,19]. Finally, an analog VLSI circuit, silicon retina, inspired by these computational studies is introduced [7]. The silicon retina developed here carries out a real time image preprocessing in a natural illumination and therefore meets wide fields of engineering applications.

2 Computational Theory of Early Vision

Previous theoretical studies pointed out that early vision can be interpreted as a set of ill-posed problems [18]. Accordingly, some of early vision problems, e.g., edge detection, and stereo vision, were found to be formulated in terms of standard regularization theory [18,19]. For instance, these problems are attributed to minimization of a cost function,

$$J(y) = \|y - d\|^2 + \lambda_r \int \sum_{r=1}^{p} (\frac{d^r y(\xi)}{d\xi^r})^2 d\xi, \quad \lambda_r > 0. \tag{1}$$

Here, d and y correspond to a raw image and a corresponding filtered image to be interpolated, respectively. The second term of cost function ensures uniqueness and convergence of the solution. This formulation corresponds to a constraint that the (noiseless) image is smooth because of a band-limited nature of the image and has therefore bounded derivatives. The first term of (1) requires y being close to d. On the other hand, the second term of (1) requires y being smooth. The emphasis put on either the former or the later is determined by λ_r.

It is convenient to replace the derivative operations in (1) by the difference operations to find a relation between the function of analog circuits and above formulation [9], *i.e*,

$$\frac{dy(\xi)}{d\xi} \rightarrow y_k - y_{k-1},$$

$$\frac{d^2y(\xi)}{d\xi^2} \rightarrow y_{k-1} - 2y_k + y_{k+1}.$$

Here $k = 0, 1...n$. These operations can, therefore, be expressed by

$$\frac{dy(\xi)}{d\xi} \rightarrow \mathbf{Dy}, \quad \frac{d^2y(\xi)}{d\xi^2} \rightarrow \mathbf{Ly}$$

where

$$\mathbf{y} = \begin{pmatrix} y_1 \\ y_2 \\ \vdots \\ y_n \end{pmatrix}, \mathbf{D} = \begin{pmatrix} 1 & 0 & 0 & \ldots & 0 \\ -1 & 1 & 0 & 0 & \ldots \\ 0 & -1 & 1 & 0 & \ldots \\ \vdots & \ddots & \ddots & \ddots & 0 \\ 0 & \ldots & 0 & -1 & 1 \end{pmatrix},$$

$$\mathbf{L} = \begin{pmatrix} -2 & 1 & 0 & \ldots & 0 \\ 1 & -2 & 1 & 0 & \ldots \\ 0 & 1 & -2 & 1 & \ldots \\ \vdots & \ddots & \ddots & \ddots & 1 \\ 0 & \ldots & 0 & 1 & -2 \end{pmatrix}.$$

Note that

$$\mathbf{D}^T\mathbf{D} = -\mathbf{L}. \tag{2}$$

Here ^{T} denotes the transpose. Consequently, the regularization problem is formalized in discrete case by the minimization of

$$J(\mathbf{y}) = \|\mathbf{y} - \mathbf{d}\|^2 + \sum_{r=1}^{p} \lambda_r \begin{cases} \|\mathbf{L}^{\frac{r}{2}}\mathbf{y}\|^2 & \text{r:even} \\ \|\mathbf{D}\mathbf{L}^{\frac{r-1}{2}}\mathbf{y}\|^2 & \text{r:odd,} \end{cases}$$

which is referred to as p-th order regularization problem in discrete case [9]. The unique solution to minimize $J(\mathbf{y})$ is obtained by the Euler equation,

$$\mathbf{y} - \mathbf{d} + \sum_{r=1}^{p} (-1)^r \lambda_r \mathbf{L}^r \mathbf{y} = \mathbf{0}. \tag{3}$$

Here (2) was used. Note that if $J(\mathbf{y})$ has the r-th order smoothness term, *i.e.*, $\lambda_r \neq 0$, then the equation (3) necessarily contains the $2r$-th order derivative. For the 2*nd* order regularization problem, which will be shown to be relevant to characterize the function of the outer retinal circuit later, the cost function is expressed as

$$\begin{aligned} J(\mathbf{y}) = (\mathbf{y} - \mathbf{d})^T(\mathbf{y} - \mathbf{d}) \ &- \lambda_1(\mathbf{D}\mathbf{y})^T(\mathbf{D}\mathbf{y}) \\ &+ \lambda_2(\mathbf{L}\mathbf{y})^T(\mathbf{L}\mathbf{y}). \end{aligned} \tag{4}$$

And the corresponding Euler equation is

$$\mathbf{y} - \mathbf{d} - \lambda_1\mathbf{L}\mathbf{y} + \lambda_2\mathbf{L}^2\mathbf{y} = \mathbf{0}. \tag{5}$$

2.1 Analog circuit model of the outer retina

The vertebrate retina is one of the few tissues of the nervous system in which electrical properties and structural organization of neurons are well correlated. Five principal cell types of neurons have been identified in the retina (see for review [4]). Fig.1 is a schematic illustration showing the gross structure of the vertebrate retina. Each of these principal cell types can be classified into several subtypes, which are not shown in the Fig. to avoid complexities. In Fig.1, the bottom side corresponds to the frontal surface of the retina from which the light comes through the optical apparatus of the eye (arrows). The light passes through the transparent retina to reach the photoreceptor array. The light-sensitive pigment catches photons and a chemical reaction cascade transduces light to a voltage response in the photoreceptor. The voltage signal is transmitted to the second order neurons, which are the horizontal cell and the bipolar cell. Photoreceptors, horizontal cells and bipolar cells interact with each other in the outer plexiform layer (OPL), which is an anatomically identifiable lamina seen in the cross section of the retina.

We refer to the neuronal circuit consisting of these three types of neurons as the outer retinal circuit.

Bipolar cells transmit the outputs of the outer retinal circuit to amacrine cells and ganglion cells. Interplexiform cells (IP cells) are located at the amacrine cell layer and provide feedback pathways to OPL. A possible function of the IP cell will be discussed later in relation to the adaptive change of the receptive field.

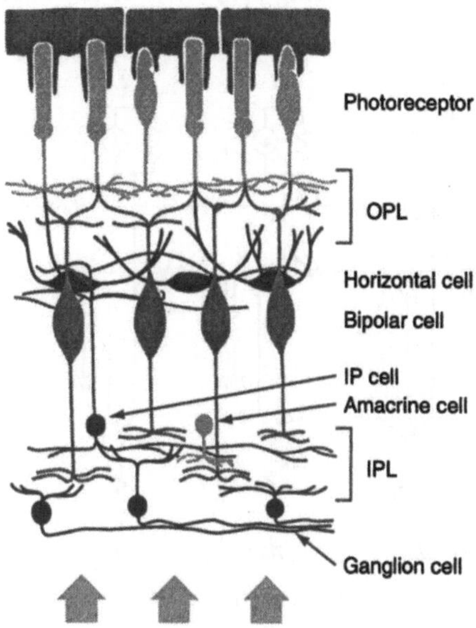

Fig. 1. A schematic drawing of the cross section of the retina

The outer retinal circuit was described by an analog circuit based on physiological and anatomical observations as shown in Fig.2 [17]. $U_k(k = 1, ..., n)$ is the light-induced current which is an alteration of current generated in the photoreceptor in response to light. k is a number to specify the location of the cone, a class of the photoreceptor. P_k and H_k are the voltage responses of the k-th cone and horizontal cell measured from the membrane potential in the dark, respectively. All the currents and voltages here designate the light-induced change measured from the level in the dark. Each cone is represented by the membrane conductance g_{m1}. Each horizontal cell is represented by the membrane conductance g_{m2}. It is well known that neighboring cones are coupled electrically by gap-junctions [2], a typical electrical synapse, as well as horizontal cells [15,22]. The coupling conductance connecting neighboring cones is represented by g_{s1} and that of horizontal cells by g_{s2}. The strength of synaptic input from the cone to the horizontal cell is expressed by t_1. The strength of the synaptic input is defined by a ratio of the postsynaptic current to the voltage of presynaptic neuron, *i.e.*

$$t = \frac{U_{pos}}{V_{pre}}.$$

Here, U_{pos} is the postsynaptic current and V_{pre} is the voltage response of presynaptic neuron. The synaptic strength has the unit of conductance (*Siemens*). The voltage and the current are time-independent in the present study, since we consider only the spatial filtering properties of the outer retinal circuit.

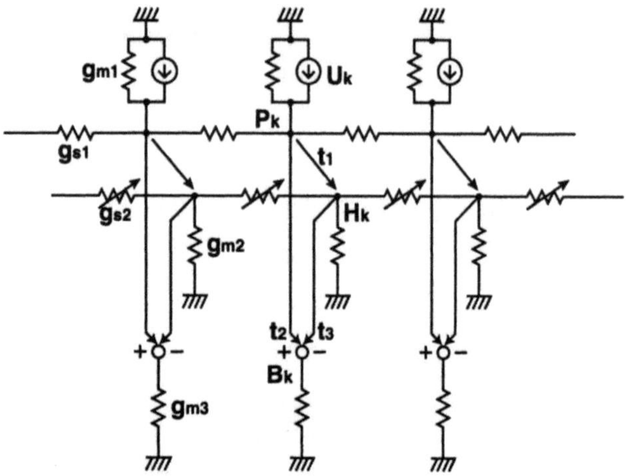

Fig. 2. An analog circuit model of the outer retina

We denote the light-induced current of the cones with **u** and the light-induced voltage responses of the cones, horizontal cells and bipolar cells with **p**, **h** and **b**, respectively:

$$\mathbf{u} = \begin{pmatrix} U_1 \\ U_2 \\ \vdots \\ U_n \end{pmatrix}, \mathbf{p} = \begin{pmatrix} P_1 \\ P_2 \\ \vdots \\ P_n \end{pmatrix}, \mathbf{h} = \begin{pmatrix} H_1 \\ H_2 \\ \vdots \\ H_n \end{pmatrix}, \mathbf{b} = \begin{pmatrix} B_1 \\ B_2 \\ \vdots \\ B_n \end{pmatrix}.$$

Applying Kirchhoff's Current Law at each node of Fig.2, we obtain a set of matrix equations,

$$\mathbf{C}_1\mathbf{p} = -\frac{1}{g_{s1}}\mathbf{u}, \tag{6}$$

$$\mathbf{C}_2\mathbf{h} + \frac{t_1}{g_{s2}}\mathbf{E}\mathbf{p} = \mathbf{0}. \tag{7}$$

Here, $\mathbf{E}$ is the identity matrix and $\mathbf{0}$ is the null vector. $\mathbf{C}_1$ and $\mathbf{C}_2$ are

$$\mathbf{C}_1 = \begin{pmatrix} c_1+1 & 1 & 0 & \dots & 0 & 0 \\ 1 & c_1 & 1 & 0 & \dots & 0 \\ 0 & 1 & c_1 & 1 & \ddots & \vdots \\ \vdots & \ddots & \ddots & \ddots & \ddots & 0 \\ 0 & \dots & 0 & 1 & c_1 & 1 \\ 0 & 0 & \dots & 0 & 1 & c_1+1 \end{pmatrix},$$

$$\mathbf{C}_2 = \begin{pmatrix} c_2+1 & 1 & 0 & \dots & 0 & 0 \\ 1 & c_2 & 1 & 0 & \dots & 0 \\ 0 & 1 & c_2 & 1 & \ddots & \vdots \\ \vdots & \ddots & \ddots & \ddots & \ddots & 0 \\ 0 & \dots & 0 & 1 & c_2 & 1 \\ 0 & 0 & \dots & 0 & 1 & c_2+1 \end{pmatrix}.$$

Here

$$c_1 = -(2 + \frac{g_{m1}}{g_{s1}}), \quad c_2 = -(2 + \frac{g_{m2}}{g_{s2}}).$$

Combining (6) and (7), we obtain equations to calculate the voltage responses of cones and horizontal cells, *i.e.*,

$$(\mathbf{C}_2\mathbf{C}_1)\mathbf{p} = -\frac{1}{g_{s1}}\mathbf{C}_2\mathbf{u}, \tag{8}$$

$$(\mathbf{C}_1\mathbf{C}_2)\mathbf{h} = \frac{t_1}{g_{s2}g_{s1}}\mathbf{u}. \tag{9}$$

Solving above equations, we will obtain voltage distributions of cones and horizontal cells as a function of physiological mechanisms, *i.e.*, membrane conductance, gap-junctional conductance and strength of synaptic input [17]. We do not refer to the solutions further here, since it is not necessary to explain the function of the present analog circuit describing the outer retinal circuit.

The bipolar cell exhibits the center-surround antagonistic receptive field [8]. It is widely believed that the inhibitory surround receptive field of the bipolar cell is generated by the horizontal cell [13,21]. There are two possible pathways for the horizontal cell response to generate the receptive field surround of the bipolar cell. One possible pathway is the direct input from the horizontal cell to the bipolar cell [5]. In that case polarity of the synaptic input from the horizontal cell to the bipolar cell (t_3 in Fig.2) is opposite from that of the cone to the bipolar cell (t_2 in Fig.2). Another possibility of inhibitory pathway is the negative feedback from the horizontal cell to the cone. The horizontal cell is known to send a feedback signal to the cone by the sign-inverting synapse [2]. And this feedback signal can be transmitted to the bipolar cell through the cone to generate the receptive field surround of the bipolar cell. More recently, these two pathways were shown to contribute simultaneously to generate the receptive field surround of the bipolar cell in the salamander retina [26]. Here we deal with only the former input for simplicity.

The voltage distribution of the bipolar cell response, **b**, is expressed simply by a difference between the cone response and the horizontal cell response [6],

$$\mathbf{b} = \frac{t_2}{g_{m3}}\mathbf{p} + \frac{t_3}{g_{m3}}\mathbf{h}.$$

Here, g_{m3} is the membrane conductance of the bipolar cell. t_2 and t_3 are strengths of synaptic inputs to the bipolar cell from the cone and from the horizontal cell, respectively. Here, we assume

$$\frac{t_2}{g_{m3}} = 1, \frac{t_3}{g_{m3}} = -1. \tag{10}$$

2.2 Spatial filtering properties of bipolar cell

We will interpret the spatial filtering properties of the bipolar cell receptive field in terms of the standard regularization theory using the analog circuit model introduced in the previous section.

Let us first consider how the spatial filtering properties of the bipolar cell response are characterized with a quadratic cost function. Combining (8), (9) and (10), we obtain the equation to find the voltage distribution of the bipolar cell response:

$$(\mathbf{C}_1\mathbf{C}_2)\mathbf{b} = -\frac{t_2}{g_{m3}g_{s1}}\mathbf{C}_2\mathbf{u} + \frac{t_1 t_3}{g_{m3}g_{s1}g_{s2}}\mathbf{u}. \tag{11}$$

Eliminating the boundary effect and substituting $\mathbf{C_1}$ and $\mathbf{C_2}$ of (11) by $(\mathbf{L} - g_{m1}/g_{s1})\mathbf{E}$ and $(\mathbf{L} - g_{m2}/g_{s2})\mathbf{E}$, respectively, we find that the response of the bipolar cell holds an equation similar to the Euler equation (5). Namely,

$$\mathbf{b} - \mathbf{d} - \lambda_1\mathbf{L}\mathbf{b} + \lambda_2\mathbf{L}^2\mathbf{b} = \mathbf{0}. \tag{12}$$

Here

$$\lambda_1 = \frac{g_{m1}g_{s2} + g_{m2}g_{s1}}{g_{m1}g_{m2}}, \lambda_2 = \frac{g_{s1}g_{s2}}{g_{m1}g_{m2}}.$$

And

$$\mathbf{d} = \nu R_0\mathbf{u} - R_0\mathbf{L}\mathbf{u}, \tag{13}$$

where

$$R_0 = \frac{g_{s2}t_2}{g_{m1}g_{m2}g_{m3}}, \quad \nu = \frac{1}{g_{s2}}(g_{m2} - t_1).$$

As was defined, $\mathbf{u}$ is the light-induced current of cones. R_0 has a unit of resistance (*ohm*) and ν is a constant. Therefore, $\nu R_0 \mathbf{u}$ designates the spatial voltage distribution which is proportional to the raw image, provided that the light-induced current is proportional to the intensity of illumination. Similarly, $R_0 \mathbf{L}\mathbf{u}$ designates the spatial voltage distribution which is proportional to the second order difference of the raw image. Note that the second order difference operation enhances the contrast of image as well as noise. The similarity between (5) and (12) indicates that the voltage response of bipolar cell distributes in such a way to minimize the cost function,

$$J(\mathbf{b}) = (\mathbf{b} - \mathbf{d})^T(\mathbf{b} - \mathbf{d}) + \lambda_1(\mathbf{D}\mathbf{b})^T(\mathbf{D}\mathbf{b}) + \lambda_2(\mathbf{L}\mathbf{b})^T(\mathbf{L}\mathbf{b}). \tag{14}$$

Using this cost function, the spatial filtering properties of the bipolar cell receptive field is explained as follows. The first term of the right hand side of (14) requires the response of the bipolar cell being close to $\mathbf{d}$. Note that $\mathbf{d}$ is composed of the raw image and the contrast-enhanced image as shown in (13). The second and the third term of the right hand side of (14) are the smoothness constraints.

It is easier to find the meaning of cost function when we simply rewrite (14) to

$$\begin{aligned} J(B_k) = &\sum_k [B_k - R_0(-U_{k-1} + 2U_k - U_{k+1}) - \nu R_0 U_k]^2 \\ &+ \lambda_1 \sum_k (B_k - B_{k-1})^2 \\ &+ \lambda_2 \sum_k (B_{k-1} - 2B_k + B_{k+1})^2. \end{aligned} \tag{15}$$

Now it becomes clear to infer the spatial filtering properties of the bipolar cell in early vision. The first term of right hand side,

$$\sum_k [B_k - R_0(-U_{k-1} + 2U_k - U_{k+1}) - \nu R_0 U_k]^2,$$

decreases as the voltage distribution of bipolar cell becomes closer to either the profile of the raw image, $\nu R_0 u_k$, or the contrast-enhanced image $R_0(-U_{k-1} + 2U_k - U_{k+1})$. R_0 is an amplification factor. As R_0 increases, the amplitude of the bipolar cell response increases to minimizes $J(B_k)$. The constant ν is relevant to the degree of emphasis put on the raw image relative to the contrast-enhanced image. As ν becomes larger, the spatial distribution of the bipolar cell response becomes more sensitive to the raw image. As ν becomes smaller, the spatial distribution of the bipolar cell response becomes more sensitive to the contrast-enhanced image or in other words the contour of image rather than the raw image. It is noticeable that νR_0 does not include g_{s2} and therefore the sensitivity to the raw image does not change even when the gap-junctional conductance connecting the horizontal cell, g_{s2}, changes (see the section 2.3).

The second and the third term of the cost function (15) correspond to stabilizers which smooth the image to filter the noise. These stabilizers are important since the bipolar cell enhances the contrast by differentiating the raw input image, which in return amplifies the noise and degrades the signal to noise ratio of the output image seriously [18]. The degree of smoothing is changed by regularization parameters, λ_1 and λ_2. The optimum values of the regularization parameters are to be determined corresponding to the signal to noise ratio of the input image.

2.3 Variable receptive field

The voltage distribution of the bipolar cell responses to a point stimulus, which corresponds to the receptive field, was calculated with the analog circuit (Fig.3). The horizontal axis measures the position of the retina. The stimulus position is indicated by an arrow. The vertical axis measures the response voltage of the bipolar cell at each position. As shown by the solid curve, the polarity of response reverses as the distance from the stimulus increases. The calculated receptive field approximates $\nabla^2 G$ function very well [23].

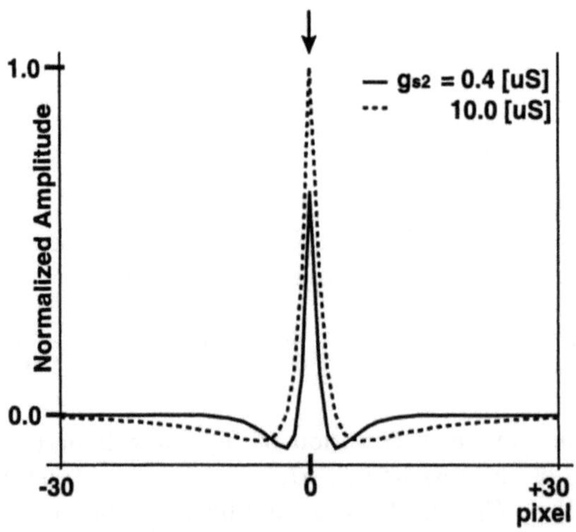

Fig. 3. Distribution of voltage responses of the bipolar cell calculated by the model. The calculation was conducted with PSPICE. $g_{m1} = g_{m2} = g_{m3} = 0.1$uS , $g_{s1} = 0.167$uS , $g_{s2} = 0.4$uS (solid line), 10.0uS (dotted line), $t_1 = t_2 = 0.1$uS, $t_3 =$ -0.1uS

The size of the receptive is controlled by the regularization parameters, which are expressed by physiological parameters of the retinal circuit model. Among the physiological parameters included in the model, the conductance of the gap-junction connecting neighboring horizontal cells, g_{s2}, is interesting. There are accumulating lines of physiological evidence indicating that the gap-junctional conductance connecting neighboring horizontal cells is controlled by IP cell [16,20]. Since the receptive field surround of the bipolar cell is generated by the horizontal cell regardless of the inhibition pathway, receptive field properties of the bipolar cell are also modulated by IP cell. The effect of g_{s2} is unambiguously explained from the cost function (15). When g_{s2} increases, λ_1 and λ_2 also increase and therefore the smoothing on the image is enhanced.

The effect of g_{s2} on spatial filtering properties of the bipolar cell was examined with the analog circuit model. The receptive field of the bipolar cell was calculated with different values of g_{s2} and plotted by the dotted line in Fig.3. As shown in Fig.3, the receptive field of the bipolar cell was broadened when g_{s2} increases. The width of both center and surround receptive field becomes wider as g_{s2} increases. This indicates that the smoothing effect is emphasized. Moreover, it is important to point out that the peak amplitude of receptive field center becomes larger as g_{s2} increases. This indicates that the response gain of the bipolar cell increases as g_{s2} increases. This is considered to be a suitable modulation mechanism for transition from the dark-adapted stage to the light-adapted stage of the retina [17].

3 Silicon Retina

The previous section elucidated how the analog circuit model inspired by the outer retinal circuit solves the early vision problem. In recent years, VLSI analog CMOS circuits, silicon retinas, have been fabricated to mimic the function of the retinal circuit (for outlines refer to [3,11,12,14]. Silicon retinas are expected to carry out real time image computation with a low power consumption and a compact hardware. As an early attempt, Mead et.al. developed a silicon retina inspired by a simple resistive circuit model of the outer retina [14]. Then, based on the retinal circuit model with detailed physiological structures[25], Laplacian-Gaussian ($\nabla^2 G$)-like receptive field chips were also designed [9]. Those silicon retinas, however, have not reached a real application stage, since the outputs of these chips are not sufficiently accurate to be used for image recognition. The photosensors used in those silicon retinas have a wide dynamic rang but the sensitivity of light intensity is low instead. The intrinsic noise due to the statistic mismatch of transistor characteristics is another serious problem. The quality of signal processing from all-analog processors has usually been degraded by transistor offsets and fixed pattern noise of photosensors. We have improved such problems by using charge accumulation type photosensors and sample/hold circuits.

The chip fabricated is a one-dimensional 100 pixel line sensor. Fig.4 shows the block diagram for one pixel of the 100 pixel chip. The photosensor is an active pixel sensor that consists of a photodiode and a source-follower circuit.

The parasitic capacitor of the photodiode is charged by P_{SW} and then the light discharges the capacitor with the photo-current. After a certain time (we refer to this time as an accumulation time), the voltage between the photodiode is read through a source-follower circuit. The photosensor circuit is connected to the first layer resistive network which represents the cone network. Here, neighboring pixels are connected by MOS resistors [14]. The output voltage is fed to a sample/hold buffer circuit (BUF) where the fixed pattern noises of photosensor and transistor offsets are compensated. The output of BUF is fed to another sample/hold buffer circuit (DSB) either directly or via the second layer resistive network, which corresponds to the horizontal cells. Neighboring horizontal cells are connected by MOS resistors again. The difference between voltages of these two pathways is taken by the sample/hold circuit of DSB, corresponding to the bipolar cell response. The offset of DSB is also compensated by the sample/hold circuit simultaneously during the subtraction. Therefore, the set of voltages from the DSB buffers has an extremely low noise. Detailed operations of the chip were described elsewhere [7,25].

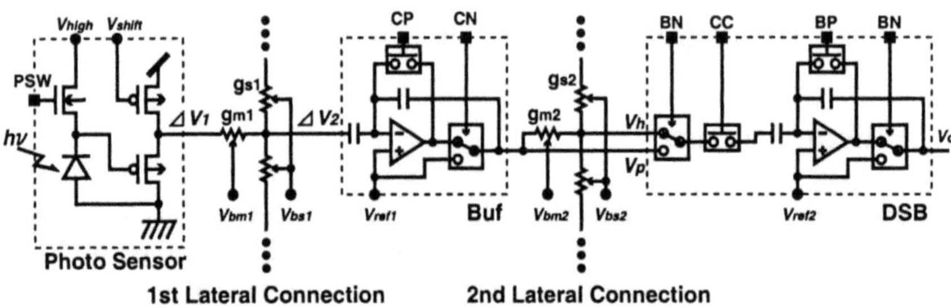

Fig. 4. Circuit design of a single pixel of the chip

3.1 Receptive field properties of the silicon retina

The spatial filtering properties of the silicon retina were tested by mounting a camera lens to focus an image on the photosensor array. A slit of white image was placed one meter from the lens. The width of the slit was 0.5 cm. The experiment was carried out under indoor illumination (about $4.0W/m^2$). The accumulation time of the photosensors, which is equal to a read out time of one frame, was 0.53 ms. The horizontal axis measures the pixel position and the vertical axis is the output voltages of each pixel. The spatial distribution of the output voltages of the chip showed a Laplacian-Gaussian like receptive field (Fig.5, solid line), which is similar to the receptive field of Fig.3.

The receptive field size was modulated by changing the conductance of the second layer MOS resisters, which correspond to the coupling conductance of horizontal cells. The dotted curve in Fig.5 shows the broadened receptive field. In this case, the bias voltage (V_{bs2}) was varied from 0.5V (solid line) to 2.0V (dotted line) to increase the coupling conductance. The broadened receptive field indicates that the filtered image will be blurred by a smoothing effect. It is noticeable that the response to a constant light intensity portion, which is the base line of the response (indicated by arrow), was not affected even though the size of the receptive field changes. The amplitude of the response becomes larger as g_{s2} increases as was expected from the calculation of Fig.3.

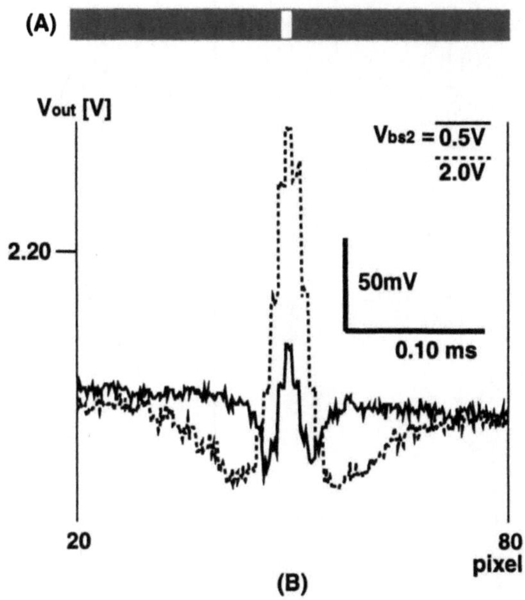

Fig. 5. Response of the chip to a slit image, (A) Presented pattern, (B) Output voltage

3.2 Application to image processing

The one-dimensional silicon retina fabricated here can be used for real time image acquisition of moving objects. When objects move along a direction perpendicular to the line sensor, two-dimensional images of objects are obtained. In the experiment shown in Fig. 6(a), a hand was moved in front of the lends with a velocity of about 14.6cm/sec. The outputs of the chip were fed to a computer through a A/D converter at a rate of 102 flame (100pixels)/sec. Then 140 flames, each of which corresponds to a sliced portion of the hand, were aligned to display a two-dimensional image. As shown in Fig., a Mach-band like effect is clearly seen in the image. The image of the hand was filtered with different receptive field size by changing g_{s2} (Fig.6(c)). As shown in the Fig., the boundary between the hand and background became blurred.

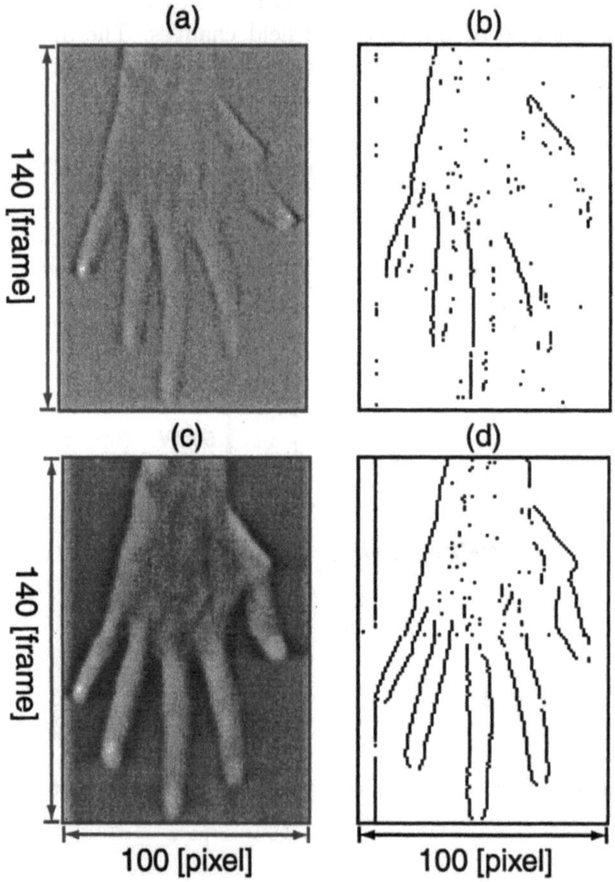

Fig. 6. Scaned image of a hand, (a),(c) Response of the chip, (b),(d) Output of the zero-crossings detector, (a),(b) narrow receptive field (c),(d) wide receptive field

Adding a zero-crossing detector at the output of the chip [1], an edge of the image can be obtained in real time. Fig.6(b) and (d) show the edge obtained with the images (a) and (c), respectively. The edge of the image is not accurately extracted in this illumination condition in (b). However, the edge is almost completely extracted using (c). This observation indicates, as expected, that the regularization parameters are critical to carry out an appropriate image processings. Further computational studies are needed to formulate an algorithm to find an optimal regularization parameters.

4 Conclusion

The retina computes the image with a completely different architecture from the one that most engineers are familiar with. Because of this unique architecture, the retina can perform real-time image processing with very low power dissipation. In the present study, we elucidated the computational significance of the architecture. Namely, the electrical properties of the analog circuit model describing the outer retina are characterized with the standard regularization theory, which was proposed from a purely computational point of view to solve early vision problems [18,19]. The cost function accompanying the analog circuit corresponds, in fact, to the total power dissipated as heat from the circuit and therefore the regularization parameters of the cost function is naturally expressed by parameters of the circuit which are directly relevant to physiological mechanisms underlying the image processing of the outer retina.

An analog VLSI, silicon retina, with variable receptive field was designed inspired by the architecture of the outer retinal circuit. We applied the silicon retina for a real time image processing under a natural illumination. The silicon retina is useful not only for engineering applications but also for gaining an insight of computational principle underlying sensory information processing of the neuronal circuit.

5 Acknowledgments

This work was partially supported by the Japan Society for the Promotion of Science, grant-in-aid for Research for the Future Program, JSPS-RFTF 97I00101 (PI: T. Yamakawa).

References

1. Bair, W., Koch, C. (1991) : An analog VLSI chip for finding edges from zero-crossings. NIPS 3, 399-405
2. Baylor, D.A., Fuortes, M.G.F., O'Bryan, P.M. (1971) : Receptive fields of cones in the retina of the turtle. J Physiol Lond 214, 256-294
3. Douglas, R., Mahowald, M., Mead, C. (1995) : Neuromorphic analogue VLSI. Annual Review Neuroscience 18, 255-258
4. Dowling, J.E. (Ed.) (1987) : The retina: an approachable part of the brain. Belknap Press of Harvard University Press, Cambridge Massachusetts
5. Dowling, J.E., Werblin, F.S. (1969) : Organization of the retina of the mudpuppy. Necturus maculosus I. Synaptic structure. Journal of Neurophysiology 32, 315-338
6. Hare, W.A., Owen, G.W. (1990) : Spatial organization of the bipolar cell's receptive field in the retina of the tiger salamander. Journal of Physiology London 421, 223-245

7. Kameda, K., Honda, A., Yagi, T. (1999) : Real time image processing with an analog vision chip systems. International Journal of Neural Systems 9, 423-428
8. Kaneko, A. (1973) : Receptive field organization of bipolar and amacrine cells in the goldfish retina. Journal of Physiology London 235, 133-153
9. Kobayashi, H., Matsumoto, T., Yagi, T., Shimmi, T. (1993) : Image processing regularization filters on layered architecture. Neural Networks 6, 327-350
10. Koch, C., Marroquin, J., Yuille, A. (1986) : Analog "neuronal" networks in early vision. Proceedings of National Academy of Science USA 83, 4263-4267
11. Koch, C., Li, H. (Eds.) (1995) : Vision chips: implementing vision algorithms with analog VLSI circuits. IEEE computer society press
12. Mahowald, M. (Ed.) (1994) : An analog VLSI system for stereoscopic vision. Kluwer Academic Publishers
13. Marchiafava, P.L. (1978) : Horizontal cells influence membrane potential of bipolar cells in the retina of the turtle. Nature 275, 141-142
14. Mead, C. (1989) : Analog VLSI and Neural Systems. Addison-Wesley
15. Naka, K.-I., Rushton, W.A.H. (1967) : S-potential from luminosity units in the retina of the fish (Cyprinidae). Journal of Physiology 192, 437-461
16. Negishi, K., Drujan, B. (1979) : Reciprocal changes in center and surrounding S-potentials of fish retina in response to dopamine. Neurochemical Research 4, 313-318
17. Ohshima, S., Yagi, T., Funahashi, Y. (1995) : Computational studies on the interaction between red cone and H1 0 cell. Vision Research, 149-160
18. Poggio, T., Koch, C. (1985) : Ill-posed problems in early vision: from computational theory to analogue networks. Proceedings of Royal Society London B 226, 303-323
19. Poggio, T., Torre, V., Koch, C. (1985) : Computational Vision and Regularization Theory. Nature 317, 314-319
20. Teranishi, T., Negishi, K., Kato, S. (1983) : Dopamine modulates S-potential amplitude and dye-coupling between external horizontal cells in carp retina. Nature 301, 234-246
21. Toyada, J.-I., Tonosaki, K. (1978) : Effect of polarization of horizontal cells on the on-center bipolar cell of the carp retina. Nature 276, 399-400
22. Yagi, T. (1986) : Interaction between the soma and the axon terminal of retinal horizontal cells in Cyprinus carpio. Journal of Physiology London 375, 121-135
23. Yagi, T., Ohshima, S., Funahashi, Y. (1997) : The role of retinal bipolar cell in early vision: an implication with analogue networks and regularization theory. Biological Cybernetics, 163-171
24. Yagi, T., Matsumoto, T., Kobayashi, H. (1998) : Parallel analog image processings: solving regularization problems with architecture inspired by the vertebrate retinal circuit. In: Leondes, C.T.(Ed.) : Neural Network Systems Techniques and Applications, Academic Press, 201-285
25. Yagi, T., Kameda, S., Iizuka, K. (1999) : A parallel analog intelligent vision sensor with a variable receptive field. Systems and Computers in Japan 30, 1, 60-69
26. Yang, X.L., Wu, S.M. (1991) : Feedforward lateral inhibition: input-output relation of the horizontal cell to bipolar cell synapse in the tiger salamander retina. Proceedings of National Academy Science USA 88, 3310-3313

Extending the Decision Accuracy of a Bioinformatics System

A. Chong, T.D. Gedeon and K.W. Wong

School of Information Technology
Murdoch University
South St.
Murdoch, Western Australia 6150
Australia

Summary: We introduce a simple fuzzy technique to improve the prediction decision accuracy of a bioinformatics neural network system from the literature for protein structure prediction. We also describe an unsound assumption made by the authors of the neural network system, and propose a fuzzy hybrid solution, which eliminates the need for this assumption and can further enhance performance.

Keywords: Bioinformatics, Decision accuracy, Protein structure prediction, Hybrid system, Neural network, Fuzzy logic

1. Introduction

Protein is the machinery of life. It is required in all organisms for the structure, function, and regulation of the body's cells, tissues, and organs. Each protein has unique functions. For example, one type of protein, known as an enzyme, helps in our body's digestion system. The structure of a particular protein determines its function. The techniques to experimentally determine the 3D structure of proteins are complicated and time consuming. Determining a structure can take from one to several years.

A protein is formed by a chain of amino acids (hereafter known as a protein sequence). Over the years, many new proteins have been identified by large-scale genome sequencing projects. While the protein sequence of the new protein can be identified, the protein structure is often not known. As an attempt to narrow the gap between the number of known protein sequences and the number of experimentally determined protein structures, methods for protein structure prediction have been studied (Defay and Cohen, 1996; Fischer and Eisenberg, 1996; Flockner et. al, 1995; Lathrop and Smith, 1996).

In general, the protein structure prediction is performed by observing the protein sequence combined with our prior knowledge on a set of homologous proteins whose structure has been determined. At the time of writing, the prediction of a protein's three-dimensional structure from its amino acid chain (protein sequence) remains an unsolved problem. A review of the literature suggests that most of the research in this problem domain addresses the prediction of protein secondary structure.

Most of the work done in this problem domain attempts to predict a protein sequence to be one of the following: Helix (H), Extended (E) and Loop (L) (Zhang et. al, 1992). The problem can be viewed as a simple classification problem. Given a protein sequence, some algorithms are applied to classify the protein as Helix, Extended or Loop. Artificial neural networks are one of the predominant classifiers used in this problem domain (Qian and Sejnowski, 1988; Baldi et. al, 1999; Rost and Sander, 1993; Zhang et. al, 1992). Fuzzy logic and genetic algorithms have also been tried (Zhang et. al, ; Vivarelli et. al, 1995).

2. Research Goal

In this research, we explore the use of a fuzzy inference system (a.k.a. fuzzy system) to improve the protein secondary structure prediction accuracy of a successful neural network protein prediction system.

Since a fuzzy set allows for the degree of membership of an item in a set to be any real number between 0 and 1, this allows human observations, expressions and expertise to be modelled more closely. Once the fuzzy sets have been defined, it is possible to use them in constructing rules for fuzzy expert systems and in performing fuzzy inference. Fuzzy system can produce more accurate results based on the basic idea of the defuzzification. A defuzzification technique is used to calculate the conclusion by evaluating the degree of matches from the observation that triggered one or several rules in the model. This will lead to a better result by handling the fuzziness in the decision making. Thus, the fuzzy technique can improve the neural network prediction in certain cases.

Among the neural networks used for protein structure prediction, the PHD (Profile-Based Network from Heidelberg) (Rost and Sander, 1993) was one of the first to claim to have achieved an accuracy of more than 70%. For this reason, the PHD has been chosen to be our base system to implement our fuzzy improvements.

This research aims to improve the accuracy of the PHD prediction by using a fuzzy system. While the fuzzy system is used in conjunction with the PHD networks in this study, it is reasonable to generalise that the technique can be used with other neural networks for protein structure prediction. The main emphasis here is the development of a technique that improves the performance of neural network-based protein structure prediction tools using fuzzy logic.

3. Materials and Methods

An overview of the PHD network is presented in the section below. Complete details can be obtained from the original paper (Rost and Sander, 1993).

3.1 The prediction process

The PHD protein secondary structure prediction is carried out based on the following steps:

1. For each input protein sequence, the SWISSPROT data bank is searched for protein sequences that are homologous to the input sequence. This is done by using a program called BLAST, which is based on a well-known fast alignment method. The output of the program is a list of protein sequences in the SWISSPROT data bank as well as their similarity (in percentages) to the input protein sequence.
2. The list of homologues identified by BLAST is then fed into a more sensitive profile-based multiple alignment program, known as MaxHom.
3. The multiple sequence alignment produced by MaxHom is then refined by applying a filter. Only sequences with a similarity to the input protein sequence higher than a threshold are selected for the prediction process.

3.2 The PHD neural network

Figure 1 summarizes the operation of the PHD Neural Networks. The PHD network is composed of multiple 3-layer neural networks (i.e. networks with 1 hidden layer). The entire architecture consists of 3 levels.

The first level is a neural network, known as the sequence-to-structure net (SQSCN). The network takes a protein sequence as input and predicts the protein secondary structure as output, thus the name sequence-to-structure. For the SQSCN, the input is given a window of 13 basic cells. Given a protein sequence, 13 amino acids within the sequence are examined at a time and the secondary structure of the amino acids in the middle (position 7) is predicted. The input to each basic cell in the SQSCN is the profile computed from the multiple sequence alignment (as discussed previously). The network produces 3 real numbers representing the probability of the prediction being Helix, Extended and Loop respectively (more on this later).

The structure-to-structure net (SCSCN) is found in the second level of the PHD Network. The input of this network is given by the output of the first level network (SQSCN). The underlying theory is that the secondary structure at position N of a protein is affected by the structures at nearby positions, such as N-2, N-1, N+1, N+2 ...etc. The input of SCSCN is given by a window of 17 basic cells, each cell takes in 3 real numbers (Helix, Extended and Loop) produced by the first level network (SQSCN). The output of SCSCN is again a set of 3 real numbers (the probabilities).

The SQSCN and SCSCN are trained separately, using the Backpropagation algorithm. There are altogether 2 SQSCNs and 9 SCSCNs produced by using slightly different types of approach and training data. For each input protein sequence, different outputs are computed by the differently trained neural networks.

At the third level of the PHD architecture, the outputs of all the second level networks (SCSCN) are averaged to product the final output. This level is called the jury level.

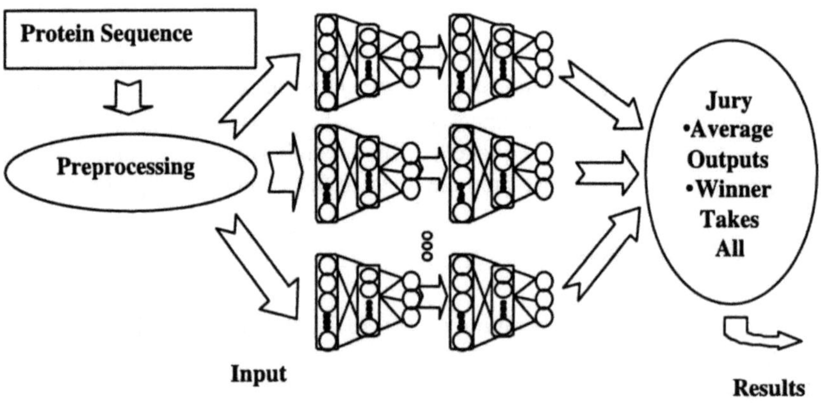

Figure 1: PHD Neural Networks in Operations

3.3 Research method

There is a major over-simplification in the PHD model, in that the network outputs are used as probabilities. The use of the outputs from neural networks which have been trained for classification as probabilities is not sound. The use of network outputs as probabilities is sound if the network has been trained using probabilities. This is not the case in the PHD system, however this usage is understandable given the lack of availability of probabilistic data.

A further problem is the use of averaging in the jury layer which eliminates much of the dynamics of the predictions made by individual networks. Our methodology is to use a fuzzy system instead of the jury layer, and we also propose a fuzzy enhancement of the decision rules to improve the soundness of the decision making. This derives from the nature of fuzzy systems as possibilistic systems as opposed to probabilistic systems.

3.4 Fuzzy systems

Most fuzzy systems can be classified into three types (Jang et. al, 1997):

Mamdani style fuzzy system A fuzzy rule, with two input *(X,Y)* and one output *(Z)*, comes in the form *if X is **A** and Y is **B** then Z is **C***, where ***A, B*** and ***C*** are fuzzy sets. Since the overall output of the system is a fuzzy set, a defuzzication process is normally performed to compute a crisp value out of the resulting fuzzy output.

Sugeno style fuzzy system A fuzzy rule, with two input *(X,Y)* and one output *(Z)*, comes in the form *if X is **A** and Y is **B** then* Z = $pX+qY+r$, where ***A, B*** and ***C*** are fuzzy sets, *p, q, r* are parameters that are used in conjunction with the inputs to compute the output. No defuzzication is required in this type of inference system.

Tsukamoto style fuzzy system The fuzzy rules used in this type of fuzzy system are similar to those of Mamdani style with the exception that the fuzzy set in the consequent part is characterized by a monotonical membership function. As a result, the output of each rule is defined as a crisp value induced by the rule firing strength.

It is noted that the three types of fuzzy systems differ only in the consequent part.

3.5 Fuzzy system as jury level

Fuzzy sets allow human expertise and decisions to be modelled more closely, thus it is suggested that we can replace the third level of the PHD architecture, the jury level, by a fuzzy system. Each of the nine neural networks in PHD produces three outputs representing the probabilities of Helix, Extended and Loop respectively. Hence, the number of inputs to the jury level is 3 x 9 = 27.

Fuzzy systems are well known for their "curse of dimensionality". In general, the number of fuzzy rules grows exponentially with the number of input variables and number of fuzzy terms per input variable. Even for moderate number of inputs, the number of rules involved in the fuzzy system can be very large. As a result, the operation and training of the fuzzy system become very slow.

As an effort to overcome this problem, we propose the use of multiple cooperative fuzzy systems as shown in figure 2. The task of secondary structure prediction is segregated to three independent fuzzy systems. The first fuzzy system, called the Helix fuzzy system, receives the Helix output from all the neural networks and output a value representing the probability that the final prediction is Helix. The second and third fuzzy systems perform similar tasks on the Extended and Loop predictions respectively. In this design, each fuzzy system deals with minimal amount of inputs. For simplicity, three out of the nine PHD neural networks are selected for the purpose of this research. That is, the fuzzy systems receive a total of 3 x 3 = 9 inputs.

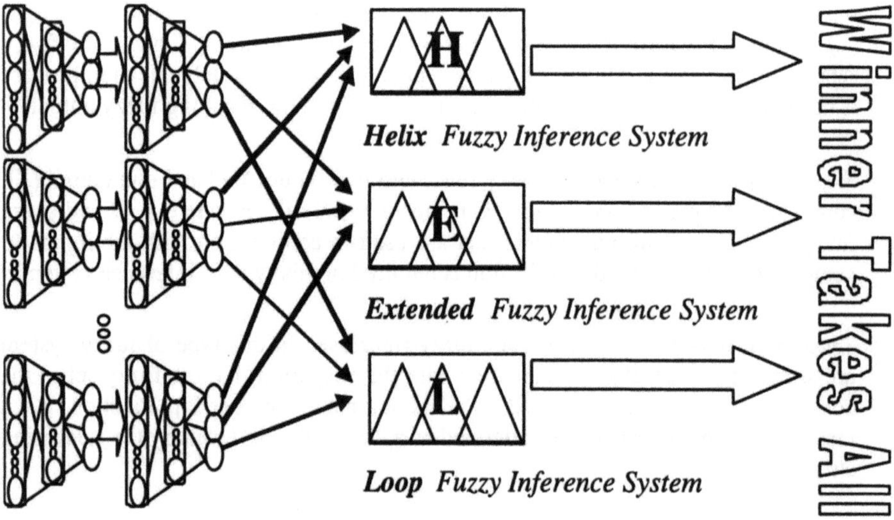

Figure 2: The use of Multiple Corporative

Pairs of input-output data are used to train the fuzzy systems. The data comes from the 126 protein sequences used in Rost and Sander (1993) are used to train the PHD neural networks. In the following sections, we would describe the training data and procedures used in this study.

3.6 Data

The input-output pairs come in the form:

$$(X_1^H, X_1^E, X_1^L, X_2^H, X_2^E, X_2^L, X_3^H, X_3^E, X_3^L; Y \in \{H, E, L\})$$

where X_i^T is the output of the ith neural network representing the membership of class T, H = class Helix, E = class Extended, L = class Loop, and Y is the desired output. When training the three independent but corporative fuzzy systems, each input-output pair is split into three individual pairs:

Helix Set: $(X_1^H, X_2^H, X_3^H; f(y, H))$

Loop Set: $(X_1^L, X_2^L, X_3^L; f(y, L))$

Extended Set: $(X_1^E, X_2^E, X_3^E; f(y, E))$

where $f(s,t) = 100$ when $s = t$; 0 otherwise. In other words, we suppress the desired output of the individual pair for a fuzzy system if the desired output of the original pair is of the type which the fuzzy system predicts. By suppressing and depressing the proper individual pair, we allow the three fuzzy systems to learn the correct overall output in a corporative manner.

3.7 Adaptive-network-based fuzzy inference system (ANFIS)

The Adaptive-Network-Based Fuzzy Inference System (ANFIS) technique has been used in this research to generate the fuzzy rules from the training data (Jang, 1993). In this section, we outline the procedures involved in the technique.

ANFIS is a multi-layer feedforward network as shown in figure 3. Each node in the network is associated with a function. Some nodes have a set of parameters that are used in conjunction with the function to compute the output based on the input. Each node receives incoming signals from the previous layer and passes its output to the nodes in the next layer.

One point to take note is that unlike some of the common neural networks such as the multilayer perceptron, the links in an adaptive network do not have weights. They merely indicate the flow direction of signals between nodes.

It is interesting to note that ANFIS can mimic the function of a fuzzy system. By carefully designing the network structure, ANFIS can operate like any of the three types of fuzzy systems described in section 3.4. For the purpose of this study, only the Sugeno style fuzzy system will be discussed.

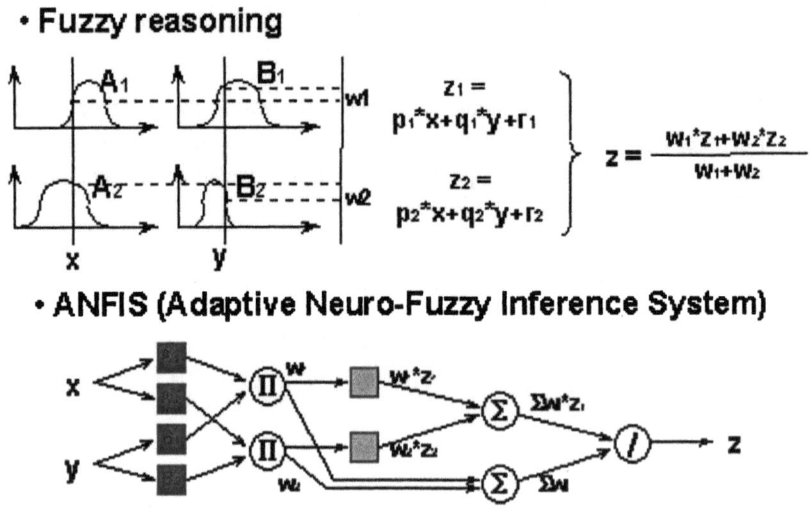

Figure 3: Sugeno Fuzzy System and ANFIS architecture

A typical ANFIS has five layers. Layer 1 is the membership layer. The functions associated with the nodes in this layer are membership functions used in fuzzy systems. The output of each node in this layer is the membership degree of the input. Common choices of the membership functions include the Triangular, Bell, Trapezoidal, and Gaussian function. Layer 2 is the rule layer. Each node output of this layer represents the firing strength of a rule. The function of each node can be any T-norm operators that performs fuzzy AND.

Layer 3 is the normalizing layer. The ith node computes the ratio of the ith rule's firing strength to the sum of all rules' firing strengths. Layer 4 is a layer that models the consequent part of a Sugeno fuzzy rule. We call it the consequent layer. It computes the output based on the normalized weights computed by the previous layer, system input (i.e. x, y, in the figure) as well as the three parameters (i.e. p,q,r) used in the consequent part of the Sugeno fuzzy rule. Layer 5 consists of only one node that sums up all the incoming signals and yields the output.

Hence, the network constructed operates similarly to the Sugeno fuzzy system. The adaptive structure enables the network to be trained using pairs of input-output data. Optimization techniques such as Backpropagation can be applied. However, it is well known that some of the common disadvantages of the gradient descent techniques are that they tend to be slow and become trapped in local minima. As a better alternative, the Hybrid Learning Algorithm has been proposed in Jang (1993), and is used in this study.

3.8 Training the fuzzy systems

The fuzzy rules of the Helix, Loop, and Extended fuzzy systems are learnt using ANFIS. The structure of the ANFIS used for this purpose is illustrated in figure 4.

There are three inputs (Helix, Extended, Loop) to the system. Three membership functions (Low, Medium, High) are used giving 3 x 3 = 9 nodes in the membership layer. The number of fuzzy rules involved in the fuzzy system is $3^3 = 27$. This led to the 27 nodes in the rules and consequent layer. The lower node with the summation sign (∑) in the figure represents the normalization factor (i.e. the sum of all firing strengths from the rule layer). This slight variation in design (as compared to figure 3) allows the normalization to be performed at the last layer. The upper summation sign node computes the weighted sum of the outputs from the previous layer. The individual node in the last layer divides the weighted sum by the normalization factor.

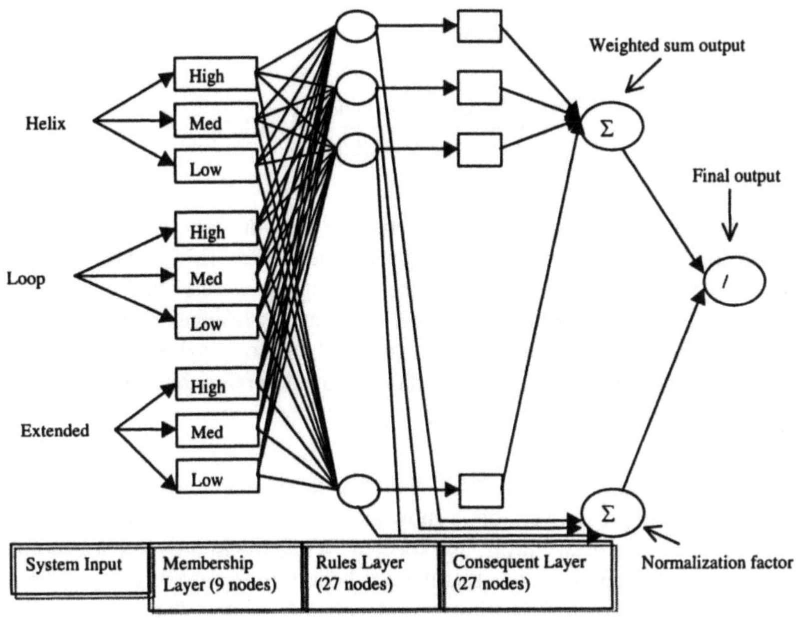

Figure 4: The ANFIS architecture of the Helix, Loop, and Extended fuzzy systems

The Hybrid Learning algorithm has been used to train the ANFIS for Helix, Extended and Loop. 4000 residues/entries are used for training. Training stops when one of the following occurs:

- The least mean squared (LMS) error is sufficiently small. In this case, the fuzzy system is considered completely trained and expected to achieve satisfactory performance.
- Over-fitting is sighted. To detect overfitting, a test set that is separate from the training set is prepared prior to the training. At the end of each epoch, the errors of both the training set as well as the test set are computed. In the beginning of the training, the errors of both sets are expected to decrease. The possibility of over-fitting can be observed when the error of the test set increases continuously.
- The decrease of LMS error is not significant over each epoch.

4. Results and Discussions

In general, accuracy can be computed:

$$\frac{\textit{Correctly predicted residues}}{\textit{Total number of residues}} \times 100\%$$

Each protein sequence contains, on average, a few hundred residues. Residues were randomly selected from different protein sequences as test data. Altogether 8000 residues that were distinct from the training set were used for testing. The test data was organized into 4 sets. Each set contains residues of several protein sequences. The test data were used to test both the original PHD network that operates with the jury layer as well as the improved version that operates with the three fuzzy systems.

The accuracy of the original PHD Network is measured to be 74.92%. The accuracy of the improved version is shown in table 1. The fuzzy system improves the performance of the jury layer, as expected. While the improvement is not large the result is useful, and indicates that there is benefit of the use of fuzzy systems in a hybrid manner with the existing neural network model. The next step is the extension to the individual networks.

Apart from the accuracy of the prediction system, other issues have also been investigated. It is observed that by segregating the jury task to three corporative fuzzy system, the training process has been sped up. A fuzzy system that accepts 9 inputs (from three neural networks), takes more than 40 minutes to complete one epoch. The decrease in least mean square error over each epoch is also insignificant suggesting the possibility of slow convergence. This is in contrast with the use of three small fuzzy systems where each system requires less than 5 minutes to compete an epoch and converges, on average, in approximately 50 epochs. Table 2 shows the number of epochs required by each fuzzy system to complete training in this study. The benefits of using small and cooperative fuzzy systems over large fuzzy systems are obvious. By small fuzzy systems, we mean fuzzy systems that have small number of inputs. At the time of writing, designing fuzzy systems that requires large amount of inputs is still a difficult task. This is because fuzzy systems suffer from rules explosion. The use of multiple cooperative fuzzy systems could be a simple yet efficient technique to address part of this problem.

Table 1: Accuracy of PHD network with three fuzzy systems that replaces the jury layer

Test Set	**Accuracy**
1	*75.50%*
2	*74.75%*
3	*75.35%*
4	*74.45%*
Average	***75.01%***

Table 2: Epochs required by each fuzzy system to complete training

Fuzzy System	**Epoch**
Helix	*40*
Extended	*33*
Loop	*76*

Table 3: Several Fuzzy Rules of the Helix Fuzzy System

If (H is Low) and (L is Low) and (E is Medium) then O = -48.68H + 105.8L + 1.013E - 48.74
If (H is Low) and (L is Low) and (E is Low) then O = 89.43H + 8.439L + 1.14E + 5.863
If (H is Low) and (L is Medium) and (E is Low) then O = 71.92H + 8.105L - 49E – 375.8

AGENDA:
H = Helix
L = Loop
E = Extended
O = probability of the final prediction output to be Helix

There is another advantage to the use the fuzzy systems as compared to the jury layer. Fuzzy systems operate using fuzzy rules. Table 3 shows 3 out of the 27 fuzzy rules generated for the Helix fuzzy system. The use of fuzzy rules allows for user manipulation and interaction. It allows domain experts to adjust each fuzzy rule to improve the overall system performance. As a result, a system becomes more adaptable to changes over time. Domain experts can encode new knowledge into the system from time to time as well.

5. Conclusions

We proposed in this paper the use of individual fuzzy systems for each neural network's outputs in previous layers. We have noted earlier that neural network conclusions cannot safely be used as probabilities if the network has not been trained using probability information in the training set, but has been trained (as is usually the case) for classification using 0 and 1 (or 0.1 and 0.9) values.

The use of hybrid neural network and fuzzy systems can improve decision accuracy in a practical application in bioinformatics for protein structure prediction. This improvement derives from the more sophisticated defuzzification techniques available as compared to simple averaging. Using this technique, no probabilistic assumption is necessary, instead the network outputs are transformed into fuzzy membership values, which is expected to further enhance the accuracy of the overall system. We have also proposed the use of the fuzzy logic's possibilistic nature to replace the unsound probabilistic assumption often made for neural network conclusions. The latter is ongoing work which will be reported elsewhere.

As part of the study, we have also addressed another issue, the use of multiple cooperative small fuzzy systems, as opposed to large fuzzy systems. This can be beneficial. The amount of time that is required for training the three-input fuzzy systems is small.

The use of fuzzy systems in place of the jury layer of the neural networks also results in a more user-friendly system. Domain experts can encode knowledge into the fuzzy systems by adjusting the fuzzy rules. The fuzzy rules also help to explain the steps taken by the prediction system in producing a particular output.

References

Baldi, P., Brunak, S., Frasconi, P., Soda, G., and Pollastri, G. (1999): Exploiting the past and the future in protein secondary structure prediction. Bioinformatics, 15(11), 937-946.

Defay, T.R., and Cohen, F.E. (1996): Multiple sequence information for threading algorithms. J. Mol. Biol., 262, 314-323.

Fischer, D., and Eisenberg, D. (1996): Protein fold recognition using sequence-derived predictions. Protein Sci., 5, 947-955.

Flockner, H., Braxenthaler, M., Lackner, P., Jariz, M., Ortner, M., and Sippl, M.J., (1995): Progress in fold recognition. Proteins: Struct Funct. Genet, 23, 376-386.

Jang, J-S.R., (1993): ANFIS: Adaptive-Network-Based Fuzzy Inference System. IEEE Transaction on Systems, Man, and Cybernetics, 23(3).

Jang, J-S.R., Sun, C-T., and Mizutani, E. (1997): Neuro-Fuzzy and Soft Computing: A Computational Approach to Learning and Machine Intelligence. Prentice Hall, US.

Lathrop, R.H., and Smith, T.F. (1996): Global Optimum protein threading with gapped alignment and empirical pair score functions. J. Mol. Biol., 255, 641-665.

Qian, Ning., and Sejnowski, T.J. (1988): Predicting the Secondary Structure of Globular Proteins Using Neural Network Models. J. Mol. Biol., 202, 865-884.

Rost, B., and Sander, C. (1993): Prediction of Protein Secondary Structure at Better than 70% Accuracy. J. Mol. Biol., 232, 584-599.

Vivarelli, F., Giusti, G., Villani, M., Campanini, R., Fariselli, P., Compiani, M. and Casadio, R. (1995): LGANN: a parallel system combining a local genetic algorithm and neural networks for the prediction of secondary structure of proteins. Bioinformatics, 11, 253-260.

Zhang, X., Mesirov, J., P., and Waltz, D.L. (1992): Hybrid System for protein secondary structure prediction. J. Mol. Biol., 225, 1049-1063.

Zhang, Chun-Ting, Chou, Kuo-Chen, and Maggiora, C.M.: Predicting protein structural class from amino acid composition: application of fuzzy clustering.

On Fuzzy Controllers Having Radial Basis Transfer Functions*

Domonkos Tikk[1,2], László T. Kóczy[1,2], and Tamás D. Gedeon[2]

[1] Department of Telecommunication & Telematics
Budapest University of Technology and Economics
H-1117 Budapest, Pázmány sétány 1/d, Hungary
[2] School of Information Technology, Murdoch University
South Street, Murdoch, 6148 W.A., Australia

Abstract. This works establish a link between certain fuzzy controllers and radial basis function networks. We deal with the two soft computing techniques and compare them from the approximation capability point of view. The fact that the approximation behaviour of these systems are similar establishes the ground of analysing their possible functional equivalence. We show that certain fuzzy systems can be considered as radial basis function (RBF) approximation scheme, having the same transfer function. This allows to implement those fuzzy systems by RBF networks, and hence, establishing a learning algorithm for them and improving their approximation capabilities. We also points out that the classes of the corresponding fuzzy system can be generalized to some extent if we use general basis function with weighted norm.

1 Introduction

The approximation capabilities of soft computing techniques has been investigated intensively in the last 10–15 years. The approximation behaviour of these techniques was found to be similar: most of the general fuzzy systems and neural networks are universal approximators on a compact subset of continuous functions with respect to the supremum or L_p norm $p \in [1, \infty)$.

These results reveal that the set of transfer functions which can be approximated by fuzzy systems or neural networks more or less the same. However, the way they captures the underlying transfer functions is different. Fuzzy systems operating with if-then rules have the advantage of easy linguistic interpretability, while neural networks can adapt learning methods to improve their performance according to a training data set. Neuro-fuzzy systems, which link these techniques, enable us to combine their advantageous properties, and hence, we can obtain better behaving tools. This can be backed from the practical side by the numerous successful applications (see e.g. [1–3]).

* This research was supported by the Australian Research Council, the Hungarian Ministry of Education (MKM) FKFP 0235/1997 and FKFP 0422/1997 and the National Science Research Fund (OTKA) 019671 and 030655.

In this paper we will show that several fuzzy controllers implement one of the typical neural networks (having radial basis activation functions). One type of those fuzzy systems is the KH interpolative controller to be recalled in Section 2. We will briefly summarize the mathematical background of (RBF) network in Section 3. We present how the implementing fuzzy systems vary if we use weighted norm in the network.

2 Interpolation in Sparse Fuzzy Rule Bases

The classical approaches of fuzzy control deal with dense rule bases where the universe of discourse is α-covered (at any point x from the input set, there exists a fuzzy set A, for which $A(x) \geq \alpha$ holds, usually $\alpha \geq 0.5$) by the antecedent fuzzy sets of the rule base in each dimension. If for given α the antecedents form no α-cover, the rule base is α-sparse. If they do not form a cover for any $\alpha > 0$, the rule base is sparse. The classical fuzzy reasoning methods (Zadeh's CRI [4], Mamdani [5], Larsen [6], and Takagi–Sugeno [7]) and all other fuzzy control approaches based on some form of degree of matching between observation and antecedents cannot be used for sparse rule systems.

Instead, a more general notion of the degree of similarity was introduced, based on the family of *α-distances between two convex and normal fuzzy sets* (CNF sets), which is defined as

$$\begin{aligned}\widetilde{d}(A_1, A_2) &= \{d_{\alpha,C}(A_1, A_2),\ \alpha \in (0,1], C \in \{L, U\}\} \\ &= \{\|A_{1\alpha L} - A_{2\alpha L}\|, \|A_{1\alpha U} - A_{2\alpha U}\|, \alpha \in (0,1]\}\end{aligned} \tag{1}$$

subscripts L and U denoting the minimum and maximum of the respective α-cuts. For simplicity, the distance belonging to fixed and L or U will be denoted by $d_{\alpha,C}(A_1, A_2)$. The conditions for the existence of the fuzzy distance set is that both fuzzy sets are CNF, and that they are comparable in the sense of $\prec$.

Using this notion of distance (1), the fuzzy similarity set can be defined as

$$\begin{aligned}\widetilde{s}(A_1, A_2) &= \{s_{\alpha,C}(A_1, A_2),\ \alpha \in (0,1], C \in \{L, U\}\} \\ &= \left\{\frac{1}{d_{\alpha,C}(A_1, A_2)}, \alpha \in (0,1]\right\}\end{aligned} \tag{2}$$

the elements of the similarity degree set being the reciprocals of the elements of the distance set.

The basic idea of the rule interpolation is formulated in the *Fundamental Equation of Rule Interpolation* (FERI):

$$D(A^*, A_1) : D(A^*, A_2) = D(B^*, B_1) : D(B^*, B_2). \tag{3}$$

In this equation A^* and B^* denote the observation and the corresponding conclusion, while $R_1 = A_1 \to B_1$, $R_2 = A_2 \to B_2$ are the rules to be interpolated, such that $A_1 \prec A^* \prec A_2$ and $B_1 \prec B_2$. If D denotes the Euclidean distance between two symbols, the solution for B^* results in simple linear interpolation. If $D = \widetilde{d}$ (the fuzzy distance family), linear interpolation between corresponding α-cuts is performed.

A more general form of (3) gives

$$B_\alpha^* = \sum_{j=1}^{r} s_\alpha(A^*, A_j) B_{j\alpha}^*, \quad \alpha \in (0, 1]$$

where s_α is some α-cut related similarity degree, e.g., the fuzzy similarity (2) obtained from the reciprocal distances of the α-cuts. This similarity can be considered as an extended "degree of matching", and these similarity degrees replace the degrees $A_j(\underline{x}^*)$ or height$(A_j \wedge A^*)$ (height$(A) = \max_x A(x)$ is the height of a fuzzy set; $\wedge$ denotes an appropriate t-norm) which are used in the Mamdani, Larsen, (Takagi–) Sugeno and other classical fuzzy reasoning algorithms. These methods can be substituted by fuzzy interpolation (KH interpolation [8]), which also functions in sparse rule bases. Hence the basic formula of the Mamdani type algorithm,

$$B^* = \max_j \left\{ \min \left\{ \text{height}(A^*(\underline{x}), A_j(\underline{x})), B_j(y) \right\} \right\}, \tag{4}$$

is replaced by

$$B_\alpha^* = \frac{\sum_{j=1}^{r} s_\alpha(A^*, A_j) B_{j\alpha C}}{\sum_{j=1}^{r} s_\alpha(A^*, A_j)} = \frac{\sum_{j=1}^{r} \frac{1}{d_{\alpha,C}(A^*, A_j)} B_{j\alpha C}}{\sum_{j=1}^{r} \frac{1}{d_{\alpha,C}(A^*, A_j)}} \tag{5}$$

where the normalized degree of similarity for fixed α and C is the reciprocal distance of the observation from the corresponding antecedent, divided by the sum of all these distances. It must be noted, however, that B^* reconstructed directly from the above α-cuts does not always exist, as various abnormalities in the shape of the conclusion might necessitate some transformations, which eventually result in obtaining sub-normal conclusions (cf. [8]). It should be noted that in practical applications, it is enough to do calculations for $\alpha \in \mathcal{B}$, the breakpoint set of the membership functions, which is four points altogether [9] in the case of the most widely applied trapezoidal functions $\mathcal{B} = \{0, 1; L, U\}$.

In all these approaches, the approximated mapping is "fuzzy set to fuzzy set":

$$\widetilde{\Psi} : \widetilde{P}(X) \to \widetilde{P}(Y)$$

where $\widetilde{P}(Z)$ denotes the fuzzy power set of Z, i.e. all fuzzy sets of the universe of discourse Z, so that a fuzzy observation is always mapped into a fuzzy conclusion by the respective rule base and inference engine: $\widetilde{\Psi}(A^*) = B^*$. In the special case where the observation is a crisp singleton, we have $\widetilde{\Psi}(\underline{x}^*) = B^*$. In control applications the conclusion set should always be defuzzified, so that $y^* = \text{defuzz}\,(B^*)$. With both crisp singleton inputs and outputs, we get $y^* = \text{defuzz}\left(\widetilde{\Psi}(\underline{x}^*)\right) = \Psi(\underline{x}^*)$. (Some of the most important Ψ-s are explicitly given in [10].) In this sense, fuzzy rule based controllers can be considered as multivariate real function approximators.

The respective interpolative extension for the equation for the Takagi–Sugeno algorithm is

$$y^* = \frac{\sum_{j=1}^{r} s_\alpha(A^*, A_j)(b_{j_1}\underline{x}^* + b_{j_0})}{\sum_{j=1}^{r} s_\alpha(A^*, A_j)}$$

where $\forall j \in \{1, r\} : b_{j_1} = 0$ in the case of the Sugeno controller [11]. Here, the similarity degree is an overall one that can be derived from the reciprocal of an overall distance of the two fuzzy sets in question, e.g. by defining it as

$$s(A^*, A_j) = \frac{1}{\int\limits_{\alpha_L \in (0,1] \wedge \alpha_u \in (0,1]} d_{\alpha,C}(A^*, A_j)\mathrm{d}\alpha}. \tag{6}$$

In the case of trapezoidal membership functions (6) can be simplified to

$$s(A^*, A_j) = \frac{4}{d_{0,L}(A^*, A_j) + d_{1,L}(A^*, A_j) + d_{0,U}(A^*, A_j) + d_{1,U}(A^*, A_j)}.$$

because both the left and right pairs of flanks of the antecedents define trapezoidal areas. This latter interpolation has the advantage of obtaining directly defuzzified consequents, which fact eliminates the problem of the abnormal conclusion shape. This interpolation method will be investigated in more detail in the future.

The investigation of the KH interpolation has shown that using a more radical similarity degree improves the convergence properties tremendously. If s is defined as

$$s_{\alpha,C}(A^*, A_j) = \frac{\frac{1}{d^n_{\alpha,C}(A^*,A_j)}}{\sum_{j=1}^{r} \frac{1}{d^n_{\alpha,C}(A^*,A_j)}}, \tag{7}$$

it will be guaranteed that when the consequent membership functions are known ε with accuracy, the conclusion function will be also calculable with $\delta(\varepsilon)$ accuracy [12]. Due to this result, the input-output function of the general KH interpolation obtained by means of the radical similarity defined in (7),

$$B^*_{\alpha,C} = \frac{\sum_{j=1}^{r} \frac{1}{d^n_{\alpha,C}(A^*,A_j)} B_{j\alpha C}}{\sum_{j=1}^{r} \frac{1}{d^n_{\alpha,C}(A^*,A_j)}} \tag{8}$$

possesses the universal approximation property for every fixed α, i.e., every continuous function defined on a compact set can be approximated arbitrarily accurately with respect to the norm L_p ($p \in \{[1,\infty)\} \cup \infty$) with functions of this form. By this, the mathematical stability of the method can be formulated simply as follows: If the antecedents or the observation of a rule base change only slightly, it does not cause a significant change in the conclusion.

General KH interpolation has strong connections with the so-called Balázs–Shepard interpolator operator well known in approximation theory (see e.g. [13–15]). This operator is also capable of approximating real valued continuous functions arbitrarily well, where the convergence of the approximation depends on the modulus of continuity (a sort of smoothness factor) of the approximated real function. It has been shown that for every fixed value of α, the function generated by the general KH interpolation is a Balázs–Shepard operator [12]. On the other hand, for practically important piecewise linear membership functions it can be proved that it is sufficient to calculate the conclusion for the elements of the breakpoint set of the membership functions, and in the regions between the breakpoint levels the conclusion can be approximated well by linear functions. Using these results, general KH interpolation can be considered as a generalized function approximator being able to approximate fuzzy-to-fuzzy mappings on the Cartesian product of compact universes of discourse.

There are further methods for fuzzy rule interpolation that we do not discuss here in detail (cf. [16–18]).

3 Radial Basis Functions in Neural Networks

3.1 Mathematical background

An alternative approach in soft computing field is the artificial neural network model. Three-layer feedforward neural networks using *radial basis functions* are similarly function approximators as are rule based fuzzy systems. Park and Sandberg proved [19] that any continuous function f could be arbitrarily well approximated on a compact subset $X \subset \mathbb{R}^n$ in L_∞ norm:

$$f(\underline{x}) = \sum_{i=1}^{r} \lambda_i \varphi(\|\underline{x} - \underline{\xi}_i\|) \tag{9}$$

where $\varphi : \mathbb{R}^+ \to \mathbb{R}$, $\underline{\xi}_i$ are called the centres of the functions and are real numbers. If there are r known points of the function to be approximated, $\{y_i = F(\underline{x}_i),\ i = 1, \ldots, r\}$, by substituting them into (9), a set of linear equations for is obtained [20]:

$$\begin{pmatrix} y_1 \\ \vdots \\ y_r \end{pmatrix} = \begin{pmatrix} \varphi_{11} & \cdots & \varphi_{1r} \\ \vdots & \ddots & \vdots \\ \varphi_{r1} & \cdots & \varphi_{rr} \end{pmatrix} \begin{pmatrix} \lambda_1 \\ \vdots \\ \lambda_r \end{pmatrix} \tag{10}$$

where

$$\varphi_{ij} = \varphi(\|\underline{x}_i - \underline{\xi}_j\|)$$

By denoting $\underline{\underline{\Phi}} = \begin{pmatrix} \varphi_{11} & \dots & \varphi_{1r} \\ \vdots & \ddots & \vdots \\ \varphi_{r1} & \dots & \varphi_{rr} \end{pmatrix}$, $\underline{\lambda} = \begin{pmatrix} \lambda_1 \\ \vdots \\ \lambda_r \end{pmatrix}$, and $\underline{y} = \begin{pmatrix} y_1 \\ \vdots \\ y_r \end{pmatrix}$, (10) is simply written as

$$\underline{y} = \underline{\underline{\Phi}}\underline{\lambda}$$

From here, the solutions for λ_i can be obtained by

$$\underline{\lambda} = \underline{\underline{\Phi}}^{-1}\underline{y}$$

if $\underline{\underline{\Phi}}$ is regular. For a rather large set of various functions it has been proved that $\underline{\underline{\Phi}}$ is always regular (see e.g. [21]), if $\underline{\xi}_i$ $(i = 1, \dots, r)$ are distinct vectors. For example, x, $x^2 \log x$, x^3, $(x^2 + c)^\alpha$, $(x^2 + c)^{-1/2}$, and e^{-cx^2} $(c > 0, 0 < \alpha < 1)$. These functions belong to two different groups: *localized* functions which satisfy $\varphi(\underline{z}) \to 0$, when $z \to 0$, and so $\underline{\underline{\Phi}}$ is positive definite, while *non-local* functions $\varphi(z)$ become unbounded, when $z \to 0$, and so $\underline{\underline{\Phi}}$ is not positive definite, having $r - 1$ negative eigenvalues and only one positive one [22].

The RBF approach can be extended to multiple output mappings, and more general approximation, where the number of known points rather larger than the number of functions φ_i, using the Moore–Penrose pseudo inverse (cf. [20,23]):

$$\underline{\underline{\Phi}}^+ = (\underline{\underline{\Phi}}^T\underline{\underline{\Phi}})^{-1}\underline{\underline{\Phi}}^T,$$

where $\underline{\underline{\Phi}}^+\underline{\underline{\Phi}} = \underline{\underline{I}}_r$, the $r \times r$ identity matrix.

3.2 HyperBF approximation with weighted norm

The usually Euclidean norm function used in the RBF approximation (9) is suitable when the elements of the vectors to calculate with belong to the same classes. However, when this is not the case, it is more appropriate to consider a general *weighted norm*, which can be defined by its square as [24]:

$$\|\underline{x}\|_C^2 = (\underline{\underline{C}}\underline{x})^T(\underline{\underline{C}}\underline{x}) = \underline{x}^T\underline{\underline{C}}^T\underline{\underline{C}}\underline{x} \tag{11}$$

where $\underline{\underline{C}}$ is the *norm weighting matrix* with size $n \times n$, n being the dimension of $\underline{x}$.

We can generalize the RBF approximation scheme by using the weighted norm defined in (11) as

$$f(\underline{x}) = \sum_{i=1}^{r} \lambda_i \varphi(\|\underline{x} - \underline{\xi}_i\|_C) \tag{12}$$

which scheme are called *generalized radial*, or *hyper basis function* approximation scheme [24].

The use of the weighted norm can be interpreted as a simple *affine transformation* of the input space. As a special case, the approximation (12) contains the original RBF scheme (9), when $\underline{\underline{C}} = I_n$, the n dimensional identity matrix. We will return to HyperBF approximation scheme in the next section.

3.3 The RBF network

Three-layer feedforward neural networks using radial basis functions and with fully interconnected layers are similarly suitable function approximators as the fuzzy systems. The number of units in the input layer should be n, the number of input variables. Each connection between the input layer and the hidden layer should be assigned a value ξ_{ij}, where $i \in \{1, \ldots, r\}$ indicates the subscript of the corresponding unit centre, and $j \in \{1, \ldots, n\}$ refers to the respective component. (In the basic case $r = n$.) The ith hidden unit computes the norm $\|\underline{x}^* - \underline{\xi}_i\|$, and its mapping by φ. The output layer consists of a single unit in the one output variable case, and this unit is connected to the ith hidden unit by a weight λ_i, and is linear, so it generates the value $f(\underline{x}^*) = \sum_{i=1}^{r} \varphi(\|\underline{x}^* - \underline{\xi}_i\|)$ (see Figure 1). Multiple unit output layers can similarly implement multiple output variable cases.

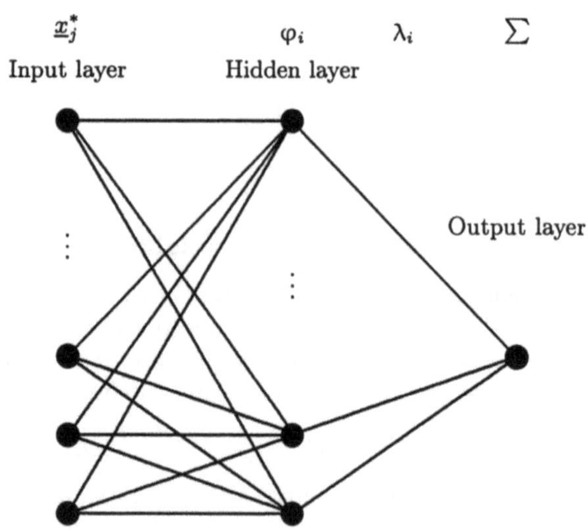

Fig. 1. Three-layer feedforward neural network with RBF activation functions in the hidden layer

4 Fuzzy Controllers Implementing RBF Networks

4.1 Approximation of fuzzy-to-fuzzy mappings

Comparing the formulae of various fuzzy interpolation techniques and RBF based feedforward networks, some similarities can be detected. Fuzzy reasoning and control systems are approximators on two different levels. They approximate primarily fuzzy-to-fuzzy mappings in the following sense:

$$S : \widetilde{P}(X) \to \widetilde{P}(Y),$$

S is the generalization of the notion of a set-to-set function. As every fuzzy set is described by a membership function, for every given $\alpha \in (0,1]$ and fixed C („lower or upper end"), S realizes a function

$$S_{\alpha,C} : X \to Y,$$

and fuzzy interpolation means the solution of an ordinary interpolation problem for every $S_{\alpha,C}$. The r rules define r vectors $(A_{i\alpha C}, B_{i\alpha C})$, so that the interpolation problem is equivalent to fitting a function to these r control points. Fuzzy interpolation means finding a family of functions $y = F(\underline{x}; \alpha, C)$, $\alpha \in (0,1]$, $C \in \{L, U\}$interpolating these families of points. For a fixed α-cut and C, the problem to solve is identical with that of ordinary interpolation.

4.2 Classical controllers and RBF schemes

On the other hand, if a fuzzy system is considered with crisp observations and defuzzified outputs, the whole system is equivalent to a single approximator. Because of the different defuzzification techniques are rather complicated to be expressed in terms of the approximation error involved, in this context we refer to the explicit formulae of the most frequently used fuzzy controllers [10], but will not discuss this aspect here in detail.

Getting back to the first type of problem, it can be recognized that Mamdani control has a similarity to RBF approximation, but is less specific concerning the type of functions used.

Obviously, the original Mamdani method is rather different from the RBF approach, as not addition but a max operator combines the rule premises and the observation (i.e. functions height$(A^*(\underline{x}), A_j(\underline{x}))$, and also the "weighting" is done by min rather than multiplication. So, the formula (4) can be written like:

$$F(\underline{x}; \alpha, C) = \max_{j=1}^{r}\{\min\{\lambda_j(\alpha, C), A_j(\underline{x})\}\}, \tag{13}$$

where $\lambda_j(\alpha, C) = B_{j\alpha C}(y)$ if B^* is calculated point by point. There is a striking analogy with the formulae of the RBF-s, especially as in the theory of possibility measures, the max operator replaces the addition and min might play a similar role as multiplication in probability theory. So, the above equation can be interpreted as a "*possibilistic linear combination*" of the consequent function points. An additional essential difference is however the lack of radial behavior.

According to many results, Mamdani type approximators are universal [25–27], at least in the sense that their full versions including defuzzification can arbitrarily well approximate an arbitrary continuous function over a compact domain. It is well worth investigating in the future whether this statement remains true for arbitrary continuous fuzzy-to-fuzzy mappings as well.

If Larsen interpolation is applied with centre of gravity defuzzification, the equivalent conclusion can be written point by point as

$$F(\underline{x}; \alpha, C) = \sum_{j=1}^{r} \lambda_j(\alpha, C) A_j(\underline{x}), \tag{14}$$

where $\lambda_j(\alpha, C)$ is the same as in (13) for the Mamdani formula. Here, the only difference to the RBF case is the non-radial nature of the general antecedent function.

Proposition 1. *If in the Larsen algorithm all antecedent functions are symmetric around the centre of the kernel:* $A_i(\underline{\xi}_i - \delta) = A_i(\underline{\xi}_i + \delta)$*, and further, if* $A_i(\underline{x})$ $(i = 1, \ldots, r)$ *differ from each other only in a constant factor, this special fuzzy controller implements an RBF approximator for every fixed* α*.*

It has been shown in [10], that the transfer function of symmetric Sugeno controllers [11] is identical with the transfer function of Larsen algorithm using the centre of gravity method. Exploiting this functional equivalence, we can have the following

Corollary 1. *Sugeno controllers with symmetric antecedent sets with respect to the centre of the kernel:* $A_i(\underline{\xi}_i - \delta) = A_i(\underline{\xi}_i + \delta)$*, and further, if* $A_i(\underline{x})$ $(i = 1, \ldots, r)$ *differ from each other only in a constant factor, this special fuzzy controller implements an RBF approximator for every fixed* α*, as well.*

Considering the results of [27], it would be interesting to investigate whether in general, any convex and normal fuzzy (CNF) membership function could generate a suitable RBF with positive definite matrix $\underline{\underline{\Phi}}$, where the only necessary condition for the radial basis function is that $\varphi(0) = 1$, $\varphi(z) \to 0$, when $z \to 0$ and is monotone decreasing. We conjecture that the answer of this question is rather negative.

4.3 Fuzzy KH interpolators and RBF schemes

Considering now the formulae of the linear KH interpolation,

$$F(\underline{x}; \alpha, C) = \sum_{j=1}^{2} \lambda_j(\alpha, C) \varphi(\underline{x}),$$

where $\varphi(x) = \dfrac{1}{d_{\alpha,C}(A^*(\underline{x}), A_j(\underline{x}))}$ and $\lambda_j(\alpha, C) = \dfrac{B_{j\alpha C}}{\sum_{j=1}^{2} \frac{1}{d_{\alpha,C}(A^*(\underline{x}),A_j(\underline{x}))}}$, is obviously an RBF approach for every fixed level and end point.

Similar constructions apply for the general KH interpolation, where

$$F(\underline{x}; \alpha, C) = \sum_{j=1}^{r} \lambda_j(\alpha, C)\varphi(\underline{x}),$$

$\varphi(x) = \dfrac{1}{d_{\alpha,C}(A^*(\underline{x}), A_j(\underline{x}))}$ and $\lambda_j(\alpha, C) = \dfrac{B_{j\alpha C}}{\sum_{j=1}^{r} \frac{1}{d_{\alpha,C}(A^*(\underline{x}),A_j(\underline{x}))}}$, for every fixed level and end point.

Analogously, for the stabilized KH interpolator also similar constructions apply, where we take the sth power of the distance function in the previous expressions, where $s \geq n$ and n stands for the number of variables.

Hence we have the following result:

Proposition 2. *The general KH interpolation implements an RBF approximator for every fixed* α. *Here, due to the construction of* $\varphi(\underline{x})$ *and* $\lambda_j(\alpha, C)$, *the vectors* $\underline{\xi}_i$ $(i = 1, \ldots, r)$ *need not to be distinct to assure the regularity of the coefficient matrix* $\underline{\underline{\Phi}}$.

A very similar statement can be said about the stabilized version of the KH interpolation, where however the distance function is raised to the power s ($s \geq n$ and n is the number of variables). Consequently we have:

Proposition 3. *The stabilized general KH interpolation also implements an RBF approximator for every fixed* α. *Here, due to the construction of and* $\varphi(\underline{x})$ *and* $\lambda_j(\alpha, C)$, *the vectors* $\underline{\xi}_i$ $(i = 1, \ldots, r)$ *need not to be distinct to assure the regularity of the coefficient matrix* $\underline{\underline{\Phi}}$.

Observe that KH interpolators implement RBF network straightforwardly: no restrictions apply on the shape or location of the involved membership functions.

4.4 Generalization possibilities with HyperBF networks

Let us now investigate how the use of HyperBF schemes can enhance the equivalent fuzzy controller type, i.e. how it generalize the design parameters of the Larsen (or Sugeno) controllers.

Let us examine which condition of Proposition 1 can be relaxed with the use of more general weighted norm. Clearly, the condition on the symmetry of the antecedent fuzzy sets can not be omitted, because the weighted norm (11) is symmetric, as well. The antecedents satisfying Proposition 1 are symmetric, and in each dimension, their projections has a fixed shape, termed as *base shape*, from which the antecedents can differ only in a constant factor. In

Proposition 1 the only degree of freedom, once a base shape of the antecedents is chosen, is the stretching (or shrinking) factor of λ_j in (14), which can be different for each n-dimensional antecedents. This only allows the projections of an arbitrary antecedent A_j to be different from the base shape of the dimension with the fixed stretching/shrinking factor defined by the coefficient λ_j.

If we use weighted norms instead of the regular Euclidean one, the antecedents have two degrees of freedom. One which enables them to be different to a constant extent for each n-dimensional antecedents (λ_j), and the other which enables them to be different to a constant extent in each dimension (determined by the weighting matrix C). It means that the antecedents should be symmetric, in each dimension there is a base shape of the antecedents' projection, but from this shape the projections can differ with a constant factor for each antecedents, and in each input dimension a further stretching/srinking factor can be given which enables the input spaces to be different by this scaling factor. The scaling factors of the input spaces are determined with the weighting matrix C (see 11). Finally, we can state that with the HyperBF approximation scheme, the generalization of the RBF approximation scheme, we can implement slightly more general Larsen and Sugeno type fuzzy controllers. This is summarized in the next

Proposition 4. *If in the Larsen or the Sugeno algorithm all antecedent functions are symmetric around the centre of the kernel:* $A_i(\underline{\xi}_i - \delta) = A_i(\underline{\xi}_i + \delta)$, *and further, if each projection* $A_{ij}(\underline{x}_j)$ *of* $A_i(\underline{x})$ $(i = 1, \ldots, r; j = 1, \ldots, n)$ *differ from each other only in a constant factor for any fixed* $i \in [1, r]$, *where* $A_i(\underline{x})$ $(i = 1, \ldots, r)$ *can also differs from each other in a constant factor, this special fuzzy controller implements an HyperBF approximator for every fixed* α.

4.5 On the advantages of neuro-fuzzy techniques

According to these result, general Larsen and KH-controllers can be implemented by RBF or HyperBF networks. With a neuro-fuzzy method, we can exploit the advantageous properties of both neural and fuzzy techniques as described in the next paragraphs.

In the case where one has certain knowledge about the modelled system, e.g. interpreted as fuzzy rules, this information can be used to determine the initial parameters of a neural network. By means of the training data and the learning algorithm applied to the network, the corresponding fuzzy rule base can be tuned, this way, refining the model of the system. Such an approach results in a well-tuned model of the system (naturally to some extent depending on the reliability and appropriateness of the training data), while the tuned rule base model of the system can still provide a linguistic description of the system, thus, it assures the tracing of the system by a human operator. The linguistic description of the system gives a possible tool for quality control in certain cases: if the training data set is noisy or degenerated, a human operator can discover this kind of mistakes much easier based on a linguistic description than merely on the transfer function calculated by the network.

In the case when only input-output sample data are available, a combined neuro-fuzzy approach helps in the extraction of fuzzy rules from the data. Once the rules are extracted from the training data set the rule based description of the system is available.

We emphasize that the advantage of using fuzzy rule interpolation approaches compared to the classical fuzzy reasoning approach is not solely that the former is capable to handle sparse rule base, but — as it has been shown in the previous sections — also less restriction applies concerning the implementation by RBF networks. We remark that sparse rule bases may occur extensively when automatic rule extraction and generation is used. In such cases when the input-output sample data set does not cover the whole range of the input space, the appropriate output value may remain unknown for large uncovered input regions. As a consequence, the extracted rule base is sparse, so fuzzy controllers operating on dense rule bases can not be applied. A practical example which uses fuzzy interpolation (different from the KH method) to solve this problem (applied for well-log analysis in petroleum industry) is presented in [28].

5 Conclusion

In this paper new examples have been given for the approximate functional equivalence of fuzzy and neural models. We pointed out that under certain conditions Larsen type controllers and Sugeno type controllers implement RBF approximation scheme.

We have also shown that the (stabilized) general fuzzy KH interpolation and the RBF approximation scheme (usually implemented by RBF networks); both being considered as universal approximators, can be connected straightforwardly. The general fuzzy KH interpolators can be considered as a special RBF approximation scheme for fixed α-cuts $\alpha \in (0, 1]$, and $C \in \{L, U\}$.

We also pointed out to what extent the use of the weighted norm can generalize the implementing Larsen controllers. Finally, some of the possible advantages of the combination of fuzzy and neural techniques were outlined.

Acknowledgement

The authors want to thank to János Levendovszky for his valuable suggestions on the connection of KH interpolation and RBF functions.

References

1. J.-S. R. Jang, C.-T. Sun and E. Mizutani. (1997) Neuro Fuzzy and Soft Computing. Prentice Hall, Upper Saddle River, NJ.
2. P. M. Wong, K. W. Wong, C. C. Fung, and T. D. Gedeon. (1997) A neural-fuzzy technique for interpolating spatial data via the use of learning curve. Biological and Artificial Computation: From Neuroscience to Technology (eds: J. Mira, R. Moreno-Diaz, J. Cabestany), Springer-Verlag, 323–329.
3. K. W. Wong, D. Myers and C. C. Fung. (1999) A generalised neural-fuzzy well log interpretation model with a reduced rule base. The Sixth International Conference of Neural Information Processing ICONIP, Perth, Australia, 188-191.
4. L. A. Zadeh. (1973) Outline of a new approach to the analysis of complex systems and decision processes. IEEE Trans. on SMC **3**, 28–44.
5. E. H. Mamdani and S. Assilian. (1975) An experiment in linguistic synthesis with a fuzzy logic controller. Int. J. of Man Machine Studies **7**, 1–13.
6. P. M. Larsen. (1980) Industrial application of fuzzy logic control. Int. J. of Man Machine Studies **12**(4), 3–10.
7. M. Sugeno and T. Takagi. (1985) Fuzzy identification of systems and its applications to modeling and control. IEEE Trans. on SMC **15**, 116–132.
8. L. T. Kóczy and K. Hirota. (1993) Approximate reasoning by linear rule interpolation and general approximation. Internat. J. Approx. Reasoning **9**, 197–225.
9. Y. Shi and M. Mizumoto. (1995) Some considerations on Kóczy's interpolative reasoning method. Proc. of the FUZZ-IEEE/IFES'95, Yokohama, 2117–2122.
10. L. T. Kóczy and M. Sugeno. (1996) Explicit functions of fuzzy control systems. International Journal of Uncertainty, Fuzziness and Knowledge-Based Systems **4**, 515–535.
11. M. Sugeno. (1985) An introductory survey of fuzzy control. Information Science **36**(1–2), 59–83.
12. D. Tikk, I. Joó, L. T. Kóczy, P. Várlaki, B. Moser and T.D. Gedeon. Stability of interpolative fuzzy KH-controllers. To appear in Fuzzy Sets and Systems.
13. D. Shepard. (1968) A two dimensional interpolation function for irregularly spaced data. Proc. of the 23rd ACM International Conference, 517–524.
14. J. Szabados. (1991) Direct and converse approximation theorems for Shepard-operator. J. Approx. Theory and its Appl. **7**, 63–76.
15. B. Della Vecchia and G. Mastroianni and J. Szabados. (1996) Balázs–Shepard operators on infinite intervals. Annales Uni. Sci. Budapest, Sectio Comput. **16**, 93–102.
16. T. D. Gedeon and L. T. Kóczy. (1996) Conservation of fuzziness in rule interpolation. Proc. of the Symp. on New Trends in Control of Large Scale Systems, Herlany, vol. 1, 13–19.
17. P. Baranyi and T. D. Gedeon. (1996) Rule interpolation by spatial geometric representation. Proc. of the IPMU'96, Granada, 483–488.
18. D. Tikk and P. Baranyi. Comprehensive analysis of a new fuzzy rule interpolation method. To appear in IEEE Trans. on Fuzzy Systems.
19. J. Park and I. W. Sandberg. (1991) Universal approximation using radial-basis-function networks. Neural Computation **3**, 246–257.

20. K. Hlavačková and R. Neruda. (without year) Radial basis function networks. Inst. of Information and Computer Science, Czech Academy of Sciences, Prague, report.
21. A. C. Faul and M. J. D. Powell. (1999) Krylov subspace methods for radial basis function interpolation. DAMPT 1999/NA11, University of Cambridge, 25p.
22. C. A. Micchelli. (1986) Interpolation of scattered data: Distance matrices and conditionally positive definite functions. Constructive Approximation **2**, 11–22.
23. D. S. Broomhead and D. Lowe. (1988) Multivariable functional interpolation and adaptive networks. Complex Systems **2**, 321–355.
24. T. Poggio and F. Girosi. (1990) Networks and the best approximation property. Biological Cybernetics **63**, 169–176.
25. L. X. Wang. (1991) Fuzzy systems are universal approximators. Proc. of the IEEE Int. Conf. on Fuzzy Systems, San Diego, 1163–1169.
26. B. Kosko. (1991) Fuzzy systems as universal approximators. Proc. of the IEEE Int. Conf. on Fuzzy Systems, San Diego, 1153–1162.
27. J. Castro. (1995) Fuzzy logic controllers are universal approximators. IEEE Trans. on SMC **25**, 629–635.
28. K. W. Wong and T. D. Gedeon. Petrophysical properties prediction using self-generating fuzzy rules inference system with modified α-cut based fuzzy interpolation. Submitted to Intelligent Systems and Application (ISA'2000, Sydney).

Evolutionary Scene Recognition and Simultaneous Position/Orientation Detection

Mamoru Minami[1], Julien Agbanhan[2], and Toshiyuki Asakura[1]

[1] Fukui University, 3-9-1, Bunkyo, Fukui 910-8507, JAPAN
[2] Graduate School of Engineering, Fukui University, 3-9-1, Bunkyo, Fukui 910-8507, JAPAN

Abstract. This paper presents a new method of scene recognition for manipulator real-time visual servoing, which utilizes a hybrid genetic algorithm in combination with a model shaping a target of known shape, and the unprocessed gray-scale image of a scene, termed here as raw-image. The scene recognition method presented here is concerned with the simultaneous recognition of the shape and detection of the position and orientation in the two-dimensional raw-image, of a three-dimensional target being imaged. This approach of scene recognition is applied for the recognition of either a non-self moving target as well as for the a self-moving target such as a living animal, in the presence of noises like lighting condition variations and other objects in the scene, considered as noises. The raw-image is employed since it does not increase the original noise, thereby being more transportable, and moreover contrary to a binary image processing, does not require any filtering processing time. In fact here, the problem of an object recognition in the raw-image is changed to an optimization problem of a model-based evaluation function, named surface-strips model-based fitness function. This fitness function possesses information about the shape of a target, and consists in the computation of the brightness difference between an internal surface and a contour-strips. In this research, in order to find a target object in every newly input raw-image to the recognition system, the highest peak of the distribution of the surface-strips model-based fitness function, which corresponds to the recognition results of the designated target, is searched by a hybrid genetic algorithm, which employs a population of potential solutions to perform the search of the target. This hybrid genetic algorithm employs the "global"search feature of a two-point crossover genetic algorithm (GA), to search a target, together with a GA-based localized search technique that focuses on the target of interest found so far, in order to perform an intensive localized search. The localized GA search technique relies on mutation of bits on the lower portion of genes in positional and orientational binary strings, in order to improve the GA-based scene recognition performance, in terms of fast and reliable recognition of the target. This generational scene recognition by the hybrid genetic algorithm can be designated as "evolutionary scene recognition and position/orientation detection". In order to appraise the proposed scene recognition method, experiments by a hand-eye camera of a robot manipulator have been conducted to show its robustness and reliability with respect to various disturbing objects and lighting condition changes, and its effectiveness to recognize a natural fish swimming in a pool. These results have shown the suitableness of the method for manipulator real-time visual servoing.

1 Introduction

It has long been recognized [1], [2] that integrating a vision sensing into a machine is fundamental for making its action more effective and efficient, and moreover, the existence of vision might increase its versatility and application domain. For a robot to be much smarter than just a mechanical device, vision is required so that it can adapt itself to a changing working environment and recognize objects that are there in its surrounding. In recent years [3], [4], a new trend of machine intelligence, that is, a new concept of Artificial Intelligence (AI), that differs from the classical AI is being used intensively in the field of robotics and also in other research areas. In that new machine intelligence trend, "new"computational methods such as fuzzy logic, neural networks, genetic algorithm (GA) and so on, or a hybrid systems of them, are employed so that a kind of intelligence is implemented to the machine. GA, due to its inductive learning feature [5] is related to AI as an example of AI programs. A task in which visual information is used to direct the end-effector of a manipulator toward a target object is referred to in the literature as visual servoing. We think that to apply some degree of intelligence into a manipulator for real-time visual servoing purposes, it is critical that the machine performs visual tracking actions similar to gazing actions done by human beings. Robotics, biomedical applications, and a host of other areas are potential beneficiaries of breakthroughs in machine-vision. Expectational knowledge, that is, expected characteristics such as color, shape, size, texture, and so on, plays an important part in the human, as well as the machine image-understanding process. For computers to truly "see", and more importantly understand what they "see", they must be programmed with knowledge about what should be expected to "see". Many researches concerning scene recognition [6], [7] employ some image segmentation techniques that necessitates several processing levels. Such approaches seem to be computationally intensive. Also, in most of the researches that deal with object recognition [8]-[11], a binary image is used for recognition purposes. However, in most cases, obtaining a binary image necessitates several preprocessing of an original gray-scale image, which may generate [12] annoying noise due to thresholding problems. The recognition of an object in the image, that is, its shape recognition together with the detection of its position and orientation, necessitates the search of a target in the input image. Traditional search methods lack efficiency for global search and moreover are computationally intensive in the sense that the whole search region must be scanned. Therefore those search methods can be seen as time consuming ones. A genetic algorithm [13] (GA) is a search method based on the natural selection, a Darwinian survival of the fittest among a population of binary strings, and genetics. This search method allows for speculation on new search points with expected improved performance, that is, an optimization process.

This paper presents a new method of scene recognition for manipulator real-time visual servoing, which utilizes a hybrid genetic algorithm in combination with a model shaping a target of known shape, and the unprocessed gray-scale image of a scene, termed here as raw-image. This approach is basically a model-based scene recognition method in which a model shaping a target object is employed as the knowledge base for recognition purposes. Here, "scene recognition" means the simultaneous recognition of the shape and the detection of the position and orientation in the two-dimensional raw-image, of a three-dimensional target being imaged. This approach of "scene recognition" is applied for the recognition of either a non-self moving target or a self-moving one such as a living animal, in the presence of noises like lighting condition variations and other unexpected objects in the scene, considered as noises. The raw-image is employed since it does not increase the original noise, thereby being more transportable, and moreover contrary to a binary image processing, does not require any filtering processing time.

The GA is used to perform the search of a target in the raw-image, since contrary to the traditional search techniques, it can efficiently explore the search space, without need to scan whole the 2-D image area, but only need to check the model-based fitness function in the selected areas designated by genes, and exploit the best solutions obtained so far by this search method. Hence, GA will provide faster recognition performance to the vision system. In the GA process, each individual in the population of potential solutions scattered in the raw-image, is characterized by genes that describe the position and orientation of a searching model in current image. Hence, a GA allows for speculation on new search points with expected improved performance toward the target in the input raw-image. Here, a hybrid GA is employed for optimization of the evaluation function and the optimization means simultaneous recognition of a target and detection of its position/orientation. In fact here, the problem of an object recognition in the raw-image is changed to an optimization problem of a model-based evaluation function, named surface-strips model-based fitness function. In this research, we consider the shape of an object as the expectational knowledge for recognition purposes in the raw-image. This fitness function possesses information about the shape of a target, and consists in the computation of the brightness difference between an internal surface of the target and a background.

To assess the performance of the proposed scene recognition method, three object models, namely a frame model, a surface model, and a surface-strips model are investigated in order to determine which one is the best for scene recognition in a noisy environment. Also in order to appraise the recognition performance of each model, a comparative study is performed by analyzing the answers to the following criteria questions: *sensitivity*, that is, how does a model based evaluation function react to slight translational and orientational changes? *Reliability*, that is, how does a model based evaluation function react in case of high noise in the image? *Speed*, that is, does a model based evaluation function influence the GA convergence speed? In other words, we want to investigate to what extent that a model based evaluation function affects the recognition results. The GA-based scene recognition method described here can be designated as "evolutionary scene recognition method", since for every step of the GA's evolution, it struggles to perform the recognition of a target in the newly input raw-image to the recognition system. To recognize a target input by CCD camera in real-time, and to avoid time lag waiting for the convergence to a target, we used GA in such manner that only one generation is processed to newly input image. The GA searching process and the convergence to the target does not consist in one image but the convergence is achieved in the sequence of the input image to recognize it in the continuously input images. Next, the hybrid GA comprises on the one hand, the "global"search feature of a two-point crossover-based GA, to search a target in the raw-image. Based on a fitness-threshold value criteria and on the existence of a target, the global GA search shifts to the local search, *i.e.*, GA-based restricted search technique that focuses on the target of interest found so far, in order to perform an intensive search in a restricted image area. In order to provide for fast recognition and reliable results, we are interested in investigating the fundamental interaction between mutation and selection. The restricted GA search technique relies on a mutation technique. In this technique, mutation is performed on the bits on the lower portion of genes in positional binary strings and on mutation of bits in orientational binary strings. This is done in order to improve the GA-based scene recognition performance, in terms of fast and reliable recognition of a target, like a human action of gazing a target of interest. In order to evaluate the effectiveness of the proposed scene recognition method concerning noises such as lighting condition variations and other unexpected objects in the scene, experiments have been performed using real world raw-images, and results have shown that the method is suitable for a manipulator real-time visual servoing.

The remainder of this paper is organized as follows: section 2 deals with the raw-image and section 3 is concerned with object recognition problem as an optimization problem. Next, section 4 presents the models and evaluation function. Then section 5 discusses about sensitivity and reliability analysis, followed by the description of the GA-based scene recognition method in section 6. Next, a concept of 1-step-GA Evolution for Real-time Visual Servoing is explained in section 7, followed by the hybrid GA search method for fish visual tracking in section 8. Then the results of experiments are provided in section 9. Section 10 deals with the conclusion of this work.

2 Raw-image

In this paper, we are interested in using a scene description given by the image brightness corresponding to the gray-scale values which possess the best information about the scene. We want to analyze in the raw-image, the brightness values of the region of the model shaping the target, in order to

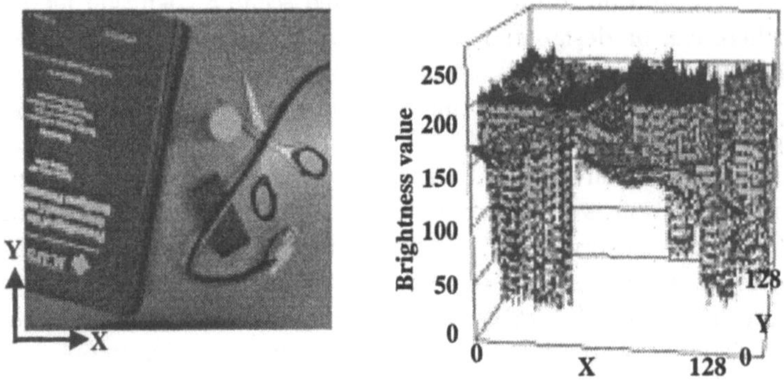

(a) Raw image (a) 3-D representation

Fig. 1. 2-D image of a scene under light condition with occluding noise

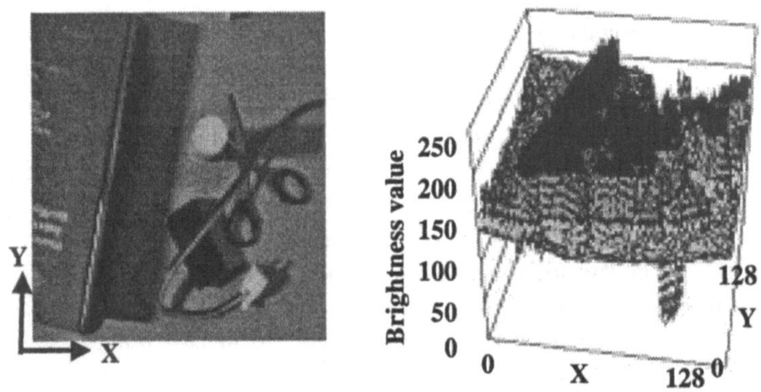

(a) Raw image (a) 3-D representation

Fig. 2. 2-D image of a scene with shadow and occluding noise

simultaneously recognize the shape of the target and find its position and orientation. The idea is that image understanding must employ raw image data such as gray-scale values, but it must also use knowledge about what to expect in a given environment.

The gray-level of a pixel is the quantized value of the measured power falling on the corresponding area in the image. We are interested in images in which many (8-bit based) brightness levels are represented. The graphs in Figs.1 to 3 show the 2-D together with the 3-D representations of a scene under different working conditions. Figures 1 (a) and (b) show respectively the 2-D and 3-D plot of a light scene including a target block with an occluding noise object. Figures 2 (a) and (b) show respectively the 2-D and 3-D plot of the scene of the target block with some shadow and other disturbing objects. Figures 3 (a) and (b) show respectively the 2-D and 3-D plot of the

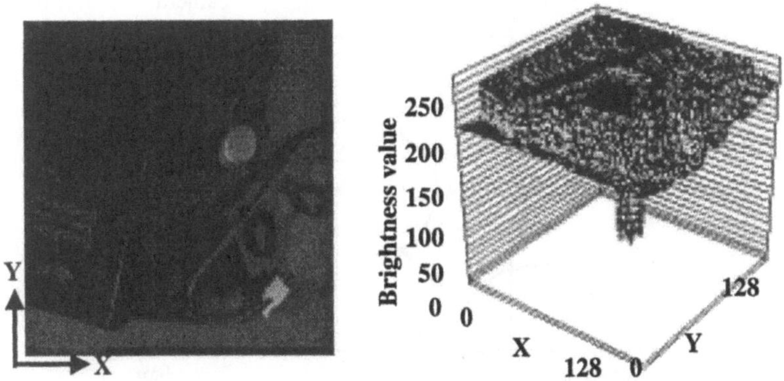

(a) Raw image (a) 3-D representation

Fig. 3. 2-D image of a scene under dark condition

scene of the target block with in dark lighting conditions. In the 3-D plots, the vertical axis represents the brightness values and the horizontal axes, the image plane. It can be seen that the brightness values vary according to the lighting conditions. Note that in Fig.3 (b), the brightness levels of the background noise and other objects is as high as the signal, that is, the brightness levels of the target block. In this paper, we want to perform recognition of a target object while keeping the original information contained in the input image. That is, we want to perform recognition of the target in spite of the noises such as lighting variations and also in the presence of other unexpected objects considered also as noises. To us, one must not cut information from the original image (case of binary images), before processing recognition. We believe that the use of the unprocessed gray-scale image, designated here as raw-image, allows us to design systems for extracting signals of interest in the presence of noises.

3 Object Recognition Problem as an Optimization Problem

One of the basic information needed to identify an object is its shape. Of course other information is still there, which makes the scene more identifiable. Here we are interested in using the knowledge of object shape for image understanding. The knowledge is used as a model of what the system should expect to see in a given environment. Thus, a typical vision system will use knowledge acquired in the past experience to drive, or control the interpretation process very much like you use past experience to control your visual perception.

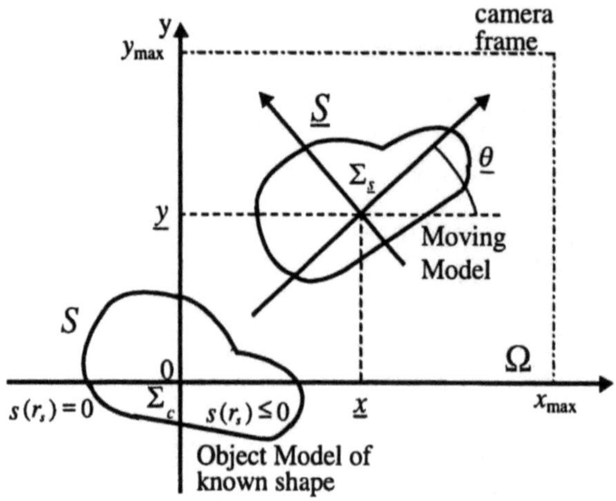

Fig. 4. Object model and camera frame

3.1 Basic Definitions

Definition of Object Model As shown in Fig.4, a camera frame Σ_c is fixed in the image plane. The domaine of the image taken by the camera is defined as Ω.

$$\Omega = \{ \boldsymbol{r} = (x, y) \mid 0 \leq x \leq x_{max}, 0 \leq y \leq y_{max} \} \tag{1}$$

Here x_{max}, y_{max} are constant values corresponding to the image plane boundaries in x and y directions. Next, let us define the contour of the object model as follows: $s(\boldsymbol{r}_s) = 0, \boldsymbol{r}_s = (x_s, y_s)$. Suppose that the representative point of the object shown in Fig.4, is coincident with the origin and that the rotation angle in Σ_c about the z axis is null. The set S inside the object model, expressed with respect to the coordinate frame Σ_c is given by following Eq.(2).

$$S = \left\{ \boldsymbol{r}_s \in \Re^2 \,\middle|\, s(\boldsymbol{r}_s) \leq 0 \right\} \tag{2}$$

Also the solid shape inside the object model S is defined by

$$m = m(\boldsymbol{r}_s), \boldsymbol{r}_s \in S \tag{3}$$

Definition of Moving Model Now let us consider the position $\underline{\boldsymbol{r}} = (\underline{x}, \underline{y})$, and orientation $\underline{\theta}$ of a moving model inside Ω. The extent of the existence of a moving model is given by:

$$\Gamma = \left\{ \underline{\phi} = (\underline{x}, \underline{y}, \underline{\theta}) \in \Re^3 \,\middle|\, \underline{\boldsymbol{r}} \in \Omega \,,\ -\pi < \underline{\theta} \leq \pi \right\} \tag{4}$$

Moreover, the inside sphere of the moving model expressed by the set $\underline{S}$ that is a function of $\boldsymbol{r}_s$ and $\underline{\phi}$, can be expressed as follows:

$$\underline{S}(\tilde{\boldsymbol{r}}(\boldsymbol{r}_s, \underline{\phi})) = \left\{ \tilde{\boldsymbol{r}} \in \Omega \subset \Re^2 \;\middle|\; \tilde{\boldsymbol{r}} = \boldsymbol{T}(\underline{\theta})\boldsymbol{r}_s + \underline{\boldsymbol{r}} \;,\; \boldsymbol{r}_s \in S \;,\; \underline{\phi} \in \boldsymbol{\Gamma} \right\} \tag{5}$$

and in the detailed form as:

$$\tilde{\boldsymbol{r}} = \boldsymbol{T}(\underline{\theta})\boldsymbol{r}_s + \underline{\boldsymbol{r}} = \begin{bmatrix} \cos\underline{\theta} & -\sin\underline{\theta} \\ \sin\underline{\theta} & \cos\underline{\theta} \end{bmatrix} \begin{bmatrix} x_s \\ y_s \end{bmatrix} + \begin{bmatrix} \underline{x} \\ \underline{y} \end{bmatrix} \quad \equiv \quad \tilde{\boldsymbol{r}}(\boldsymbol{r}_s, \underline{\phi}) \tag{6}$$

The solid shape of the model $\underline{S}$, is also expressed by the same solid shape as S, and is given by

$$\underline{m}(\tilde{\boldsymbol{r}}(\boldsymbol{r}_s, \underline{\phi})) = m(\boldsymbol{r}_s) \quad , \quad \boldsymbol{r}_s \in S \quad , \quad \underline{\phi} \in \boldsymbol{\Gamma} \tag{7}$$

Definition of Model-based Fitness Function When an image including a target object represented by the object model defined in Eqs. (2), (3) is taken by a CCD camera, the distribution of the brightness values in the input raw-image is expressed as:

$$p = p(\boldsymbol{r}) \quad , \quad \boldsymbol{r} \in \Omega \tag{8}$$

$\boldsymbol{r}$ is the position of a picture element expressed in the camera coordinate frame Σ_c and p($\boldsymbol{r}$) expresses the brightness value at that position.

A model-based fitness function F expresses the correlation between an object existing in the input image and the extent $\underline{S}$ of a moving model, and is determined as follows

$$F(\underline{\phi}) = \int_{\boldsymbol{r}, \tilde{\boldsymbol{r}} \in \Omega \cap \underline{S}} p(\boldsymbol{r}) \quad \cdot \quad \underline{m}(\tilde{\boldsymbol{r}}(\boldsymbol{r}_s, \underline{\phi}))ds \tag{9}$$

where ds means a small area in $\underline{S}$. The above equation projects in the frame $\underline{\phi}$, the degree of correlation between the raw-image $p(\boldsymbol{r})$ and the moving model $\underline{S}$.

Formulation of a Target Object Searching Problem If it can be assumed that $F(\underline{\phi})$ gives the maximum value when $\underline{S}$ coincides with the object contained in $p = p(\boldsymbol{r})$, then the problem which consists in the raw-image, in the simultaneous recognition of an object together with the detection of its position and orientation, is changed to the problem of searching $\underline{\phi}$ that maximizes F. Here, the value of $F(\underline{\phi})$ depends on the model's solid shape $m(\boldsymbol{r}_s)$ and on the raw-image $p(\boldsymbol{r})$. Let us denote the true position and orientation of an object in $p(\boldsymbol{r})$ as ϕ_p. When $m(\boldsymbol{r}_s)$ and $p(\boldsymbol{r})$ satisfy $F(\phi_p) > F(\underline{\phi})$, $\phi_p \neq \underline{\phi}$, $\underline{\phi} \in \boldsymbol{\Gamma}$, which means that the above assumption can be brought about, the problem of searching for ϕ_p can be expressed as

$$find \;\; \underline{\phi} \;\; to \;\; maximize \;\; F(\underline{\phi}) \;\; subject \;\; to \;\; \underline{\phi} \in \boldsymbol{\Gamma} \tag{10}$$

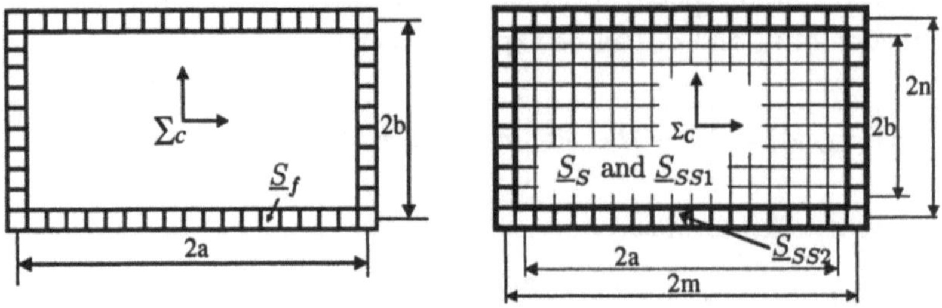

(a)Frame model (b)Surface model and Surface-strips model

Fig. 5. Frame, Surface and surface-strips models

Later on, we will confirm the validity of the assumption and will make use of a GA as a method to seek for the solution ϕ_p of the problem stated above. The detailed description of the method will be provided in next section.

4 Object Model and Evaluation Function

The selection of an appropriate fitness function is a key to the success of solving the problem of shape recognition combined with position/orientation detection. The fitness value should be a clear indication of the goodness of the solution to the problem. In this research, the fitness function is defined based on the type of object model used.

In the case of a rectangular block as the object, the 2-D representation of the frame model is shown in Fig.5(a) and the 2-D representations of the surface model and the surface-strips model are shown together in Fig.5(b). Also the solid shape of each model is shown in Fig.6 (a), (b), and (c). For each of the object models, the model's domain $\underline{S}$ defined in Eq.(5) and the model's solid shape $\underline{m}$ defined in Eq.(7), are denoted as follows: for the frame model, we have: $\underline{S}_f$, $\underline{m}_f(\boldsymbol{r}_s) = 1$, $(\boldsymbol{r}_s \in \underline{S}_f)$, for the surface model, we have: $\underline{S}_s$, $\underline{m}_s(\boldsymbol{r}_s) = 1$, $(\boldsymbol{r}_s \in \underline{S}_s)$, and for the surface-strips model, we have: $\underline{S}_{ss}$, $\underline{m}_{ss}(\boldsymbol{r}_s) = 1$, $(\boldsymbol{r}_s \in \underline{S}_{ss1})$, $\underline{m}_{ss}(\boldsymbol{r}_s) = -1$, $(\boldsymbol{r}_s \in \underline{S}_{ss2})$. Here, $\underline{S}_{ss1}$ expresses the same domain as $\underline{S}_s$. Also $\underline{S}_{ss2}$ corresponds to the contour.

4.1 Frame model

The transformation from the points in S_f to the domain of the moving model $\underline{S}_f$ is expressed as,

$$\tilde{\boldsymbol{r}} = \begin{bmatrix} \cos\underline{\theta} & -\sin\underline{\theta} \\ \sin\underline{\theta} & \cos\underline{\theta} \end{bmatrix} \begin{bmatrix} x_s \\ y_s \end{bmatrix} + \begin{bmatrix} \underline{x} \\ \underline{y} \end{bmatrix} \tag{11}$$

Here we use the values $p(\boldsymbol{r})$ of the raw-image that range from 0 to 255. The evaluation function of the frame model $F_f(\underline{\phi})$, is expressed by following

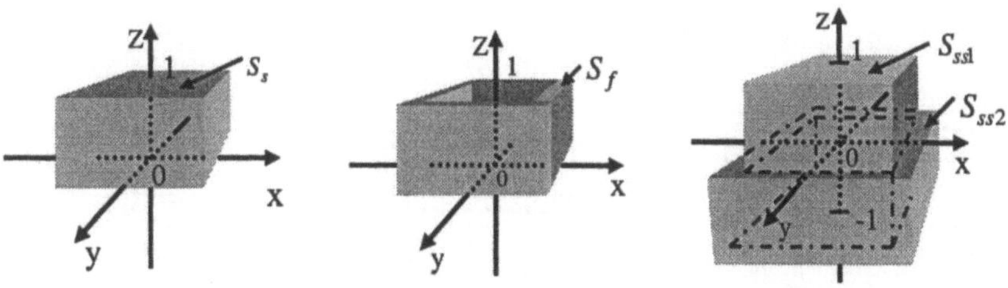

(a) Frame model (b) Surface Model (c) Surface-strips model

Fig. 6. Solid shapes of the three types of models: $m = m(\boldsymbol{r})$

Eq.(12).

$$
\begin{aligned}
F_f(\underline{\phi}) &= \int_{\boldsymbol{r},\tilde{\boldsymbol{r}} \in \Omega \cap \underline{S}} p(\boldsymbol{r}) \cdot \underline{m}(\tilde{\boldsymbol{r}}(\boldsymbol{r}_s, \underline{\phi}))ds \\
&= \int_{\tilde{\boldsymbol{r}} \in \underline{S}_f} p(\tilde{\boldsymbol{r}}) \cdot \underline{m}_f(\tilde{\boldsymbol{r}})ds \\
&= \int_{\tilde{\boldsymbol{r}} \in \underline{S}_f} p(\tilde{\boldsymbol{r}})ds \\
&= \sum_{\tilde{\boldsymbol{r}}_{i,j} \in \underline{S}_f} p(\tilde{\boldsymbol{r}}_{i,j})
\end{aligned} \tag{12}
$$

We define N_f as the number of pixels contained in $\underline{S}_f$. M corresponds to the maximum brightness value, that is 255, for the input raw-image. Then the proper expression of $F_f(\underline{\phi})$ using the maximum brightness value, is $F_f(\underline{\phi})/N_f M$. In the experiments we make use of the value of the proper expression of $F_f(\underline{\phi})$.

4.2 Surface model

The evaluation function $F_s(\underline{\phi})$ of the surface model is expressed by following Eq.(13).

$$
\begin{aligned}
F_s(\underline{\phi}) &= \int_{\boldsymbol{r},\tilde{\boldsymbol{r}} \in \Omega \cap \underline{S}} p(\boldsymbol{r}) \cdot \underline{m}(\tilde{\boldsymbol{r}}(\boldsymbol{r}_s, \underline{\phi}))ds \\
&= \int_{\tilde{\boldsymbol{r}} \in \underline{S}_s} p(\tilde{\boldsymbol{r}}) \cdot \underline{m}_s(\tilde{\boldsymbol{r}})ds \\
&= \int_{\tilde{\boldsymbol{r}} \in \underline{S}_s} p(\tilde{\boldsymbol{r}})ds \\
&= \sum_{\tilde{\boldsymbol{r}}_{i,j} \in \underline{S}_s} p(\tilde{\boldsymbol{r}}_{i,j})
\end{aligned} \tag{13}
$$

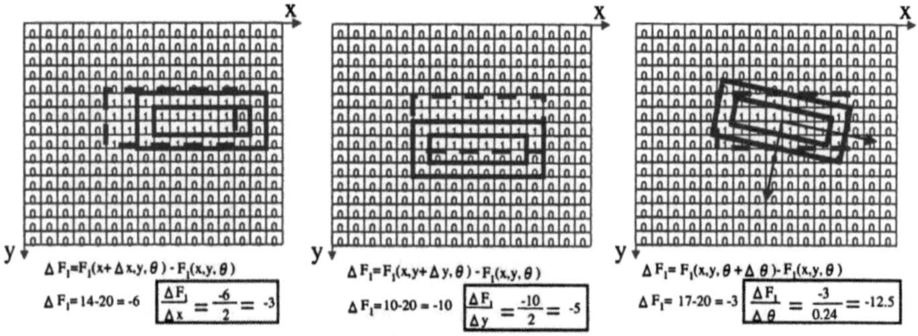

(a) x direction (b) y direction (c) θ direction

Fig. 7. Sensitivity of surface model

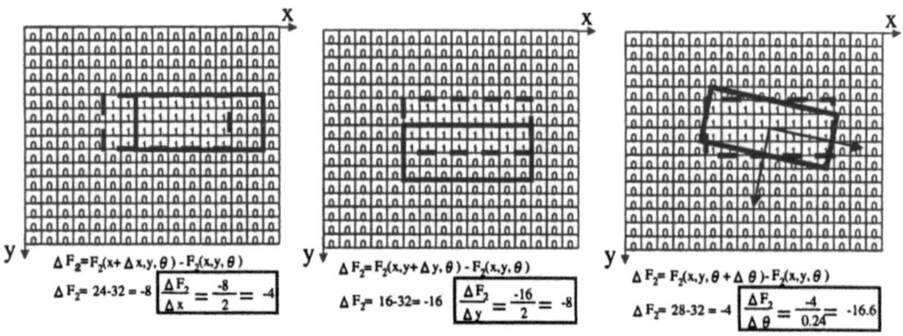

(a) x direction (b) y direction (c) θ direction

Fig. 8. Sensitivity of surface model

We define N_s as the number of pixels contained in $\underline{S}_s$ and in accordance with what was aforementioned, the proper expression of $F_s(\underline{\phi})$ using the maximum brightness value, is $F_s(\underline{\phi})/N_s M$. Similar to the proper expression of $F_f(\underline{\phi})$, in the experiments we make use of the value of the proper expression of $F_s(\underline{\phi})$.

4.3 Surface-strips model

The evaluation function $F_{ss}(\underline{\phi})$ of the surface-strips model is given as :

$$
\begin{aligned}
F_{ss}(\underline{\phi}) &= \int_{\boldsymbol{r},\tilde{\boldsymbol{r}}\in\Omega\cap\underline{S}} p(\boldsymbol{r})\cdot\underline{m}(\tilde{\boldsymbol{r}}(\boldsymbol{r}_s,\underline{\phi}))ds \\
&= \int_{\tilde{\boldsymbol{r}}\in\underline{S}_{ss}} p(\tilde{\boldsymbol{r}})\cdot\underline{m}_{ss}(\tilde{\boldsymbol{r}})ds \\
&= \int_{\tilde{\boldsymbol{r}}\in\underline{S}_{ss1}} p(\tilde{\boldsymbol{r}})\cdot\underline{m}_{ss}(\tilde{\boldsymbol{r}})ds + \int_{\tilde{\boldsymbol{r}}\in\underline{S}_{ss2}} p(\tilde{\boldsymbol{r}})\cdot\underline{m}_{ss}(\tilde{\boldsymbol{r}})ds \\
&= \int_{\tilde{\boldsymbol{r}}\in\underline{S}_{ss1}} p(\tilde{\boldsymbol{r}})ds - \int_{\tilde{\boldsymbol{r}}\in\underline{S}_{ss2}} p(\tilde{\boldsymbol{r}})ds
\end{aligned}
$$

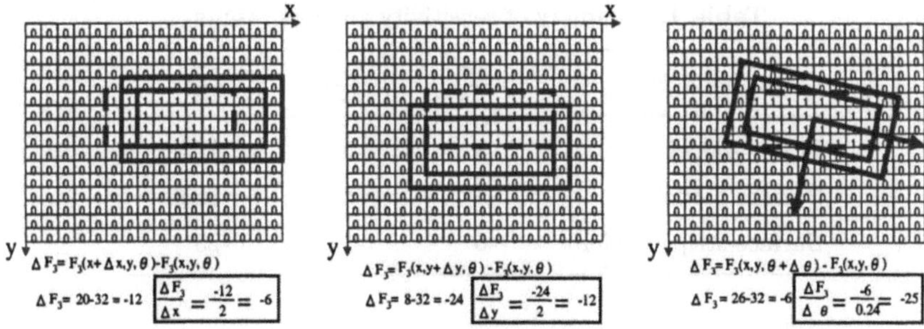

(a) x direction (b) y direction (c) θ direction

Fig. 9. Sensitivity of surface-strips model

$$= \sum_{\tilde{\boldsymbol{r}}_{i,j} \in \underline{S}_{ss1}} p(\tilde{\boldsymbol{r}}_{i,j}) - \sum_{\tilde{\boldsymbol{r}}_{i,j} \in \underline{S}_{ss2}} p(\tilde{\boldsymbol{r}}_{i,j}) \tag{14}$$

We define N_{ss1} as the number of pixels contained in $\underline{S}_{ss1}$ and N_{ss2} as the number of pixels contained in $\underline{S}_{ss2}$. Then here, the proper expression of $F_{ss}(\underline{\phi})$ using the maximum brightness value, is $F_{ss}(\underline{\phi})/N_{ss1}M$. As for $F_f(\underline{\phi})$ and $F_s(\underline{\phi})$, in the experiments we make use of the value of the proper expression of $F_{ss}(\underline{\phi})$. When S_{ss1} and the object match well, S_{ss2} coincides with the background, and $F_{ss}(\underline{\phi})$ gets its highest value. That is, $F_{ss}(\underline{\phi})$ considers the change in the brightness values of the object, that is, the differentiation information is taken into account. By using the proper expressions of the evaluation functions: $F_f(\underline{\phi})/N_fM$, $F_s(\underline{\phi})/N_sM$, and $F_{ss}(\underline{\phi})/N_{ss1}M$, a comparison among their recognition results becomes more apparent.

5 Models Sensitivity and Reliability Analysis

5.1 Sensitivity to Slight Position and Orientation Variations

An artificial binary image is used to evaluate each searching model's sensitivity to slight variations of position and orientation. The models' sensitivity to slight variations of position and orientation has been analyzed. In the experiments performed, a positional variation of 2 pixels in the x and y directions, and an orientational variation of 14 degrees have been investigated. The results are shown from Figs.7 to 9. The functions F_1, F_2 and F_3 in the figures correspond to the evaluation functions $F_f(\underline{\phi})$, $F_s(\underline{\phi})$, and $F_{ss}(\underline{\phi})$ defined previously for the frame model, the surface model, and the surface-strips model. The sensitivity analysis results are summarized in Table 1. From these results, the surface-strips model is found to be more sensitive to position/orientation variations than the frame model and the surface model. Hence we expect the most reliable recognition results with the surface-strips model.

Table 1. Summary of sensitivity analysis results

Object-models	$\frac{\Delta F_i}{\Delta x}$	$\frac{\Delta F_i}{\Delta y}$	$\frac{\Delta F_i}{\Delta \theta}$
Frame model	-3	-5	-12.5
Surface model	-4	-8	-16.6
Surface-strips model	-6	-12	-25

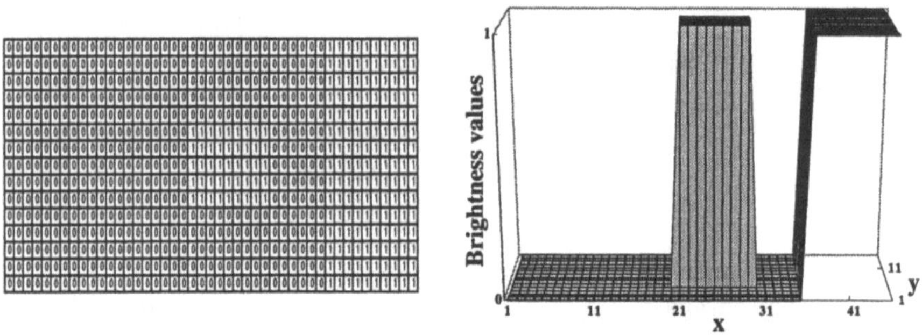

(a) Data of artificial image (b) 3-D representation

Fig. 10. Artificial noisy image

5.2 Recognition Reliability in the Presence of High Noise

Consider the artificial noisy binary image shown in Fig.10(a). In this image, the area of "1's" on the right side represent the noisy-zone, and the area of "1's" on the center part represent the target object. The 3-D plot of this image is shown in Fig.10(b). We want to evaluate the reliability of recognition of each of the models described earlier, in the presence of high noise in the image. By translating a model over all the points in this image, and computing the fitness values in accordance with Eqs.(12), (13), and (14), at each point with the condition of $\theta_p = \underline{\theta} = 0$, the distribution of the model-based fitness function in this image for each kind of searching-model has been examined and shown in Fig.11(a), (b), (c), respectively for the frame model, the surface model, and the surface-strips model. A glance at these graphs indicates that the surface-strips model shows a good *shape − based image filter* feature. It can be seen in Fig.11(c) that the surface-strips model has reduced the noisy zone in the image, by showing a good recognition result of the target with only one sharp peak in the distribution. This special feature shown by the surface-strips model is thought to come from the differentiation process inherent to this model. But the surface and frame models could not show such feature. Hence with the surface-strips model, the recognition results are found to be the most reliable against the two other searching-models. When we use $F_{ss}(\underline{\phi})$ to perform the search, the difference between the brightness

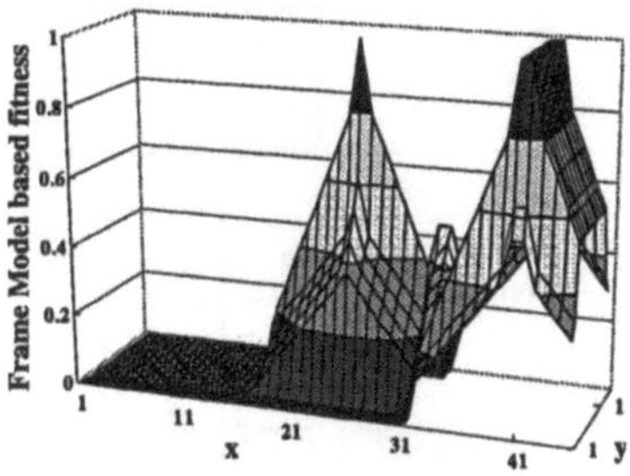

(a) Frame model

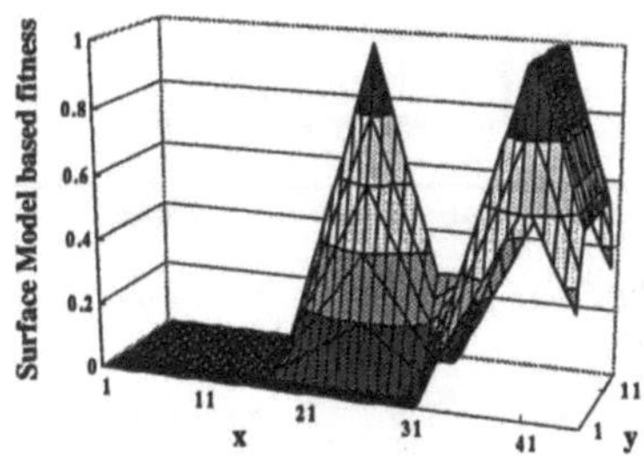

(b) Surface model

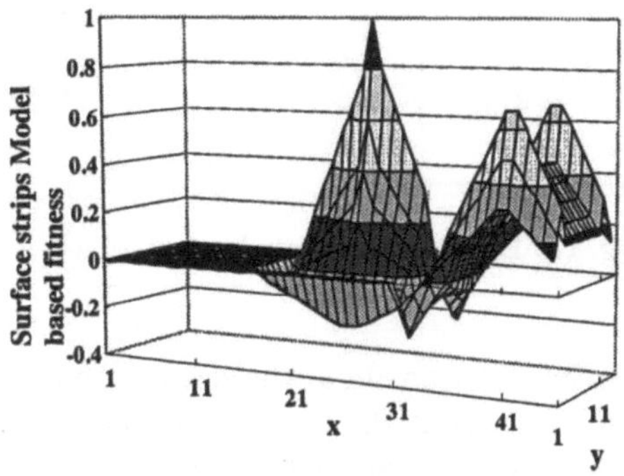

(c) Surface strips model

Fig. 11. Fitness function distribution using artificial binary image

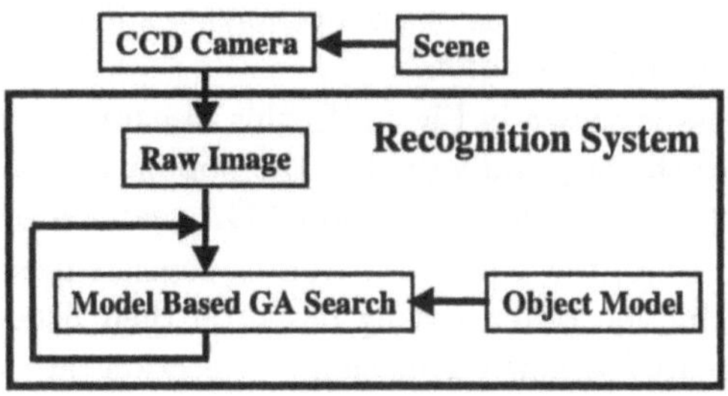

Fig. 12. Diagram of scene recognition

of the object and background is evaluated and as a result the difference of the fitness value becomes high.

6 GA-based Scene Recognition Method

The diagram describing the proposed GA-based scene recognition method is presented in Fig.12. In this vision system, a CCD camera used as the image

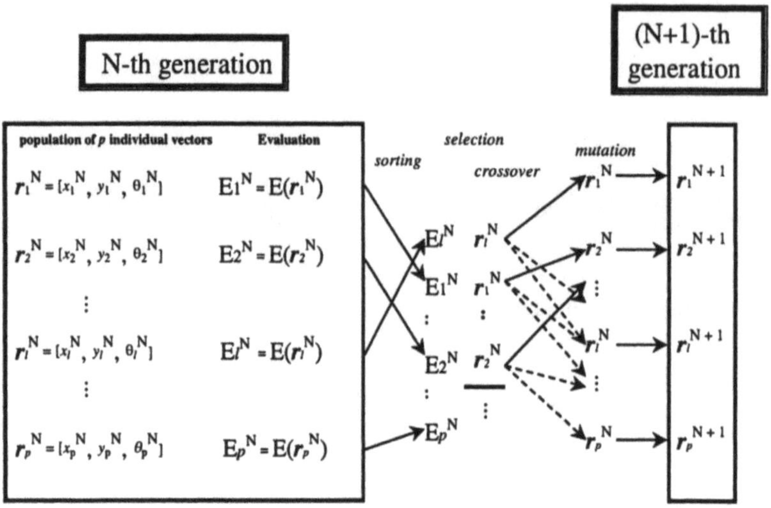

Fig. 13. Elitist model global search of a GA

input device, provides information about the scene. The image of this working space provides information about the top view surfaces of the objects in the scene and the vision system is required to recognize a designated object, that is, its shape recognition together with the detection of its position and orientation. The problem of recognizing an object description is as the process of using the shape to find the "best"account of the data. That is, we can define the recognition process as one of optimizing a description over a shape relative to goodness-of-fit-criterion.

The GA process is shown in Fig.13. In this research, an elitist model of a GA that preserves the best individual in the population at every generation is utilized. The operators used in the global GA search are described below:

Selection Here, the selection operator consists in selecting the individuals for reproduction (after sorting based on their fitness values), which will be used to create the new population. A selection rate is used to remove weak individuals based on their low-ranking fitness value (objective function value), and those individuals will be replaced by the selected top ranking individuals which undergo crossover and mutation.

Crossover A two-point crossover is employed in this research. This operation consists in crossing two individuals considered as parents to create two new ones considered as offspring. The binary strings of randomly selected two individuals are partly interchanged according to two-random crossover sites in the binary individual structure. Crossover is applied to take valuable information from both parents.

Mutation This operator consists in mutating bits. A bit of value "1" is changed to "0" and a bit of value "0" is changed "1". A random bit-by-bit

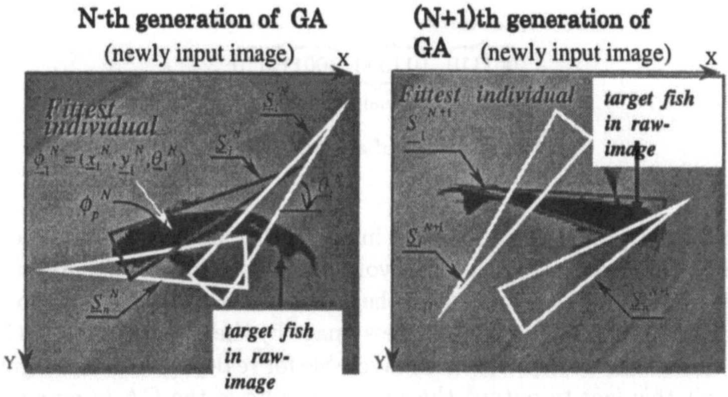

Fig. 14. Illustration of 1-step-GA evolution with fish as target

basis mutation is performed on all the bits of the intermediate population obtained after crossover.

A population composed of a set of p such individuals is used by GA process to perform the search for the target object in the current image. Since the elitist model of GA is adopted, the sorted best individual in the N-th population, designated as a vector $\boldsymbol{r}_1^N$, representing the position and orientation of the target, is selected to survive. Let us denote by $g_{kl}^N (k = 1, 2, 3)$, the $l-th$ components of an individual in the N-th generation. These components correspond to x_l^N, y_l^N, and θ_l^N, which stand respectively to g_{1l}^N, g_{2l}^N, and g_{3l}^N. The components x_l^N and y_l^N represent the position coordinates in the camera frame, of the origin of the coordinate frame a in the moving model. Likewise, θ_l^N represents the model's orientation in the camera frame. A population of models scattered in the image onto x_l^N, y_l^N, with orientation θ_l^N, is evaluated in order to determine how much each of them fits to the target object being imaged. The obtained fitness values E_1^N, $E_2^N, \cdots$, E_p^N are sorted. Based on the ranking and a selection rate to die, the weakest individuals in terms of poor fitness values are replaced by created new individuals. In creating the new individuals, first, a random two-pair-mates, random two-point crossover is performed. Next, a random bit-by-bit basis mutation (exchange of 1 by 0, or vice versa), is performed on the individuals obtained after crossover. This ends the N-th generation and the population obtained after these operations constitutes the population at the starting point of the (N+1)-th generation. Then, the preceding steps are repeated with the individual in population N+1, in order to evolve the population toward the solution.

7 1-step-GA Evolution for Real-time Visual Servoing

In this research, we adopted the elitist model of a GA, which keeps track of the best individual in the population, in every generation. For real-time vi-

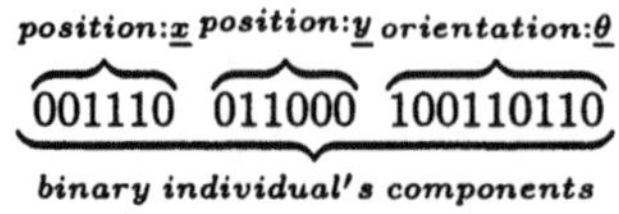

Fig. 15. Structure of an individual in GA

sual control purposes, we employ GA in a way that we denoted as 1-step-GA evolution. This means that the GA evolution is applied only one iteration to the newly input image. While using the elitist model of the GA, the location of a target can be detected in every new image by the top genes in the GA. In addition, this feature happens to be favorable for real-time visual recognition. We exploit this fact to output the current results of the GA in every newly image, to be used as input value of the manipulator controller. Thereby real-time visual servoing can be performed. This concept is illustrated in Fig.14. To recognize a target object in the detected scene, i.e., to seek for ϕ_p, a population of searching models represented in $\underline{\phi}_i$, $(i = 1, 2, \ldots, p)$, is used by GA to search for it. In order to determine the best individual in the population, i.e., the model in the population providing the best matching result, the model-based fitness function is applied to evaluate the brightness values of each model. The left side of this Fig.14 corresponds to the configuration of an N-th generation of GA's evolution. Consider p moving models $\underline{S}_i^N$, $i = 1, 2, \ldots, p$, shaping the target and scattered in the image. The position and orientation of each of them is characterized by an individual in GA process as a binary string as shown in Fig.15, which describes an individual in GA. The position and orientation parameters $\underline{\phi}_1^N = (\underline{x}_1^N, \underline{y}_1^N, \underline{\theta}_1^N)$, correspond to the ones of the fittest individual in the population. Likewise, as shown on the right side of this Fig.14, after the population has evolved to the next generation $N+1$, the position parameters $\underline{\phi}_1^{N+1} = (\underline{x}_1^{N+1}, \underline{y}_1^{N+1}, \underline{\theta}_1^{N+1})$ correspond to the ones of the fittest individual in the population. This illustration shows that an increase of the fitness value of the best individual means an improvement toward the target recognition.

8 GA-Based Evolutionary Fish Visual Tracking

Consider the 2-D raw-image of a fish in its natural scene shown in Fig. 16(a). The corresponding 3-D plot of this figure is shown in Fig.16(b). In the 3-D plots, the vertical axis represents the image brightness values. In this work, we think that giving an ability to recognize a target in real-time is the same as giving a kind of intelligence to the machine. To search for a fish in the scene being imaged, here a triangular surface-strips model as shown in Fig.17, which describes the 2-D top-surface of the target fish, is used as the expectational knowledge to search for it in the raw-image.

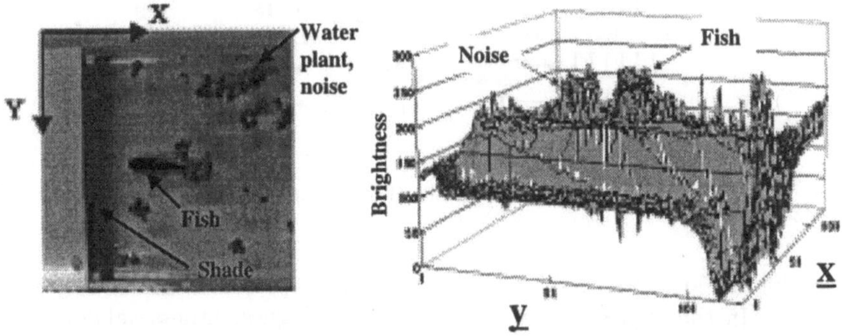

(a) Raw image (b) 3-D representation

Fig. 16. Raw-image of a fish in its natural scene and 3-D plot

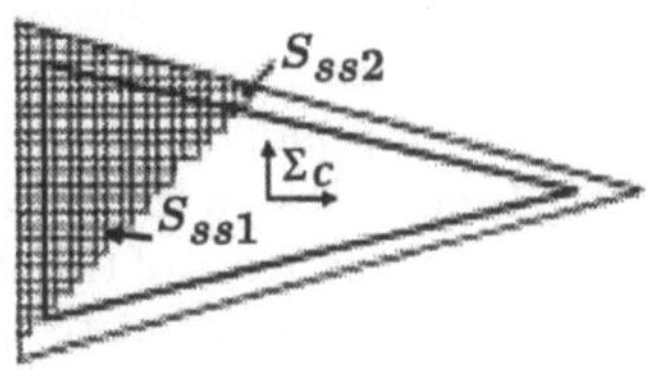

Fig. 17. A model to search for fish

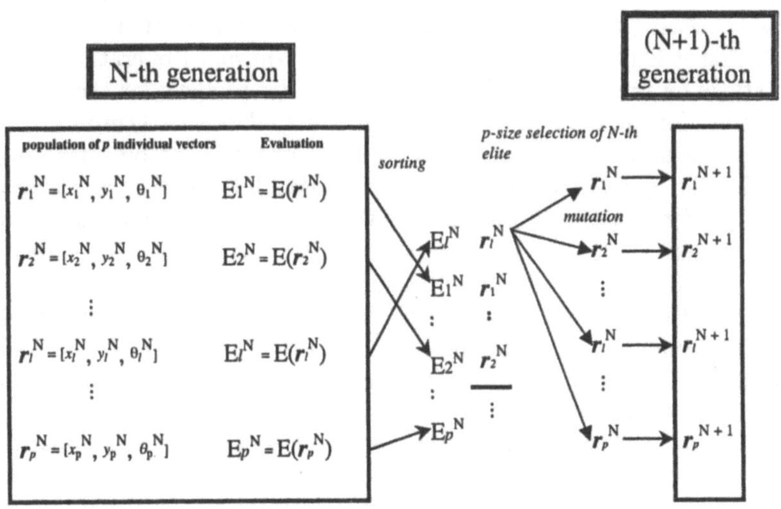

Fig. 18. Elitist model local technique search of a GA

In order to increase the tracking performance to a target fish, here we employ a localized search technique of a GA, in addition to the GA process shown in Fig.13 that we denote here as global GA search. The algorithm

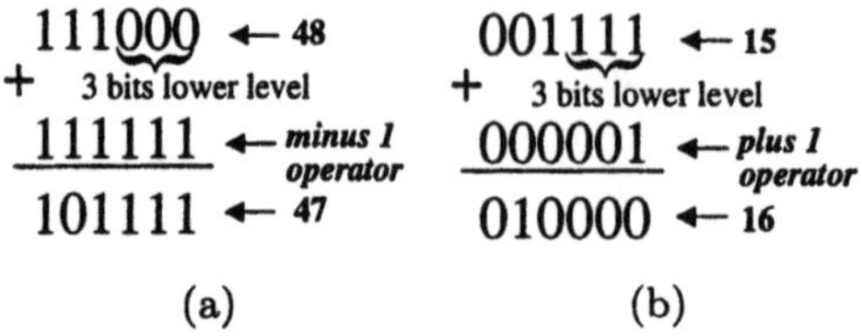

Fig. 19. Illustration of lower level mutation: case of 3 bits

describing the localized search technique of GA is represented in the diagram of Fig.18. In this figure, let us assume that the highest fitness value of an $N-th$ generation has reached a certain value where the target's position is in a range of acceptable recognition. The individual that has the highest fitness value is on the ranking 1. Then what we denoted as pure selection is performed, that is, the genes of the individual of ranking 1 are selected to be copied p times, thus making an intermediate population of p identical individuals, for a population of p individuals. Next, in the reproduction process, except that ranking 1 individual, a mutation operation is gradually performed on the lower level bits of the genes of the other individuals in order to increase the fitness value of the others and at the same time obtain better positional results. When the mutation is limited to 4 bits of the lower level, 16 reproduction patterns can be obtained for one positional direction. When the mutation is limited to 3 bits, 8 patterns can be obtained and when it is limited to 2 bits, 4 reproduction patterns can be obtained. However, we have found that it is necessary to introduce a *plus* 1/*minus* 1 operation to those limitations on the bits of the lower level, in order to allow for the search in the vicinity of the actual position of the target. This *plus* 1/*minus* 1 technique consists in adding a string of all 1 bit string to the bit string of the gene, in the case where all 0 bits for the lower bits. This is illustrated in Fig.19 (a) for a case of 3 bits lower level mutation. Figure 19 (a) shows a bit string like 111000, which is equal to 48 in decimal base. For such a bit string, it is impossible to obtain the number 47 for all possible mutation of the 3 bits. Hence, in accordance with the explanation given above, in such case, the technique is to add the string 111111. Also, when all the limited lower level bits are 1s, this operator consists in adding a bit string of value 1 to the bit string of the gene. This is also illustrated in Fig.19 (b). Figure 19 (b) shows a bit string like 001111, which is equal to 15 in decimal base. For such a bit string, it is impossible to obtain the number 16 for all possible mutation of the 3 bits. Hence, in such case, the technique is to add the string 000001. The change of the search area of the global and local search method can be illustrated as shown in Fig.20 (a). In this case, the search area of the normal GA is 64[pixels]×64[pixels]. Based on the local search mutation level, the search area is being restricted to the smaller area in accordance with the GA's evolution time. When the shape recognition together with the position/orientation detection of an object is being completed to a certain

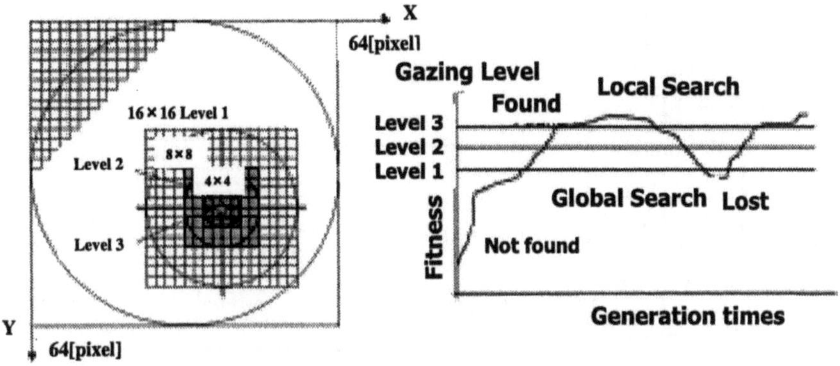

(a) Change of area (b) Switching based on fitness value

Fig. 20. Change of area and switching by local/global search method

degree, this search technique focusing on the target of interest, happen to be acted similarly by human. Thus, when compared with the global search method of the normal GA, faster and correct detection of the target will be possible.

The switching process between the global GA search and the local GA search technique can also be described by the curve shown in Fig.20 (b). This figure indicates the global search of GA that proceeds to the search in order to find a target, and once a target is found, the system switches to the local GA search as a gazing action, in order to perform an intensive localized search so as to fasten and improve the recognition results of the target.

In this gazing control, we also have to consider the time used for the restricted area searching. To maintain the real-time recognition, the restricted searching is executed in the next newly input image. A diagram describing the explanation of the hybrid (combined) global and local GA is shown in Fig.21. In practice, once the global GA has achieved the stage of the detection of a target, after reaching a certain fitness value as a threshold value, it switches to the local GA that performs the fine and fast recognition of the target. Actually, the determination of this threshold value, is environment dependent. Using the combined GA, the task of the search for a target and its tracking, when done by the machine, will be similar to the same task when done by human.

9 Experiments

9.1 Experiments With Respect to a Stationary Target

Several real images are used in order to examine the potential of the proposed method of an object recognition in the raw-image. The first scene is a man-made environment with several objects containing a rectangular block as

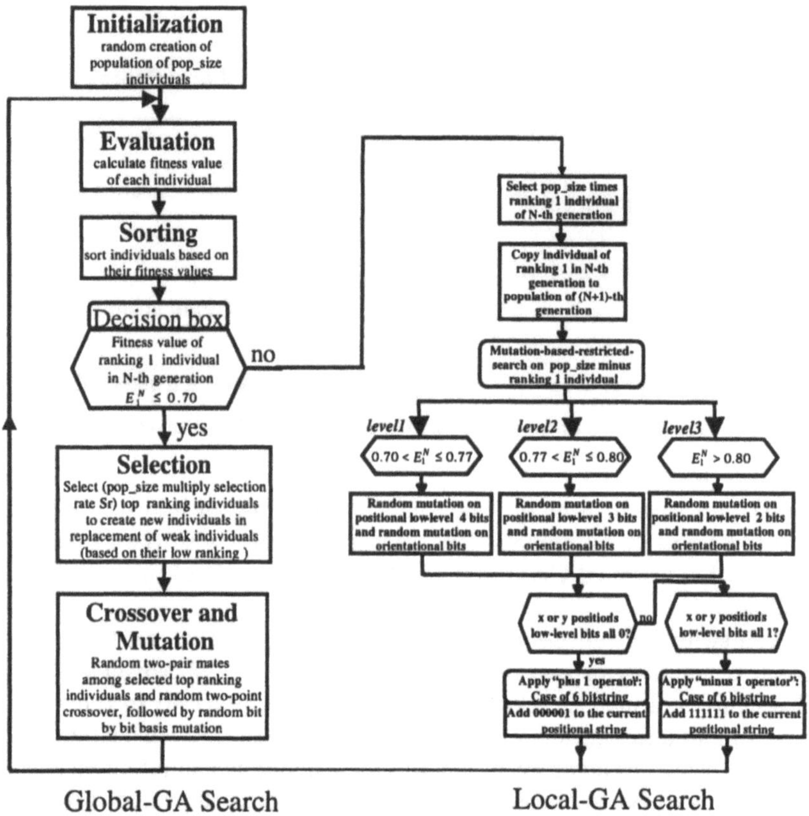

Fig. 21. Diagram of global and local GA

target. The 2-D together with the 3-D graphs of this scene under different lighting conditions are shown in Figs.1 to 3, and described in section 2.

A CCD camera system is used to obtain the image of the scene. The distance of the camera lens position to the task plane has been set to 300 [mm]. The camera system consists of a CCD camera that sees the objects from their top and an image input board. This board MT98-MN is connected to the host computer: PC-9821 Xe10/4. The real images obtained from the processor are of size 512×512 and quantized to 256 gray levels. That is, the converted pixel value will range from 0 for white to 255 for black and different shades of gray are represented by values between these two extremes. To speed up the search, the image size has been reduced to 128×128. In this process, pixels are sampled with the interval of four in both the horizontal and vertical direction.

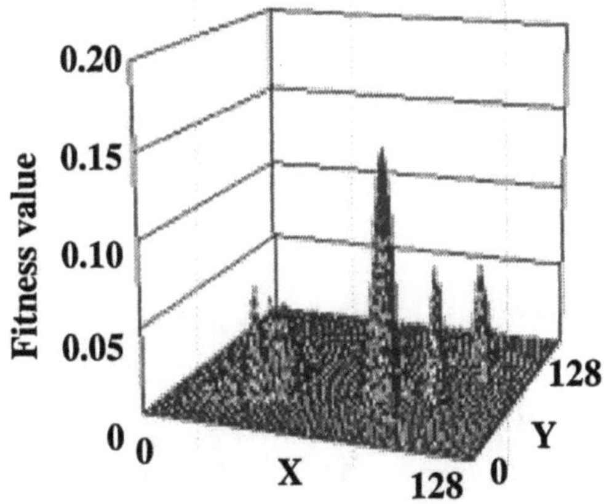

Fig. 22. Recognition results of target block in light environment with occlusion: fitness distribution with surface-strips model-based fitness function

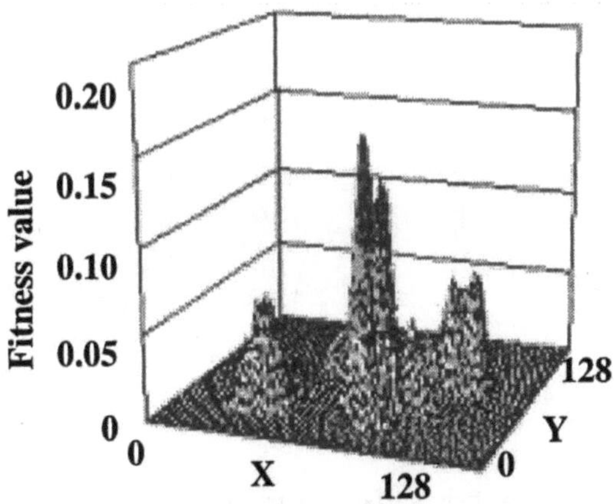

Fig. 23. Recognition results of target block in shadow environment with occlusion: fitness distribution with surface-strips model-based fitness function

Performance of Traditional Search Results with Real Images In order to evaluate the performance of the proposed GA-based scene recognition system, we will compare its recognition results with the ones obtained while scanning whole the search area while using the model-based fitness function. Obviously, a traditional search method which consists in exploring all the positional pixels in the image and which will require at each pixel, to check for the 360 degrees orientional angles (assuming 1 degree incremental angle), will lead to the optimal solution, but is computationally intensive. We have employed this search technique to detect the target block in Figs.1 to 3 under different lighting conditions. The recognition results corresponding to the highest values of the model-based fitness function, with the surface-strips

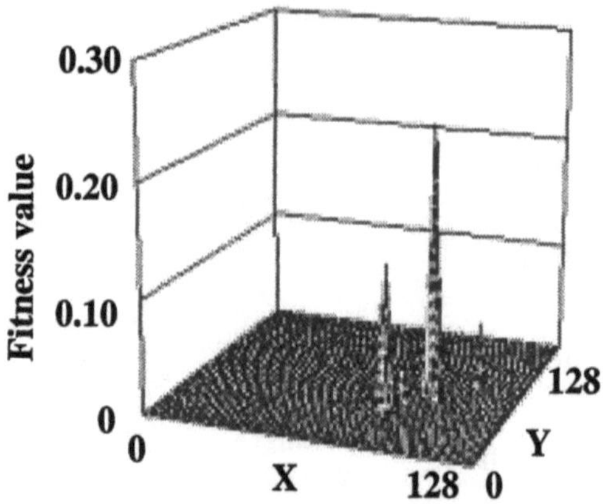

Fig. 24. Recognition results of target block in dark environment: fitness distribution with surface-strips model-based fitness function

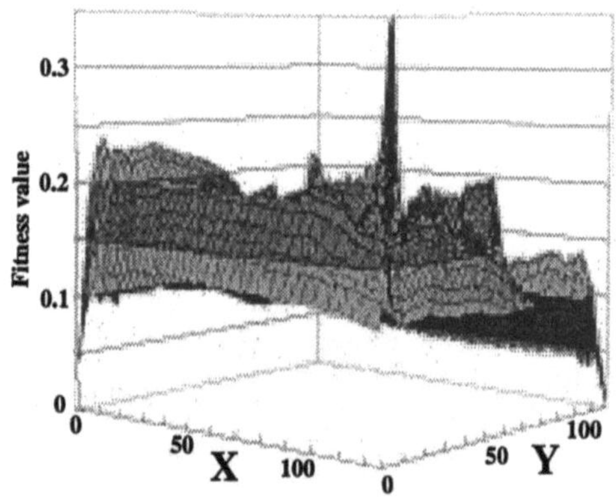

Fig. 25. Scene of target fish: surface-strips model-based fitness function plot

model, are presented in Figs.22, 23, and 24. For each graph, a sharp peak has been found, which correspond to the position and orientation of the target in the raw-image. Also, the recognition result of the target fish shown in Fig.16 is presented in Fig.25. In Fig.22, the recognition results are 68 *degrees* for the orientation, and $x = 78$, $y = 82$ for the position. In Fig.23, the recognition results are 68 *degrees* for the orientation, and $x = 72$, $y = 90$ for the position. In Fig.24, the recognition results are 248 *degrees* for the orientation, and $x = 62$, $y = 80$ for the position. In Fig.25, the recognition results are 303 *degrees* for the orientation, and $x = 71$, $y = 72$ for the position, corresponding to the one that have provided the highest fitness value to detect the target fish. From these results, it can be seen that with the surface-strips model, only one sharp peak has occurred in the distribution graph, which corresponds to the recognition result of the target block. This confirms the

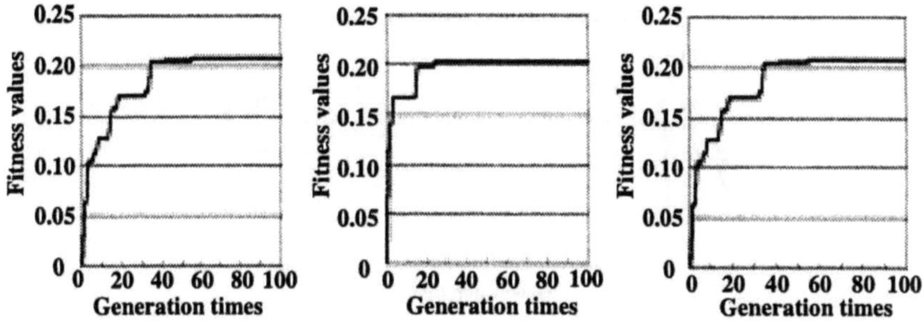

(a) light scene of Fig.1 (b) shadow-scene of Fig.2 (c) dark scene of Fig.3

Fig. 26. GA search results of target block

Table 2. GA Parameters

Population size	20 individuals
Selection rate(Two-point crossover)	0.4
Mutation rate	0.01
Length per individual	21 bits
Elitist model	*yes*

noise reduction feature of the surface-strips model. Hence, we may say that with the surface-strips model, the obtained experimental recognition results in the raw-image happen to be the most reliable ones. Further, the assumption $F(\phi_p) > F(\underline{\phi})$, $\phi_p \neq \underline{\phi}$, $\underline{\phi} \in \boldsymbol{\Gamma}$, of the recognition problem conversion, i.e., shape recognition and position/orientation detection problem into maximum value searching problem, is confirmed to be realizable, since each of the Figs. 22, 23, 24, and 25 shows a single high peak corresponding to the recognition results of the target.

GA-based Recognition Results The proposed GA-based scene recognition system using the raw-image has been applied to the raw-images of the above-mentioned scenes. In the experiment, 6 bits are used for the x positional parameter, 6 bits are used for the y positional parameter, and 9 bits are used for the orientational parameter of an individual in GA. Also, we should point out that the GA parameters are critical to a good performance of the system. The parameters shown in Table 2 provide satisfactory results. Note that this set of parameters was chosen after some preliminary studies.

The graphs provided in Figs.26 (a), (b), (c), present the fitness values of the best individual in GA, versus the generation times, during the search for the target block in the raw-image, for the three different lighting conditions early described. Also, the graphs corresponding to the positional and orien-

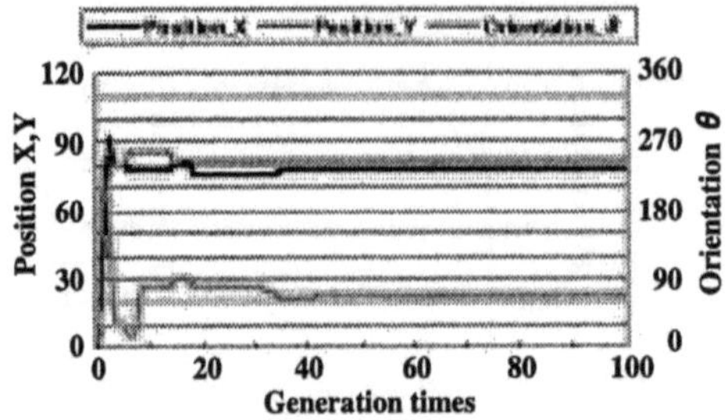

Fig. 27. Positional and oriententional results of best individual in GA with surface-strips model: results with respect to Fig.26(a)

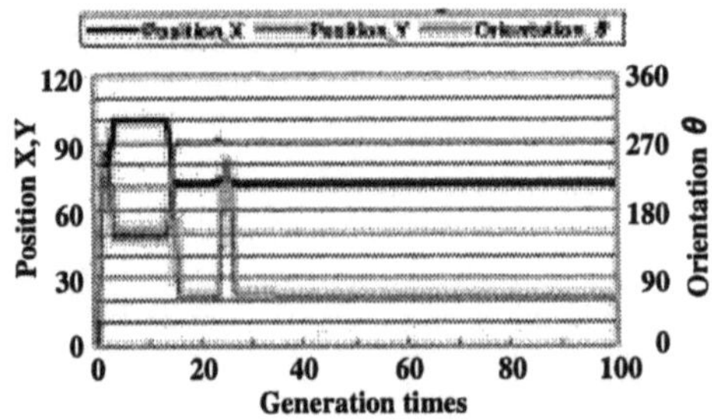

Fig. 28. Positional and oriententional results of best individual in GA with surface-strips model: results with respect to Fig.26(b)

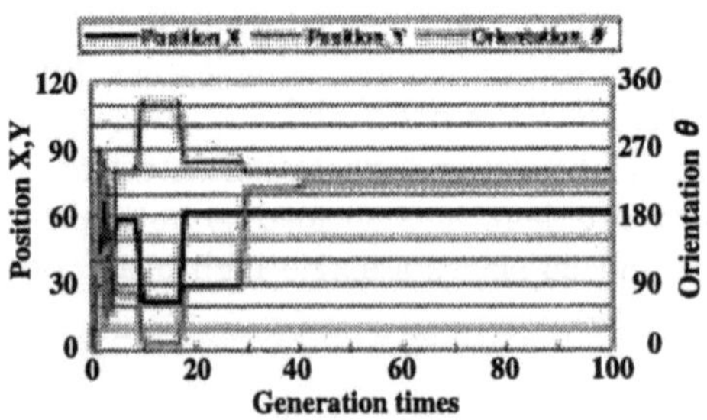

Fig. 29. Positional and oriententional results of best individual in GA with surface-strips model: results with respect to Fig.26(c)

tational values of Figs.26 (a), (b) and (c) are shown in Figs.27, 28, and 29. Regarding the surface-strips model's robustness with respect to noise, that is, its noise reduction feature and also its high sensitivity to slight translations and orientations, we suggest the use of this model for tasks requiring recognition of known objects. With the proposed object recognition method, GA can recognize the target in a short time, without need to scan all the points in the image, since GA searches for the target through an initial population of individuals scattered in the raw-image at different locations, and which

(a) (b) (c)

Fig. 30. Three fixed images

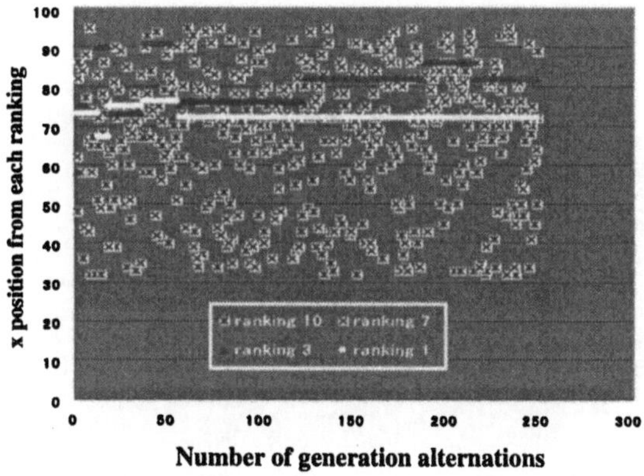

Fig. 31. Distribution in x direction with normal GA

struggle to evolve from generation to generation toward the target. When the surface-strips model is employed, the recognition method presented is shown to be robust even in the presence of noise elements in the scene.

9.2 Experiments With Respect to a Moving Fish

Computer experiments of recognition and tracking of a fish using real raw-images have been performed to appraise the performance of the proposed scene recognition method. The first experiment has considered the three fixed images shown in Fig.30. In this experiment, the images have been input into the recognition system at 0, 80 and 160 generations, in a step-input manner. Also, another experiment has considered a continuous image. In these experiments a population of 10 individuals has been used. Figure 31 presents the x positional results of four rankings (1, 3, 7, and 10), while using the global GA to track the fish shown in Fig.30 (a), (b), and (c). The wide spread of the genes at every generation is noticeable, reducing the fast detection of the target. Both GA global and local search methods have been used to perform a tracking experiment with the aforementioned fixed images. Figures 32 (a) and (b) present the detection results of the position in x and y directions. It can be seen that the global search is effective during the

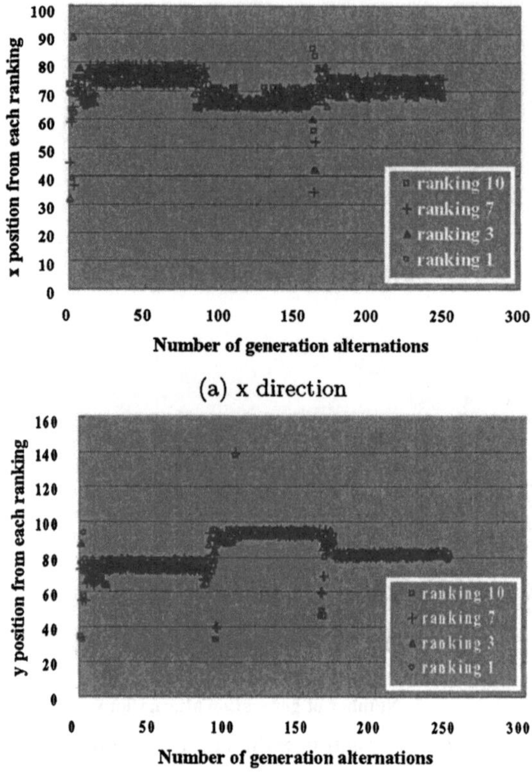

(a) x direction

(b) y direction

Fig. 32. Searching distribution with local GA

transition of images, together with the gathering of genes with the local GA search for a rapid detection of the target.

We consider the behavior during the transition of the images and recognition results inside an image. Here, it can be seen that the combined GA performs a global search together with the local search. These results indicate the expected results of the local GA search technique. It can be seen that the individuals for each ranking are gathering in the neighborhood of the point of focus. Also the results presented in Fig.33 show the intensive search for the orientation. Figure 34 shows the fitness values versus the alternation of generation number, for comparison between the results of the global GA and the results of the global/local GA that can be designated as a hybrid GA. For either image, with the hybrid GA search method, the fitness value increases quickly and higher fitness values are obtained. These results show the effectiveness of the hybrid GA search technique for real-time visual servoing purposes.

Furthermore, we employed the described hybrid GA search method, which utilizes both the global search and a local search technique of a GA, to

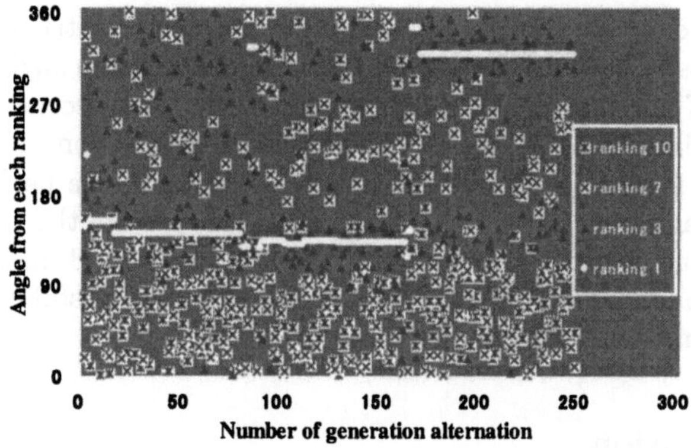

Fig. 33. Distribution of orientations with local GA

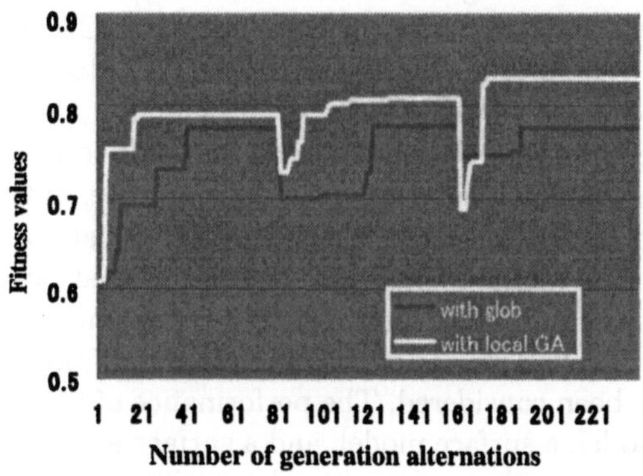

Fig. 34. Result of fitness function

Fig. 35. A sequence of recognition results of moving fish using global/local GA

perform a tracking experiment of a swimming fish in its natural scene. Figure 35 presents a sequence of the tracking results when the fish is moving. It can be seen that a high convergence speed to the position and orientation of the target could be obtained in every new input image, since the triangular model is correctly positioned with its base on the head and its sharp angle orientated on the tail of the fish. The effectiveness of evolutionary scene recognition depends on the results of the competition between the convergence speed of GA and the speed of the moving target. As a result, the performance of the evolutionary scene recognition depends on the speed of the computer. In this experiment, real-time recognition is obtained within 33 msec that is at the video rate (Japan and USA). When GA's trial is successful, this leads to tracking of a moving target in real-time.

10 Conclusion

A new method of scene recognition for manipulator real-time visual servoing has been proposed. This method utilized a hybrid GA, which comprises the global search feature of a genetic algorithm (GA), together a GA-based localized search technique. This local search technique utilizes mutation of bits in the lower portion of positional genes of an individual in GA, in order to improve the GA-based scene recognition performance, in the sense of high convergence speed toward a target object and also reliability of the search. This scene recognition method employed the unprocessed raw-image, termed here as raw-image, and a model shaping the 2-D top-surface of a target object. Also in this research, noises such as lighting conditions variations and other objects have been considered. The performance of three object models, namely a frame model, a surface model, and a surface-strips model have been investigated by analyzing the responses to three criteria questions: sensitivity, reliability, and speed. From these results, the surface-strips model has been found to be the most reliable model for object recognition in a noisy environment. The effectiveness of the proposed scene recognition method by the hybrid GA has been verified through experiments of real-time visual tracking of a moving fish using real-world raw-images. The method has shown its robustness of object recognition with the surface-strips model, in spite of the noises in the scene. The proposed GA-model-based object recognition method, which is a vision-related technique, might be applied for real-time robot navigation systems, in such case the recognition must be done in real-time. This aspect has not been developed in this paper.

References

1. D. Choi, S. Oh and K. Kim (1995) A Vehicle Steering Algorithm by Combining Computer Vision and Neural Vision. in Proc. Int. Conf. on Neural Networks, 2703-2708
2. M. Minami, J. Agbanhan and T. Asakura (1999) Manipulator visual servoing and tracking of fish using a genetic algorithm. Industrial Robot: An International Journal, **26**, 4, 278-289

3. T. Fukuda and K. Shimojima (1995) Intelligent Control for Robotics. Computational Intelligence, 202-215,
4. S. V. Ulyanov, K. Yamafuji, K. Miyagawa, T. Tanaka and T. Fukuda(1995) "Intelligent Fuzzy Motion Control of Mobile Robot for Service Use," in Proc. IEEE/RSJ Int. Conf. on Intelligent Robots and Systems, pp.486-497
5. Back T. (1996) *Evolutionary Algorithms in Theory and Practice*, OXFORD UNIVERSITY PRESS
6. A. Hanson and E. Riseman(1988) "The VISIONS Image-understanding System," in Advances in Computer Vision, C. Brown (Ed.), Lawrence Erlbaum Associates, Publishers, Hillsdale, New Jersey, 1-114
7. T. Nagata and H. Zha (1991) "Recognizing and Locating a known Object from Multiple Images," in IEEE Transaction on Robotics and Automation, **7**, 4, 434-447
8. K. Sumi, M. Hashimoto, and H. Okuda (1995) "Three-level Broad-Edge Matching based Real-time Robot Vision," in Proc. IEEE Int. Conf. on Robotics and Automation, 1416-1422
9. N. Okada and T. Nagata (1995) "A Parts Picking System with a Range Finder and a camera System," in Proc. IEEE Int. Conf. on Robotics and Automation, 1410-1415
10. G. Ao, H. Akazawa, M. Izumi, and K. Fukunaga (1996) "A Method of Model-Based Object Recognition", in Proc. Japan/USA Symposium on Flexible Automation(ASME), vol.2, 905-912
11. R. A. Brooks(1983) "Model-Based Three-Dimensional Interpretations of Two-Dimensional Images", IEEE Transaction on Pattern Analysis and Machine Intelligence, PAMI-**5**, 2, 140-150
12. M. Minami, J. Agbanhan and T. Asakura(1999) "GA-pattern matching-based manipulator control for real-time visual servoing", Advanced Robotics, **12**, 7,8, 711-734
13. D. E. Goldberg(1989)
Genetic Algorithms in Search, Optimization and Machine Learning. Reading: Addison–Wesley

Evolutionary Dynamics Identification of Multi-Link Manipulators Using Runge–Kutta–Gill RBF Networks

Thrishantha Nanayakkara[1], **Keigo Watanabe**[2], **Kazuo Kiguchi**[2], and **Kiyotaka Izumi**[3]

[1] Faculty of Engineering Systems and Technology, Graduate School of Science and Engineering, Saga University, 1-Honjomachi, Saga 840-8502, Japan.
[2] Department of Advanced Systems Control Engineering, Graduate School of Science and Engineering, Saga University, 1-Honjomachi, Saga 840-8502, Japan.
[3] Department of Mechanical Engineering, Faculty of Science and Engineering, Saga University, 1-Honjomachi, Saga 840-8502, Japan.

Abstract. This chapter discusses a method for the identification of dynamics and control of a multi-link industrial robot manipulator using Runge–Kutta–Gill neural networks (RKGNNs). RKGNNs are used to identify an ordinary differential equation of the dynamics of the robot manipulator. A structured function neural network (NN) with sub-networks to represent the components of the dynamics is used in the RKGNNs. The sub-networks consist of shape adaptive radial basis function (RBF) NNs. An evolutionary algorithm is used to optimize the shape parameters and the weights of the RBFNNs. Due to the fact that the RKGNNs can accurately grasp the changing rates of the states, this method can effectively be used for long-term prediction of the states of the robot manipulator dynamics. Unlike in conventional methods, the proposed method can even be used without input torque information because a torque network is part of the function network. This method can be proposed as an effective option for the dynamics identification of manipulators with high degrees-of-freedom, as opposed to the derivation of dynamic equations and making additional hardware changes as in the case of statistical parameter identification such as linear least-squares method. Experiments were carried out using a seven-link industrial manipulator.

1 Introduction

Developing methods to accurately grasp the full dynamics of robot manipulators is of paramount importance in judging the robustness of model-based control strategies such as computed torque controllers. The problem often encountered in the identification of dynamics of multi-link industrial manipulators is the lack of knowledge of necessary data such as input joint torques or the gains of the servo controllers. Identification of inertia parameters using the least squares algorithm has been investigated in [1], [2]. These methods need the input joint torque information, which needs special hardware changes in an industrial manipulator. Identification of dynamics using artificial neural networks (NN) has been studied in [3], [4], [5]. Yet the conventional type of direct mapping neural networks (DMNNs) are coupled with drawbacks such as the trained neural network depends on the sampling time width of the training data and the fact that the inherent structure of the dynamics of manipulators are not reflected in the structure of the NN.

Therefore the problems in hand to be solved are: to develop a method that needs only the basic reference input joint angle and its velocity, and their output data to suite a real industrial application; the model does not depend on the sampling time width; ability to grasp the structure of the dynamics of a manipulator so that the properties of dynamics are clearly reflected in the identified model; and the ability to converge to a global optimum. Therefore the motivation of this discussion has been to present solutions to all these challenging problems to construct model-based controllers for unknown industrial multi-link manipulators with high degrees-of-freedom.

The proposed method uses Runge–Kutta–Gill neural networks (RKGNNs), which is an extension version of Runge–Kutta neural networks (RKNNs), for the identification of the dynamics, because RKNNs are capable of identifying the changing rates of the states accurately due to the state space interpolation between one sampling interval [6]. Furthermore, this method only needs one state information to predict the next state and the trained NN of the function is independent of the sampling time width unlike in DMNNs [7]. In the RKGNNs, a function NN has been constructed to identify an ordinary differential equation (ODE) of the dynamics of a robotic manipulator. In [8] it has been stated that the information processing capability of an NN is determined by its architecture. Therefore the proposed method synthesizes a function NN consisting of sub-networks, so that the sub-networks represent the components of the dynamics of the manipulator [9], to catch full dynamics of a manipulator. Radial Basis Function (RBF) NNs are used for each sub-network [10]. Due to the complex structure of the dynamics of a manipulator, the number of weights of the total NN system demands for efficient optimization techniques to train the NNs. Therefore a new evolutionary optimization algorithm has been employed for the optimization of weights [11], [12]. The new evolutionary algorithm demonstrates fast convergence due to the property of monotonically reducing standard deviation [13]. In the proposed algorithm, sustaining adequate standard deviation of mutation throughout the optimization process guarantees the convergence to a global optimality [14].

To demonstrate the effectiveness of the above method, experimental studies are carried out to control a seven-link industrial robot manipulator called PA-10, manufactured by the Mitsubishi Heavy Industries Ltd. Promising results are obtained.

The rest of this chapter is organized as follows: In section 2, a brief introduction to the basic components of the dynamics of a multi-link robot manipulator and the problems associated with a practical dynamics identification study are given. Section 3 discusses the RKGNNs and the structure of the function NNs are explained. The procedure of the system identification of the PA-10 manipulator is explained in section 4. The experimental results are given in section 5.

2 Identifying the Dynamics of a Manipulator

2.1 Manipulator dynamic model and its properties

Here we will briefly discuss about the key properties of the rigid dynamics of a robot manipulator. The rigid dynamics of a manipulator is given by

$$\boldsymbol{\tau} = M(\boldsymbol{\theta})\ddot{\boldsymbol{\theta}} + V(\boldsymbol{\theta}, \dot{\boldsymbol{\theta}}) + G(\boldsymbol{\theta}) + F(\boldsymbol{\theta}) \tag{1}$$

where $\boldsymbol{\tau} \in \Re^n$ is the joint torque vector, $\boldsymbol{\theta}$, $\dot{\boldsymbol{\theta}}$, $\ddot{\boldsymbol{\theta}} \in \Re^n$ are the joint angle, angular velocity, and angular acceleration vectors, respectively, $M(\boldsymbol{\theta}) \in \Re^{n\times n}$ is the inertia matrix, $V(\boldsymbol{\theta}, \dot{\boldsymbol{\theta}}) \in \Re^n$ is the centrifugal and Coriolis force vector, $G(\boldsymbol{\theta}) \in \Re^n$ is the gravity force vector, and $F(\boldsymbol{\theta}) \in \Re^n$ is the friction vector, where n is the number of joints of the manipulator [9].

It is well known that the rigid dynamics of equation (1) has the following properties [4].

Property 1—Boundedness of the Inertia Matrix: The inertia matrix $M(\boldsymbol{\theta})$ is symmetric and positive definite, and satisfies the following inequalities:

$$m_1\|\boldsymbol{y}\|^2 \leq \boldsymbol{y}^T M(\boldsymbol{\theta})\boldsymbol{y} \leq m_2\|\boldsymbol{y}\|^2, \quad \forall \boldsymbol{y} \in \Re^n, \tag{2}$$

where m_1 and m_2 are known positive constants and $\|\cdot\|$ denotes the standard Euclidean norm.

Property 2—Skew symmetry: The inertia and centrifugal-Coriolis matrices have the following property:

$$\boldsymbol{y}^T \left(\frac{1}{2}\dot{M}(\boldsymbol{\theta}) - C(\boldsymbol{\theta}, \dot{\boldsymbol{\theta}})\right) \boldsymbol{y} = 0, \quad \forall \boldsymbol{y} \in \Re^n, \tag{3}$$

where $\dot{M}(\boldsymbol{\theta})$ is the time derivative of the inertia matrix and $V(\boldsymbol{\theta}, \dot{\boldsymbol{\theta}}) = C(\boldsymbol{\theta}, \dot{\boldsymbol{\theta}})\dot{\boldsymbol{\theta}}$.

2.2 The new concept for effective dynamics identification

It can be seen in [9] that the derivation of the dynamic equation for a manipulator with higher degrees-of-freedom such as PA-10, is very complex and time consuming. Even if the structure is known for such a manipulator, many crucial parameters such as link inertia are unknown or partially known. The identification of these unknown parameters involves data collection including the input torque sensor information if conventional least-squares method [1], [2] or DMNN approach [7] is employed. Yet in most practical situations, the torque sensors for all joints are not built-in with the manipulator. Therefore if torque sensors are to be used, some hardware changes are required that make additional trouble and expense.

The proposed method attempts to identify the dynamics involving only the reference input of the joint angle, its velocity, and their output information, using the RKGNNs trained by an evolutionary algorithm [15], [16]. This needs rearranging of the dynamic equation in (1) to form an ODE of the form $\dot{\boldsymbol{x}} = f(\boldsymbol{x})$. Given the dynamics of the manipulator as in equation (1), the state vector $\boldsymbol{x}$ can be defined by $\boldsymbol{x} = [\boldsymbol{\theta} \;\; \dot{\boldsymbol{\theta}}]^T$. Then the ODE can be formed using $\boldsymbol{x}$ and the dynamics defined by equation (1) as

$$\begin{bmatrix} \dot{\boldsymbol{\theta}} \\ \ddot{\boldsymbol{\theta}} \end{bmatrix} = \begin{bmatrix} 0 & I \\ 0 & 0 \end{bmatrix} \begin{bmatrix} \boldsymbol{\theta} \\ \dot{\boldsymbol{\theta}} \end{bmatrix} + \begin{bmatrix} 0 \\ M^{-1}(\boldsymbol{\theta})\{\tau - V(\boldsymbol{\theta}, \dot{\boldsymbol{\theta}}) - G(\boldsymbol{\theta} - F(\boldsymbol{\theta}))\} \end{bmatrix}. \tag{4}$$

The right-hand side of the above ODE gives the rate of change of the states given by the left-hand side. Therefore for this family of ODEs, RKGNNs may very accurately identify the right hand side.

3 RKGNNs and the Structure of the Function NNs

3.1 RKGNNs for the dynamics identification of manipulators

RKGNNs are an extended version of RKNNs [6]. An r-order RKNN consists of r identical function networks each with identical network structure and weights connected in such a way to realize r-order Runge–Kutta algorithm. Each function network models the right hand side of the ODE directly. The RNGNN differs from RKNNs in the way the function NNs are connected. When defining two time steps as $t_i = ih$ and $t_{i+1} = (i+1)h$, the input-output relationship of a fourth-order RKGNN is given by

$$\boldsymbol{x}(i+1) = \boldsymbol{x}(i) + \frac{h}{6}(\boldsymbol{k}_0 + e\boldsymbol{k}_1 + g\boldsymbol{k}_2 + \boldsymbol{k}_3) \tag{5}$$

where h is the sampling time width, and

$$\boldsymbol{k}_0 = N_f(\boldsymbol{x}(i))$$

$$\boldsymbol{k}_1 = N_f(\boldsymbol{x}(i) + \frac{h}{2}\boldsymbol{k}_0)$$

$$\boldsymbol{k}_2 = N_f(\boldsymbol{x}(i) + ah\boldsymbol{k}_0 + bh\boldsymbol{k}_1)$$

$$k_3 = N_f(x(i) + chk_1 + dhk_2)$$

in which $a = 1/\sqrt{2} - 1/2$, $b = 1 - 1/\sqrt{2}$, $c = -1/\sqrt{2}$, $d = 1 + 1/\sqrt{2}$, $e = 2 - \sqrt{2}$, and $g = 2 + \sqrt{2}$, with the function NN denoted by $N_f(\cdot)$. The fundamental idea of an RKGNN is depicted by **Fig. 1**. Note that if $a = 0$, $b = 1/2$, $c = 0$, $d = 1$, $e = 2$, and $g = 2$, the RKGNN becomes an RKNN.

From Fig. 1, it can be seen that the prediction of the next state involves one previous state information and the sampling width is external to the function NN model. One other feature is that all four NN models given by N_f in the RKGNN are the same. Therefore finding the optimum weights for one N_f is all that is required to identify the system.

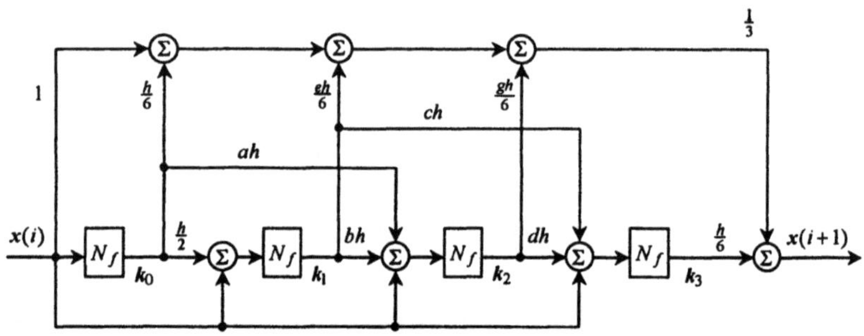

Fig. 1. Overall arrangement of one-step ahead predictive network based on the Runge-Kutta-Gill method with the function neural networks N_f.

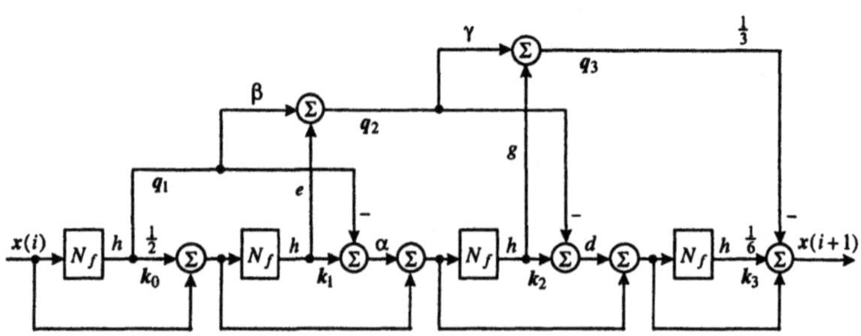

Fig. 2. Overall arrangement of one-step ahead predictive network based on the alternative Runge-Kutta-Gill method with the function neural networks N_f.

When using an RKGNN that minimizes the memory storage, we further define the following new vectors:

$$\boldsymbol{q}_1 = \boldsymbol{k}_0$$

$$\boldsymbol{q}_2 = (-2 + \frac{3}{\sqrt{2}})\boldsymbol{k}_0 + (2 - \sqrt{2})\boldsymbol{k}_1$$

$$\boldsymbol{q}_3 = -\frac{1}{2}\boldsymbol{k}_0 - (1 + \sqrt{2})\boldsymbol{k}_1 + (2 + \sqrt{2})\boldsymbol{k}_2.$$

Then, the RKGNN algorithm contains the following four steps:

Step 1:

$$\boldsymbol{k}_0 = hN_f(\boldsymbol{x}(i))$$

$$\boldsymbol{x}^{(1)}(i+1) = \boldsymbol{x}(i) + \frac{1}{2}\boldsymbol{k}_0$$

$$\boldsymbol{q}_1 = \boldsymbol{k}_0$$

Step 2:

$$\boldsymbol{k}_1 = hN_f(\boldsymbol{x}^{(1)}(i+1))$$

$$\boldsymbol{x}^{(2)}(i+1) = \boldsymbol{x}^{(1)}(i+1) + \alpha(\boldsymbol{k}_1 - \boldsymbol{q}_1)$$

$$\boldsymbol{q}_2 = \beta\boldsymbol{q}_1 + e\boldsymbol{k}_1$$

Step 3:

$$\boldsymbol{k}_2 = hN_f(\boldsymbol{x}^{(2)}(i+1))$$

$$\boldsymbol{x}^{(3)}(i+1) = \boldsymbol{x}^{(2)}(i+1) + d(\boldsymbol{k}_2 - \boldsymbol{q}_2)$$

$$\boldsymbol{q}_3 = \gamma\boldsymbol{q}_2 + g\boldsymbol{k}_2$$

Step 4:

$$\boldsymbol{k}_3 = hN_f(\boldsymbol{x}^{(3)}(i+1))$$

$$\boldsymbol{x}^{(4)}(i+1) = \boldsymbol{x}^{(3)}(i+1) + \frac{1}{6}\boldsymbol{k}_3 - \frac{1}{3}\boldsymbol{q}_3$$

$$\boldsymbol{x}(i+1) \equiv \boldsymbol{x}^{(4)}(i+1)$$

where $\alpha = 1 - \sqrt{2}/2$, $\beta = -2 + 3/\sqrt{2}$, and $\gamma = -(2 + 3/\sqrt{2})$. The alternative illustration is given in **Fig. 2**.

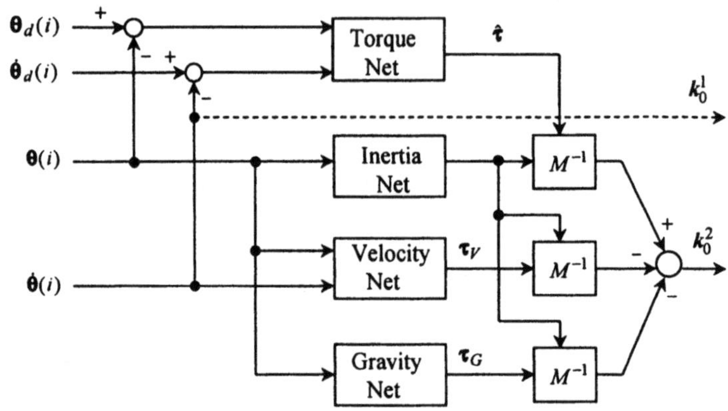

Fig. 3. Basic concept of the function NN to identify each component of the dynamic equation.

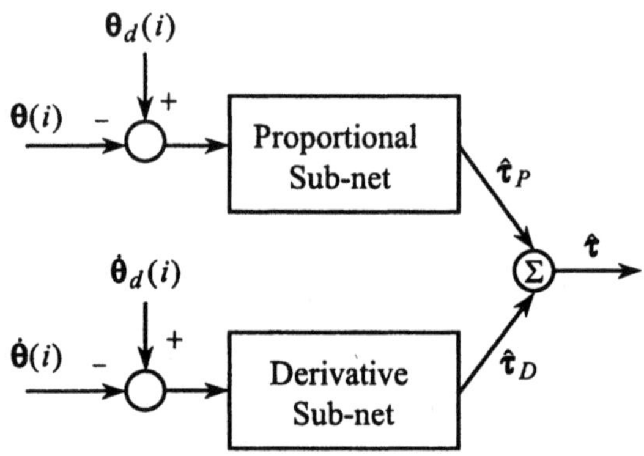

Fig. 4. Torque network for identifying the proportional and derivative gains in a P-D servo controller.

3.2 The anatomy of the function NN

The dynamics equation of a rigid link manipulator can be analyzed into the following basic components: one is the inertia matrix, the other is the velocity dynamics given by centrifugal and Coriolis force vectors, another is the gravity force vector. In the proposed method, separate networks are constructed to identify the dynamics of each of these components. In addition to these, in a practical industrial application, the information about the servo controller is unknown. Therefore a torque network is constructed that will grasp the dynamics of the servo controller. Thus, it is very important to notice that the function NN $N_f(\cdot)$ as shown in **Fig. 3** contains components for the identification of each component of the dynamic equation as shown in equation (1).

Torque network The structure of the torque network is depicted by **Fig. 4**, where two sub-networks for the identification of proportional and derivative gains of the actual controller are employed. The angular error vector given by $(\boldsymbol{\theta}_d - \boldsymbol{\theta})$ and the angular velocity error vector given by $(\dot{\boldsymbol{\theta}}_d - \dot{\boldsymbol{\theta}})$ are given as inputs to the proportional and derivative networks respectively to produce a suitable torque vector for the robot manipulator, where $\boldsymbol{\theta}_d$ is the reference angular trajectory and $\dot{\boldsymbol{\theta}}_d$ is the reference angular velocity. If detailed information of the controller structure is known, then the structure of the network can be altered accordingly. Moreover, the torque data are directly provided to the function NN, if the input torque can be measured for each joint.

Inertia network The structure of the sub-networks for the inertia network is designed such that they identify the diagonal and upper triangular components of the inertia matrix of the dynamic equation. This helps to make sure that the inertia matrix is symmetric and positive definite. The block diagram of the inertia sub-networks can be seen in **Fig. 5**, where M_{ij}^{-1} $i = 1, \cdots, n$; $j = 1, \cdots, n$ is the element in the ith row and the jth column of the inverse of the inertia matrix.

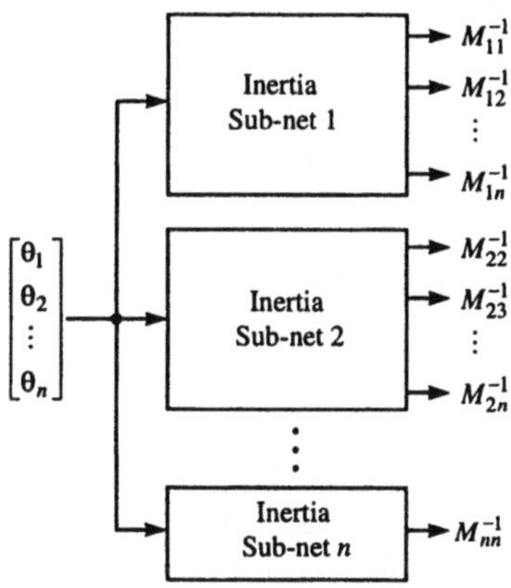

Fig. 5. Inertia network considering the symmetry of the inertia matrix.

Velocity network The velocity network set which includes centrifugal and Coriolis components is shown in **Figs. 6 and 7** respectively. Basically the output of the velocity network is given by

$$V(\boldsymbol{\theta}, \dot{\boldsymbol{\theta}}) = B(\boldsymbol{\theta})[\dot{\boldsymbol{\theta}}\dot{\boldsymbol{\theta}}] + C(\boldsymbol{\theta})[\dot{\boldsymbol{\theta}}^2] \tag{6}$$

where $B(\boldsymbol{\theta})$ is a matrix of Coriolis effect, whose size is $n \times n(n-1)/2$ and $[\dot{\boldsymbol{\theta}}\dot{\boldsymbol{\theta}}]$ is an $n \times (n-1)/2$ vector consisting of

$$[\dot{\boldsymbol{\theta}}\dot{\boldsymbol{\theta}}] = [\dot{\theta}_1\dot{\theta}_2 \quad \dot{\theta}_1\dot{\theta}_3 \quad \cdots \quad \dot{\theta}_{n-1}\dot{\theta}_n]^T. \tag{7}$$

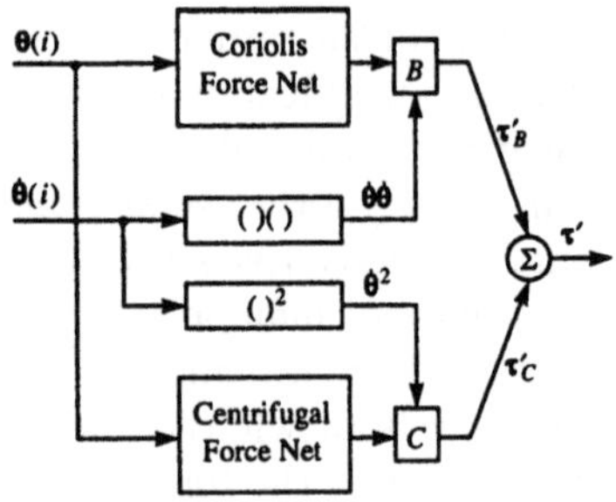

Fig. 6. Velocity network consisting of the centrifugal and Coriolis force networks.

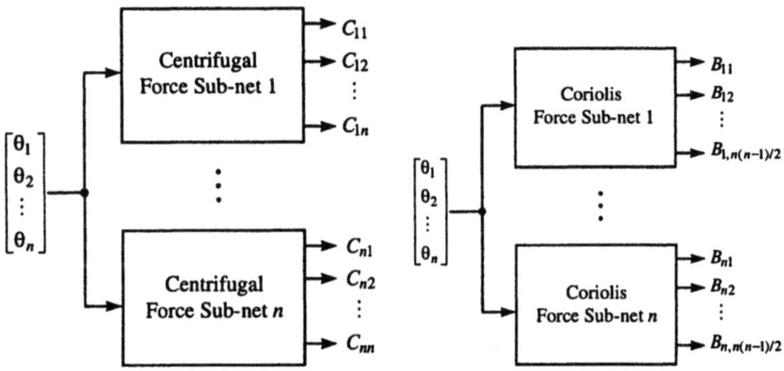

Fig. 7. Block diagrams of the centrifugal and Coriolis networks.

$C(\boldsymbol{\theta})$ is a matrix of centrifugal effect, whose size is $n \times n$, and

$$[\dot{\boldsymbol{\theta}}^2] = [\dot{\theta}_1^2 \ \ \dot{\theta}_2^2 \ \cdots \ \dot{\theta}_n^2]^T. \tag{8}$$

In Fig. 7, C_{ij} $(i = 1, \cdots, n;\ j = 1, \cdots, n))$ and B_{ij} $(i = 1, \cdots, n;\ j = 1, \cdots, n(n-1)/2)$ are the elements in the ith row and the jth column of the centrifugal and Coriolis effect vectors respectively.

Gravity network The gravity network consists of one RBF network with n RBFs and n outputs, if the manipulator will be moved in the gravity direction.

The construction of sub-networks of each of these component networks involves construction of shape adaptive RBFNNs as can be seen in **Fig. 8**.

The input-output mapping relationship of the RBF in this application is given by

$$\phi_l(\boldsymbol{y}; \bar{\boldsymbol{y}}_l, \sigma_l) = \exp\{-[(\boldsymbol{y} - \bar{\boldsymbol{y}}_l)^T(\boldsymbol{y} - \bar{\boldsymbol{y}}_l)/\sigma_l^2]\} \tag{9}$$

where $\boldsymbol{y} \in \Re^n$ and $\bar{\boldsymbol{y}}_l \in \Re^n$ are the input vector and the vector of centers of the RBF, σ_l^2 is the variance of the lth RBF. Therefore the output at the kth output node of the sub-network $N_{sub}(\cdot)$ is given by

$$[N_{sub}(\boldsymbol{y})]_k = \sum_{l=1}^{N} W_{kl}\phi_l(\boldsymbol{y}; \bar{\boldsymbol{y}}_l, \sigma_l), \quad k = 1, \cdots, m \tag{10}$$

where N is the number of RBFs that construct the sub-network and W_{kl} is the connection weight from lth RBF to kth output node. Here we make the standard assumption made in NN literature [4] given next.

Assumption: When defining the weight matrix $W = \{W_{kl}\}$, $k = 1, \cdots, n$, $l = 1, \cdots, N$, the ideal weights are bounded by known positive values so that

$$\|W\|_F \leq w_M \tag{11}$$

where $\|\cdot\|_F$ is the Frobenius norm, i.e., $\sqrt{\mathrm{tr}(WW^{\mathrm{T}})}$, and w_M is known.

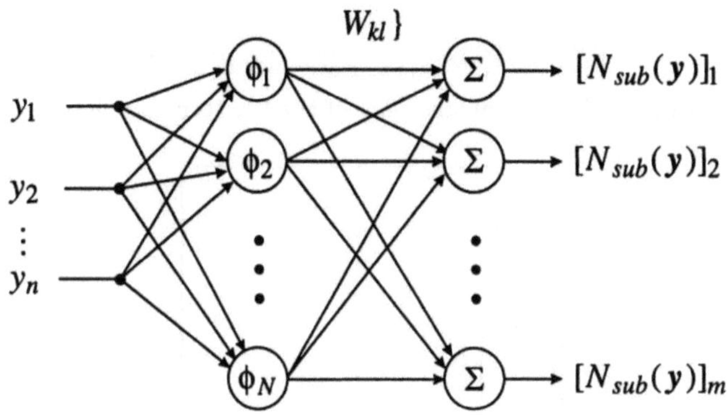

Fig. 8. Basic structure of the RBF NN.

This assumption is made in the initialization of the weights of the NN in the evolutionary algorithm employed to train the NNs, in this application.

4 Procedure of the System Identification

The PA-10 manipulator consists of seven links with seven degrees-of-freedom (three rotation axes and four pivot axes) as illustrated in **Fig. 9**. The main specifications of the PA-10 manipulator are as follows: the length of the manipulator is 1345 mm, the weight of the manipulator is 35 kgf, the maximum combined speed with all axes is 1.55 m/s, the payload weight is 10 kgf, and output torque is 9.8 Nm.

For a given trajectory, the input reference angle, its velocity and their output data of all seven joints were experimentally obtained. Then these experimental data were used to train the RKGNNs to identify the dynamics

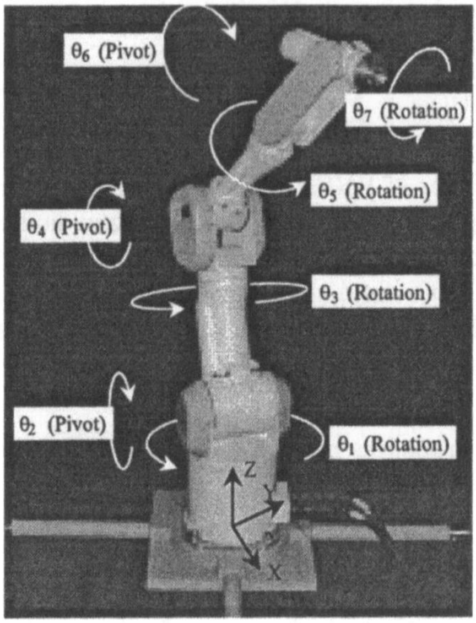

Fig. 9. Mechanical structure of the PA-10 manipulator.

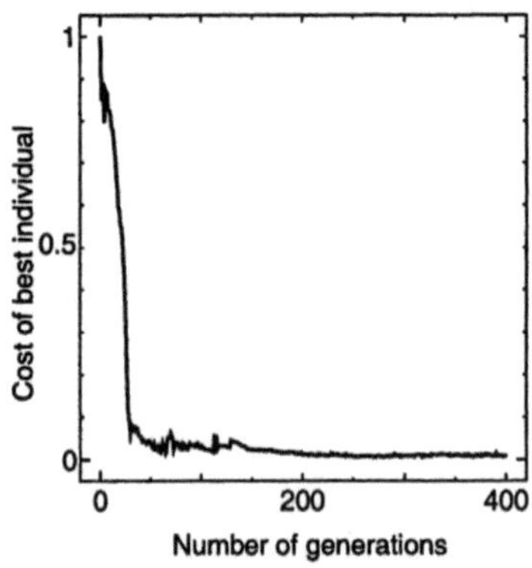

Fig. 10. Evolutionary history for the dynamics RKGNN training for the trajectory.

of the manipulator. The weights of the RKGNN and the RBF parameters, $\bar{\boldsymbol{y}}_l$ and σ_l, were found using the evolutionary algorithm described in references [11], [12], with the objective function subjected to constraints [17]. The constrained objective function is given by

$$J = \left(\frac{1}{t_N} \sum_{i=1}^{t_N} \|\boldsymbol{\theta}_d(i) - \boldsymbol{\theta}(i)\| + \|\dot{\boldsymbol{\theta}}_d(i) - \dot{\boldsymbol{\theta}}(i)\| \right) + \frac{s_t}{2} \left(\sum_{j=1}^{c_N} [g_j^+(\boldsymbol{\theta})]^2 \right) \quad (12)$$

where t_N is the total time span of the trajectory, $g_j^+(\boldsymbol{\theta}) = \max\{0, g_j(\boldsymbol{\theta})\}$ is the additional penalty function for the jth constraint satisfaction and s_t is a log penalty parameter such that $s_t = 10 + \log(i+1)$, and c_N is the number of constraints.

One constraint was considered: which is given by

$$g_1(\boldsymbol{\theta}) : -\Gamma(\boldsymbol{\theta}) \leq 0 \tag{13}$$

where $\Gamma(\boldsymbol{\theta})$ is the minimum principle minor of the inertia matrix.

Then the parameters of the function NN to be optimized were encoded in the elements of individuals of the evolutionary algorithm as real valued vectors in the following ranges. All the weights were initialized in the range $[-2, 2]$, the variances in $[0, 5]$, and centers of RBFs in $[-2, 2]$ except for the following: weights, variances, and centers of RBFs in the torque network were in $[-10, 10]$, $[0, 100]$, and $[-1, 1]$, respectively; the weights of the gravity network were in $[-5, 0]$; the weights of the velocity net were in $[-0.1, 0.1]$. The conditions of the usage of evolutionary algorithm were as follows: 100% mutation and 20% crossover were employed. The tournament size was 15 and the number of individuals was 100. One individual consists of 3080 elements. The algorithm was run for 200 generations. The size of the regression vector n_p was 2, and the strategic move percentage μ_p was 10% [12].

5 Model Based Controller

A model based controller is used to generate joint velocity commands as given by

$$\ddot{\boldsymbol{\theta}}_c = \hat{M}^{-1}(\boldsymbol{\theta})[\boldsymbol{\tau}_d + \{K_p\boldsymbol{e} + K_v\dot{\boldsymbol{e}} + K_I \int \boldsymbol{e}dt\} + \hat{V}(\boldsymbol{\theta}, \dot{\boldsymbol{\theta}}) + \hat{G}(\boldsymbol{\theta})] \tag{14}$$

where $\ddot{\boldsymbol{\theta}}_c$ is the manipulated joint acceleration; $\hat{M}^{-1}$ is the estimated inertia matrix; $\boldsymbol{\tau}_d$ is the torque command estimated by the torque network; K_p, K_v and K_I are the proportional, derivative and integral gains respectively; $\boldsymbol{e}$ is the vector of joint angle errors; $\hat{V}(\boldsymbol{\theta}, \dot{\boldsymbol{\theta}})$ is the estimated velocity force matrix; and $\hat{G}(\boldsymbol{\theta})$ is the estimated gravitational force matrix.

6 Experimental Results

The following **Figs. 11-13** show the experimental results. The experimental results encourage the use of this method for identification of dynamics of a complex robot manipulator.

7 Conclusions

The Runge–Kutta–Gill neural network (RKGNN) has been developed and it was applied for identifying the complex dynamics of manipulators with

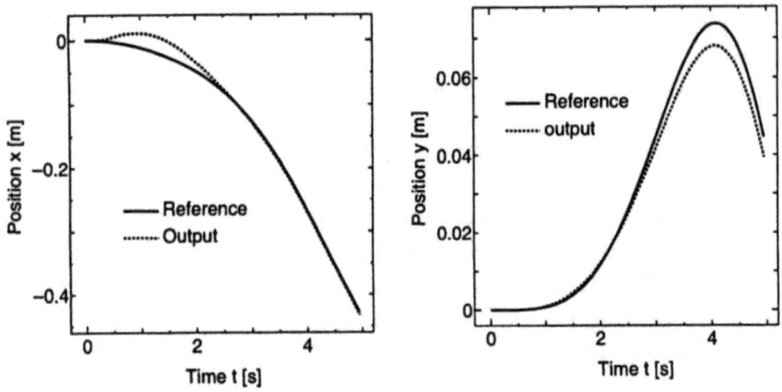

Fig. 11. The control results for position x and y

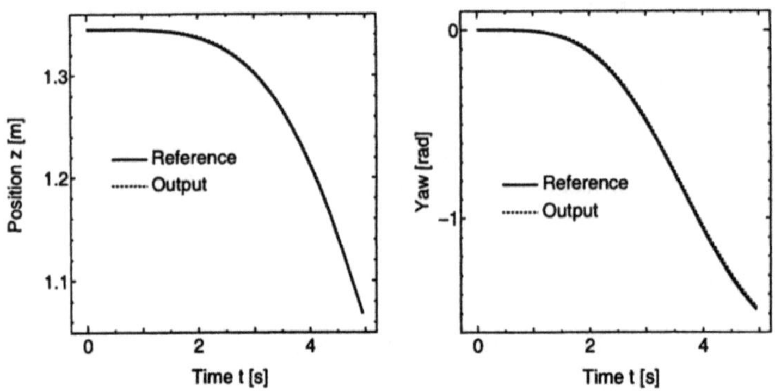

Fig. 12. The control results for position z and orientation yaw

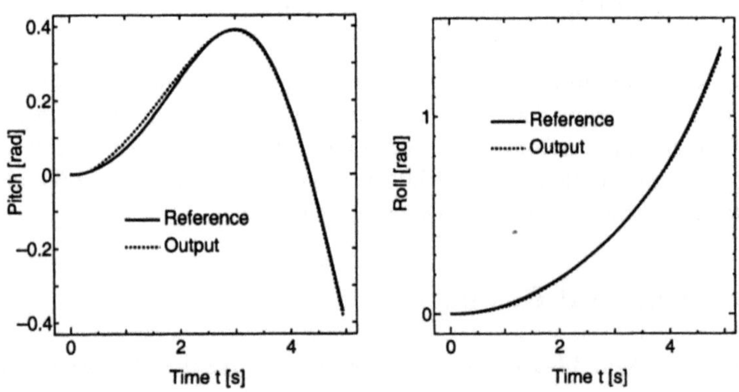

Fig. 13. The control results for orientation pitch and roll

higher degrees-of-freedom. It is very useful, because it reduces the burden of deriving the complex dynamic equations for such manipulators. The adoption of RKGNNs makes the long-term prediction of the states very accurate due to its property of interpolation of states within one sampling interval and estimating the rate of change of states accurately. This method is advantageous in the practical situation where input torque information is unknown. The key feature of using a structured function NN gives room to adopt any kind of model based controller due to the separability of the components of the dynamics such as inertia, Coriolis, centrifugal, and gravity matrices.

Experimental studies were carried out to demonstrate the ability of the proposed method to capture the complex dynamics of a multi-link manipulator, by using an industrial seven-link robot manipulator called PA-10. Promising experimental results have been obtained.

References

1. An, C. H., Atkeson, C. G., and Hollerbach, J. M. (1986): Estimation of Internal Parameters of Rigid Body Links of Manipulators. Artificial Intelligence Memo 887, MIT Artificial Intelligence Laboratory
2. Kawasaki, H. and Nishimura, K. (1998): Terminal Link Parameter Estimation of Robotic Manipulators. IEEE Trans. on Robotics and Automation. **4**(5), 485–490.
3. Wu, C.-J. and Huang, C.-H. (1996): Back-Propagation Neural Networks for Identification and Control of a Direct Drive Robot. J. of Intelligent and Robotic Systems. **16**(1), 45–64.
4. Kwan, C., Lewis, F. L., and Dawson, D. M. (1998): Robust Neural-Network Control of Rigid-Link Electrically Driven Robots. IEEE Trans. on Neural Networks. **9**(4), 481–488.
5. Behera, L., Gopal, M., and Chaudhury, S. (1996): On Adaptive Trajectory Trackingg of a Robot Manipulator Using Inversion of Its Neural Emulator. IEEE Trans. on Neural Networks. **7**(6), 1401–1414.
6. Wang, Y.-J., and Lin, C.-T. (1998): Runge-Kutta Neural Network for Identification of Dynamical Systems in High Accuracy. IEEE Trans. on Neural Networks. **9**(2), 294-307.
7. Noriega, J. R. and Wang, H. (1998): A Direct Adaptive Neural-Network Control for Unknown Nonlinear Systems and Its Application. IEEE Trans. on Neural Networks. **9**(1), 27–34.
8. Yao, X. and Liu, Y. (1997): A New Evolutionary System for Evolving Artificial Neural Networks. IEEE Trans. on Neural Networks. **8**(3), 694–713.
9. Craig, J. J. (1989): Introduction to Robotics: Mechanics and Control. 2nd edn, Reading, MA: Addison-Wesley.
10. Web, A. R. and Shannon, S. (1998): Shape Adaptive Radial Basis Functions. IEEE Trans. on Neural Networks. **9**(6), 1155–1166.
11. Nanayakkara, T., Watanabe, K., and Izumi, K. (1999): Evolving in Dynamic Environments Through Adaptive Chaotic Mutation. in Procs. of Fourth International Symposium on Artificial Life and Robotics. Oita, Japan, Jan 19–22, **2**, 520–523.

12. Nanayakkara, T., Watanabe, K., Kiguchi, K., and Izumi, K. (1999): Evolutionary Optimization with Strategic Moves Based on Historical Knowledge. in Procs. of the First SOFT Kyushu Branch Annual Conference, 21–28
13. Bäck, T. and Schwefel, H. P. (1993): An Overview of Evolutionary Algorithms for Parameter Optimization. Evolutionary Computation. **1**(1), 1–23.
14. Leung, Y. (1997): Degree of Population Diversity—A Perspective on Premature Convergence in Genetic Algorithms and its Markov Chain Analysis. IEEE Trans. on Neural Networks. **8**(5), 1165–1175.
15. Nanayakkara, T., Watanabe, K. and Izumi, K. (1999): Evolving Runge-Kutta-Gill RBF Networks to Estimate the Dynamics of a Multi-Link Manipulator. in Proc. of IEEE International Conference on Systems, Man, and Cybernetics. **2**, 770–775.
16. Nanayakkara, T., Watanabe, K., Kiguchi, K., and Izumi, K. (2000): Dynamic Identification of an Industrial Seven-Link Manipulator by Evolving Runge–Kutta–Gill RBF Networks. in Proc. of Fifth International Symposium on Artificial Life and Robotics. **2**, 669–672.
17. Kim, J.-H. and Myung, H. (1997): Evolutionary Programing Techniques for Constrained Optimization Problems. IEEE Trans. on Evolutionary Computation. **1**(2), 129–140.

Towards Reliable Sub-Division of Geological Areas: Interval Approach

David D. Coblentz[1,2], Vladik Kreinovich[1], Brian S. Penn[1,3], and Scott A. Starks[1]

[1] NASA Pan American Center for Earth and Environmental Studies, University of Texas, El Paso TX 79968, USA, contact email vladik@cs.utep.edu
[2] Institut für Geologie und Paläontologie, Universität Graz, Heinrichstr 26, A-8010 Graz, Austria
[3] Autometric, Inc., 1330 Inverness Drive, Suite 350, Colorado Springs, CO 80910, USA

Abstract. An appropriate subdivision of a geophysical area into segments enables us to extrapolate the results obtained in some locations within the segment (where extensive research was done) to other locations within the same segment, and thus, get a good understanding of the locations which weren't thoroughly analyzed.

Often, different evidence and different experts' intuition support different subdivisions schemes. For example, in our area – Rio Grande rift zone – there is some geochemical evidence that this zone is divided into three segments, but, in the viewpoint of many researchers, this evidence is not yet sufficiently convincing.

We show that if we use topographical information (this information, e.g., comes from satellite photos), then interval methods lead to a reliable justification for the tripartite subdivision of the Rio Grande rift zone.

1 Appropriate Subdivision is Important in Geophysics

In geophysics, appropriate subdivision of an area into segments is extremely important, because it enables us to extrapolate the results obtained in some locations within the segment (where extensive research was done) to other locations within the same segment, and thus, get a good understanding of the locations which weren't that thoroughly analyzed.

2 A Problem: It is Often Difficult to Produce a Reliable Subdivision

The subdivision of a geological zone into segments is often a controversial issue, with different evidence and different experts' intuition supporting different subdivisions.

For example, in our area – Rio Grande rift zone – there is some geochemical evidence that this zone is divided into three segments [5]:

- the southern segment which is located, approximately, between the latitudes $y = 29°$ and $y = 34°$;
- the central segment – from $y = 34.5°$ to $y = 38°$; and
- the northern segment – from $y = 38°$ to $y = 41°$.

However, in the viewpoint of many researchers, this evidence is not yet sufficiently convincing.

It is therefore desirable to develop new techniques for zone sub-division, techniques which would be in the least possible way dependent on the (subjective) expert opinion and would, thus, be maximally reliable.

3 Main Idea: Using Topographic Information

One reason for subjectivity is the fact that the existing subdivision is often based on the chemical and physical analysis of several samples collected throughout the area, and often, we do not have a statistically sufficient amount of thoroughly analyzed geological samples to make the conclusion about the subdivision statistically convincing.

To make this conclusion more reliable, we can use, instead of the more rare *geological* samples, a more abundant *topographical* information (this information, e.g., comes from satellite photos). We can characterize each part of the divided zone by its topography.

4 Preserving only Geophysically Meaningful Topographic Information: The Use of Spectral Values Corresponding to Long Wavelengths

In topographical analysis, we face a new problem: of too much data, most of which is geophysically irrelevant. To eliminate some of this irrelevant data, we can use the Fourier transform; indeed, it is known that while (at least some) absolute values of the map (forming a so-called spectrum) are geophysically meaningful, the phases usually are random and can be therefore ignored. So, we should only use the spectrum.

Since we are interested only in the large-scale classification, it makes sense to only use the spectrum values corresponding to relatively large spatial wavelengths, i.e., wavelengths L for which $L \geq L_0$ for some appropriate value L_0. In particular, for the sub-division of the Rio Grande rift, it makes sense to use only wavelengths of $L_0 = 1000$ km or larger.

Also, for the Rio Grande Rift, we are interested in the classification of horizontal zones, so it makes sense to do the following:

- divide the Rio Grande Rift into 1° zones $[y^-, y^+]$ (with y from $y^- = 30$ to $y^+ = 31$, from $y^- = 31$ to $y^+ = 32$, ..., from $y^- = 40$ to $y^+ = 41$);
- for each of these zones, take the topographic data, i.e., the height $h(x, y)$ described as a function of longitude x and latitude y;
- for each zone and for each y, compute the Fourier transform $H(\omega, y)$ with respect to x;
- for each zone, combine all the spectral values which correspond to large wavelength (i.e., for which $\omega \,\square\, 1/L_0$), and compute the resulting spectral value

$$S(y^-) = \int_{y=y^-}^{y^+} \int_{\omega=0}^{1/L_0} |H(\omega, y)|^2 \, d\omega \, dy.$$

We are interested in comparing the spectral values $S(y)$ corresponding to different latitudes y, so we are not interested in the *absolute* values of $S(y)$, only in *relative* values. Thus, to simplify the data, we can *normalize* them by, e.g., dividing each value $S(y^-)$ by the largest $S_{\max}$ of these values. In particular, for the Rio Grande rift, the resulting values of $y^- = y_1, y_2, \ldots$ and $s_i = S(y_i)/S_{\max}$ are as follows:

Table 1.

y_i	29	30	31	32	33	34	35	36	37	38	39	40	41
s_i	0.28	0.24	0.21	0.16	0.20	0.29	0.31	0.35	0.46	1.00	0.80	0.96	0.74

Based only on these spectral values s_i, we will try to classify locations into several clusters ("segments").

5 Traditional Clustering Techniques Do Not Lead to Reliable Classification

Most existing clustering methods (see, e.g., [2]) assume that we have some information about the *probability* distributions of the measurement errors. In real-life topographical analysis, we do not have enough data to uniquely determine these distributions, and different possible distributions lead to drastically different conclusions about the subdivision of a geological zone.

Similarly, *fuzzy* clustering techniques (see, e.g., [1]) depend on the subjective fuzzy estimates, and different expert estimates often lead to drastically different subdivisions.

Therefore, we need a new, less subjective clustering method.

6 Segments as Monotonicity Regions

How can we detect the segments based on these values s_i? In order to answer this question, let us first plot the dependence of s_i on y_i and see if there is any visible feature of this plot which can be associated with a subdivision of the area into geophysical segments.

If we plot the dependence of s_i on y_i, we will see that at first, the function $s(y)$ is (approximately) decreasing, then it is (approximately) increasing, and then it is (approximately) decreasing again. Interestingly, these "monotonicity regions" seem to be in good accordance with the empirical subdivision of the rift into segments [5]. In view of this observation, we will identify geophysical segments with the monotonicity regions of the (unknown) function $s(y)$.

7 Let's Use Intervals

The heights are measured pretty accurately, so the only errors in the values s_i come from discretization. In other words, we would like to know the values of the function $s(y) = S(y)/S_{\max}$ for all y, but we only know the values $s_1 = s(y_1)$, ..., $s_n = s(y_n)$ of this function for the points $y_1, \ldots, y_n$. For each y which is different from y_i, it is reasonable to estimate $s(y)$ as the value $s_i = s(y_i)$ at the point y_i which is the closest to y (and, ideally, which belongs to the same segment as y_i). For each point y_i, what is the largest possible error Δ_i of the corresponding approximation?

When $y > y_i$, the point y_i is still the closest until we reach the midpoint $y_{\text{mid}} = (y_i + y_{i+1})/2$ between y_i and y_{i+1}. It is reasonable to assume that the largest possible approximation error $|s(y) - s_i|$ for such points is attained when the distance between y and y_i is the largest, i.e., when y is this midpoint; in this case, the approximation error is equal to $|s(y_{\text{mid}}) - s_i|$.

If the points y_i and y_{i+1} belong to *the same* segment, then the dependence of $s(y)$ on y should be reasonably smooth for $y \in [y_i, y_{i+1}]$. Therefore, on a narrow interval $[y_i, y_{i+1}]$, we can, with reasonable accuracy, ignore quadratic and higher terms in the expansion of $s(y_i + \Delta y)$ and thus, approximate $s(y)$ by a linear function. For a linear function $s(y)$, the difference $s(y_{\text{mid}}) - s(y_i)$ is equal to the half of the difference $s(y_{i+1}) - s(y_i) = s_{i+1} - s_i$; thus, for $y > y_i$, the approximation error is bounded by $0.5 \cdot |s_{i+1} - s_i|$.

If the points y_i and y_{i+1} belong to *different* segments, then the dependence $s(y)$ should exhibit some non-smoothness, and it is reasonable to expect that the difference $|s_{i+1} - s_i|$ is much higher than the approximation error.

In both cases, the approximation error is bounded by

$$0.5 \cdot |s_{i+1} - s_i|.$$

Similarly, for $y < y_i$, the approximation error is bounded by $0.5 \cdot |s_i - s_{i-1}|$ if the points y_i and y_{i-1} belong to the same segment, and is much smaller if they don't. In both cases, the approximation error is bounded by

$$0.5 \cdot |s_i - s_{i-1}|.$$

We have two bounds on the approximation error and we can therefore conclude that the approximation error cannot exceed the smallest Δ_i of these two bounds, i.e., the value

$$\Delta_i = 0.5 \cdot \min(|s_i - s_{i_1}|, |s_{i+1} - s_i|).$$

As a result, instead of the *exact* values s_i, for each i, we get the *interval* $\mathbf{s}_i = [s_i^-, s_i^+]$ of possible values of $s(y)$, where $s_i^- = s_i - \Delta_i$ and $s_i^+ = s_i + \Delta_i$. In particular, for the Rio Grande rift, the corresponding intervals are as follows:

Table 2.

y_i	$y_1 = 29$	$y_2 = 30$	$y_3 = 31$	$y_4 = 32$
$\mathbf{s}_i = [s_i^-, s_i^+]$	[0.26,0.30]	[0.225,0.255]	[0.195,0.225]	[0.14,0.18]
$y_5 = 33$	$y_6 = 34$	$y_7 = 35$	$y_8 = 36$	$y_9 = 37$
[0.18,0.22]	[0.28,0.30]	[0.30,0.32]	[0.33,0.37]	[0.405,0.515]
$y_{10} = 38$	$y_{11} = 39$	$y_{12} = 40$	$y_{13} = 41$	
[0.80,1.10]	[0.72,0.88]	[0.88,1.04]	[0.63,0.85]	

8 How to Find Monotonicity Regions of a Function Defined with Interval Uncertainty: Idea

We want to find regions of uncertainty of a function $s(y)$, but we do not know the exact form of this function; all we know is that for every i, $s(y_i) \in \mathbf{s}_i$ for known intervals $\mathbf{s}_i$. How can we find the monotonicity regions in the situation with such interval uncertainty?

Of course, since we only know the values of the function $s(y)$ in finitely many points y_i, this function can have as many monotonicity regions between y_i and y_{i+1} as possible. What we are interested in is funding the subdivision into monotonicity regions which can be deduced from the data. The first natural question is: can we explain the data by assuming that the dependence $s(y)$ is monotonic? If not, then we can ask for the possibility of having a function $s(y)$ with exactly two monotonicity regions:

- if such a function is possible, then we are interested in possible locations of such regions;
- if such a function is not possible, then we will try to find a function $s(y)$ which is consisted with our interval data and which has three monotonicity regions, etc.

This problem was first formalized and solved in [8]. The corresponding algorithm is based on the following idea.

If the function $s(y)$ is non-decreasing, then, for $i < j$, we have $s(y_i) \Box s(y_j)$; therefore, $s_i^- \Box s(y_i) \Box s(y_j) \Box s_j^+$ and $s_i^- \Box s_j^+$. It turns out that, vice versa, if the inequality $s_i^- \Box s_j^+$ hold for every $i < j$, then there is a non-decreasing function $s(y)$ for which $s(y_i) \in \mathbf{s}_i$ for all i (e.g., we can take $s(y_i) = \max(s_1^-, \ldots, s_i^-)$ for all i and use linear interpolation to define the values $s(y)$ for $y \neq y_i$.) Thus, to check monotonicity, it is sufficient to check these inequalities.

If we have already checked these inequalities for the intervals $\mathbf{s}_1, \ldots, \mathbf{s}_k$, and we add the new interval $\mathbf{s}_{k+1}$, then, to confirm that it is still possible for a function $s(z)$ to be non-decreasing, it is sufficient to check that $s_i^- \Box s_{k+1}^+$ for all $i = 1, \ldots, k$. These k inequalities are equivalent to a single inequality $m_k \Box s_{k+1}^+$, where we denoted $m_k = \max(s_1^-, \ldots, s_k^-)$.

Similarly, in order to check that a function $s(y)$ can be non-increasing, it is sufficient to check that $s_i^+ \geq s_j^-$ for all pairs $i > j$. If we have already checked these inequalities for the intervals $\mathbf{s}_1, \ldots, \mathbf{s}_k$, and we add the new interval $\mathbf{s}_{k+1}$, then, to confirm that it is still possible for a function $s(z)$ to be non-increasing, it is sufficient to check that $M_k \geq s_{k+1}^-$, where we denoted $M_k = \min(s_1^+, \ldots, s_k^+)$.

The values m_k and M_k needed for these comparisons do not to be re-computed for every k; if we have already computed $M_{k-1} = \min(s_1^+, \ldots, s_{k-1}^+)$, then we can compute M_k as $M_k = \min(M_{k-1}, s_k^+)$ (and m_k as $m_k = \max(m_{k-1}, s_k^-)$).

9 How to Find Monotonicity Regions of a Function Defined with Interval Uncertainty: Algorithm

Thus, to find the monotonicity segments, we can use the following algorithm. In this algorithm, we process the intervals $\mathbf{s}_1, \ldots, \mathbf{s}_n$ one by one.

When we have the 1-st interval, then the only information that we have about the function $s(y)$ is that $s(y_1) \in \mathbf{s}_1$. This information is consistent with the function $s(y)$ being a constant, i.e., both non-decreasing and non-increasing. Thus, we are still consistent with monotonicity. To start the process of computing m_k and M_k, we assign $m_1 := s_1^-$ and $M_1 := s_1^+$.

If the first k intervals $\mathbf{s}_k$ are consistent with the assumption that the function $s(y)$ is non-decreasing, then when we get the new interval $\mathbf{s}_{k+1}$, we check whether $m_k \Box s_{k+1}^+$. Then:

- If $m_k \Box s^+_{k+1}$, then the new point y_{k+1} is still within the same monotonicity region. To prepare for the next interval, we compute $m_{k+1} = \min(m_k, s^-_{k+1})$.
- If $m_k > s^+_{k+1}$, this means that the new point y_{k+1} cannot be within the same monotonicity region, and so the monotonicity region must end before y_{k+1}. The point y_{k+1} itself belongs to a different monotonicity region which may expand both to the previous values y_k, y_{k-1}, ..., and to the following values y_{k+2}, ...; this new region can be traced in the same manner.

Similarly, if the first k intervals $\mathbf{s}_k$ are consistent with the assumption that the function $s(y)$ is non-increasing, then when we get the new interval $\mathbf{s}_{k+1}$, we check whether $M_k \geq s^-_{k+1}$. Then:

- If $M_k \geq s^-_{k+1}$, then the new point y_{k+1} is still within the same monotonicity region. To prepare for the next interval, we compute $M_{k+1} = \max(M_k, s^+_{k+1})$.
- If $M_k < s^-_{k+1}$, this means that the new point y_{k+1} cannot be within the same monotonicity region, and so the monotonicity region must end before y_{k+1}. The point y_{k+1} itself belongs to a different monotonicity region which may expand both to the previous values y_k, y_{k-1}, ..., and to the following values y_{k+2}, ...; this new region can be traced in the same manner.

10 How to Find Monotonicity Regions of a Function Defined with Interval Uncertainty: Example

Let us illustrate the above algorithm on the Rio Grande rift example.

We start with the 1-st interval, for which $m_1 := s^-_1 = 0.26$ and $M_1 := s^+_1 = 0.30$.

For the 2-nd interval, we check the inequalities $0.225 = s^-_2 \Box M_1 = 0.30$ and $0.255 = s^+_2 \geq m_1 = 0.26$. The first inequality holds, but the second does not hold, so we are in a non-increasing region. Thus, we compute $M_2 = \min(M_1, s^+_2) = \min(0.30, 0.255) = 0.255$.

For the 3-rd interval, the inequality $0.195 = s^-_3 \Box M_2 = 0.255$ still holds, so we are still in the non-increasing region. To prepare for the next interval, we compute $M_3 = \min(M_2, s^+_3) = \min(0.255, 0.225) = 0.225$.

For the 4-th interval, the inequality $0.14 = s^-_4 \Box M_3 = 0.225$ still holds, so we compute $M_4 = \min(M_3, s^+_4) = \min(0.225, 0.18) = 0.18$.

For the 5-th interval, the inequality $0.18 = s^-_4 \Box M_4 = 0.18$ still holds, so we compute $M_5 = \min(M_4, s^+_5) = \min(0.18, 0.22) = 0.18$.

For the 6-th interval, the inequality $0.28 = s^-_5 \Box M_5 = 0.18$ no longer holds, so the first monotonicity region cannot continue past $y_6 = 34$. Thus, the first monotonicity region must be within $[29, 34]$.

The point y_6 must belong to the new monotonicity region, where the function $s(y)$ is non-decreasing instead of non-increasing. Before we go forward and start checking on the points y_7, etc., we must go back and check which points y_5, y_4, ..., can belong to this new region. If we have already checked that $y_6, \ldots, y_k$ belong to this region, this means that $s_i^- \Box s_j^+$ for all such $i < j$. To check whether y_{k-1} belongs to this same region, we must check whether $s_{k-1}^- \Box s_i^+$ for all such i, i.e., whether $s_{k-1}^- \Box \widetilde{M}_k = \min(s_6^+, \ldots, s_k^+)$.

We start with $\widetilde{M}_6 = s_6^+ = 0.30$. Since the inequality $0.16 = s_5^- \Box \widetilde{M}_6 = 0.30$ holds, y_5 also belongs to this region, so we compute $\widetilde{M}_5 = \min(\widetilde{M}_6, s_5^+) = \min(0.30, 0.22) = 0.22$.

For the 4-th value, the inequality $0.14 = s_4^- \Box \widetilde{M}_5 = 0.22$ holds, so y_4 also belongs to this region, and we compute $\widetilde{M}_4 = \min(\widetilde{M}_5, s_4^+) = \min(0.22, 0.18) = 0.18$.

For the 3-rd value, the inequality $0.195 = s_3^- \Box \widetilde{M}_4 = 0.18$ is not true, so the new region must stop before $y_3 = 31$. Thus, the new region start at 31.

Now, we can move forward and check whether points y_7, ..., belong to the new monotonicity region.

We start with $m_6 := s_6^- = 0.28$. On the next step, we check whether $0.32 = s_7^+ \geq m_6 = 0.28$, and since this inequality holds, we compute $m_7 = \max(m_6, s_7^-) = \max(0.28, 0.30) = 0.30$. Similarly, we compute $m_8 = 0.33$, $m_9 = 0.405$, $m_{10} = 0.80$, $m_{11} = 0.80$, and $m_{12} = 0.88$. For the 13-th interval, the inequality $0.85 = s_{13}^+ \geq m_{12} = 0.88$ is no longer true, so y_{13} cannot belong to the second (non-decreasing) monotonicity region. Thus, the second region must stop before $y_{13} = 41$, and the point y_{13} must belong to the *third* monotonicity region.

To find out which other points belong to this third region, we must go back and check which points y_{12}, y_{11}, ..., can belong to this new region. If we have already checked that $y_{13}, \ldots, y_k$ belong to this region, this means that $s_i^+ \geq s_j^-$ for all such $i < j$. To check whether y_{k-1} belongs to this same region, we must check whether $s_{k-1}^+ \geq s_i^-$ for all such i, i.e., whether $s_{k-1}^+ \geq \widetilde{m}_k = \max(s_{13}^-, \ldots, s_k^-)$.

We start with $\widetilde{m}_{13} = s_{13}^- = 0.63$. Since the inequality $1.04 = s_{12}^+ \geq \widetilde{m}_{13} = 0.63$ holds, y_{12} also belongs to this region, so we compute $\widetilde{m}_{12} = \max(\widetilde{m}_{13}, s_{12}^-) = \max(0.63, 0.88) = 0.88$.

For the 11-th value, the inequality $0.88 = s_{11}^+ \geq \widetilde{m}_{12} = 0.88$ holds, so y_{11} also belongs to this region, and we compute $\widetilde{m}_{11} = \max(\widetilde{m}_{12}, s_{11}^-) = \max(0.88, 0.72) = 0.88$.

For the 10-th value, the inequality $1.10 = s_{10}^+ \geq \widetilde{m}_{11} = 0.88$ holds, so y_{10} also belongs to this region, and we compute $\widetilde{m}_{10} = \max(\widetilde{m}_{11}, s_{10}^-) = \max(0.88, 0.80) = 0.88$.

For the 9-th value, the inequality $0.515 = s_9^+ \geq \widetilde{m}_{10} = 0.88$ is not true, so the new region must stop before $y_9 = 37$. Thus, the new region starts at 37.

Thus, we have three monotonicity regions: [29, 34], [31, 41], and [37, 41]. The fact that we have discovered exactly three monotonicity regions is in good accordance with the geochemical data from [5].

11 Comment: What Happens When We Use Shorter Wavelengths

We ran the same algorithm for the spectral data corresponding to shorter wavelengths, i.e., for wavelengths 350-1000 km, 150-350 km, etc. Intuitively, the shorter the wavelength, the more small-size structures (which are irrelevant for our large-scale subdivision) influence the spectral values. In full accordance with this expectation: for 350-1000 and 150-350 km, we got a similar classification, but with a much larger uncertainty; for even shorted wavelengths, we got a subdivision to many more zones which probably reflect some short-scale structures that we ignore in our subdivision into segments.

12 From Intervals to Fuzzy

For each segment Y_i and for each point y, the algorithm tells whether a point y can (in view of the data) belong to the segment Y_i or not.

Some points belong to only one segment and therefore, are guaranteed to belong to this segment.

> In the Rio Grande rift example, such are the points from the interval $\widetilde{Y}_1 = [29, 31]$ for the first region, the points from the interval $\widetilde{Y}_2 = [34, 37]$ for the second region, and the point $\widetilde{Y}_3 = \{41\}$ for the third region.

Other points may belong to two neighboring segments Y_1 and Y_2, and the existing information is not sufficient to definitely conclude whether the given point y belongs to the segment Y_1 or to the segment Y_2.

In particular, for all the points which can belong to two neighboring segments Y_1 and Y_2, the result of the above algorithm is the same. Thus, from the interval computations viewpoint, for all points y which have not been uniquely classified, there is the same uncertainty. Intuitively, however, there is a reason to assume that we are more uncertainty about the classification of some of these points and less uncertainty about the classification of some other points.

For example, if a point y is close to the set $\widetilde{Y}_1 = [y_1^-, y_1^+]$ of all points which are proven to belong to Y_1, and it is distant from the set $\widetilde{Y}_2 = [y_2^-, y_2^+]$ of all the points which are proven to belong to Y_2, then, intuitively, it seems reasonable to conclude that the point y is more probably to be in the segment Y_1. Similarly, y is close to $\widetilde{Y}_2$ and distant from $\widetilde{Y}_1$, then, intuitively, it seems reasonable to conclude that the point y is more probably to be in the segment Y_2.

This intuitive sense can be captured if, instead of simply describing which points y can be in which segments, we produce, for each segment Y_i and for each y, not only the values "true" or "false" indicating that the point y can or cannot belong to Y_i, but also the "degree" $\mu_i(y)$ to which y can belong to the segment Y_i. For two neighboring segments Y_1 and Y_2, the degree $\mu_1(y)$ must be equal to 1 for $y \in \widetilde{Y}_1 = [y_1^-, y_1^+]$ and to 0 for $y \in \widetilde{Y}_2 = [y_2^-, y_2^+]$. It therefore makes sense to use a linear extrapolation to determine the values of the function $\mu_1(y)$ for $y \in [y_1^+, y_2^-]$. Linear extrapolation is widely (and successfully) used in applications of fuzzy techniques (see, e.g., [4,7]), and it is also known that linear extrapolation is (in some reasonable sense) the most robust extrapolation procedure (see, e.g., [6]).

As a result, for each segment Y_i, the resulting membership function $\mu_i(y)$ is *trapezoidal*:

- $\mu_i(y) = 0$ for $y \square y_{i-1}^+$;
- $\mu_i(y) = \dfrac{y - y_{i-1}^+}{y_i^- - y_{i-1}^+}$ for $y \in [y_{i-1}^+, y_i^-]$,
- $\mu_i(y) = 1$ for $y \in [y_i^-, y_i^+]$;
- $\mu_i(y) = \dfrac{y_{i+1}^- - y}{y_{i+1}^- - y_i^+}$ for $y \in [y_i^+, y_{i+1}^-]$, and
- $\mu_i(y) = 0$ for $y \geq y_{i+1}^-$.

Acknowledgments

This work was supported in part by NASA under cooperative agreement NCC5-209, by NSF grants No. DUE-9750858 and CDA-9522207, by the United Space Alliance, grant No. NAS 9-20000 (PWO C0C67713A6), by the Future Aerospace Science and Technology Program (FAST) Center for Structural Integrity of Aerospace Systems, effort sponsored by the Air Force Office of Scientific Research, Air Force Materiel Command, USAF, under grant number F49620-95-1-0518, and by the National Security Agency under Grant No. MDA904-98-1-0561.

References

1. Höppner, F. (1999) Fuzzy Cluster Analysis. J. Wiley, New York.
2. Jain, A. K., Dubes, R. C. (1988) Algorithms for Clustering Data. Prentice Hall, Englewood Cliffs, NJ.
3. Kearfott, R. B., Kreinovich, V., eds. (1996) Applications of Interval Computations, Kluwer, Dordrecht.
4. Klir, G., Yuan, B. (1995) Fuzzy Sets and Fuzzy Logic: Theory and Applications, Prentice Hall, Upper Saddle River, NJ.
5. N. J. McMillan, N. J. (1998) Temporal and spatial magmatic evolution of the Rio Grande rift. New Mexico Geological Survey Guidebook, 49th Field Conference, Las Cruces County, 107–116.

6. Nguyen, H. T., Kreinovich, V., Lea, B., Tolbert, D. (1995) Interpolation that leads to the narrowest intervals, and its application to expert systems and intelligent control. Reliable Computing **1**, No. 3, 299–316.
7. Nguyen, H. T., Walker, E. A. (1999) First Course in Fuzzy Logic, CRC Press, Boca Raton, FL.
8. Villaverde, K., and Kreinovich, V. (1993) A linear-time algorithm that locates local extrema of a function of one variable from interval measurement results. Interval Computations, No. 4, 176–194.

A Fuzzy Classifier with Pyramidal Membership Functions

Shigeo Abe

Kobe University, Kobe Japan

Abstract. In this chapter we discuss a new fuzzy classifier with hyperbox regions, namely a fuzzy classifier with pyramidal membership functions and its training method. First for each class we define a fuzzy rule with a hyperbox region that includes all the training data in the class and define a pyramidal membership function for the hyperbox. Then we tune the fuzzy rules, i.e., the locations of the hyperboxes and the slopes of the membership functions successively until there is no improvement in the recognition rate of the training data. Then if the recognition rate of the training data is not sufficient we repeat defining fuzzy rules for the misclassified training data and tuning the newly defined fuzzy rules until the sufficient recognition rate is obtained.

1 Introduction

Although training of multilayer neural network classifiers by the back-propagation algorithm is slow, the trained classifier shows good generalization ability for a wide range of applications. Therefore, to realize fast training and good generalization ability comparable to neural network classifiers, many fuzzy classifiers with a learning capability have been proposed [1]–[10]. There are three types of fuzzy classifiers:

1. a fuzzy classifier with hyperbox regions whose surfaces are parallel to one of the input variables [1]–[5];
2. a fuzzy classifier with polyhedral regions whose surfaces are expressed by a linear combination of the input variables [6], and
3. a fuzzy classifier with ellipsoidal regions [7,9,10].

There are two types of fuzzy classifiers with hyperbox regions. In [2] hyperboxes belonging to the same class are allowed to overlap but the hyperboxes belonging to different classes are not. Hyperboxes are created, expanded, or contracted according to the overlap between classes. In [3], two types of hyperboxes are introduced: activation hyperboxes that allow the existence of the class data, and inhibition hyperboxes that do not. First the activation hyperboxes are defined for each class by calculating the minimum and maximum values of the training data. Then if there is an overlap between activation hyperboxes, the overlapped region is defined as the inhibition hyperbox. And if there are data in the inhibition hyperbox, hyperbox generation in the inhibition hyperbox is iterated until there is no overlap between classes. In [5], slopes of the membership functions are tuned to improve the generalization ability.

Training time of the fuzzy classifier with hyperbox regions [3] is the shortest among the three types of classifiers because we need only to calculate the minimum and maximum values of the training data in each input variable. But the generalization ability depends on the characteristics of the data [9]; if the principal axes of the training data distribution for each class are parallel to the input variables, we can obtain good generalization ability. But if the principal axes are not, the generalization ability of the classifier is not so good compared with that of neural network classifiers.

The fuzzy classifier with polyhedral regions [6] does not have this problem since the polyhedron regions are approximated by shifting the hyperplanes extracted from the trained neural network classifier. But since the fuzzy classifier is based on the neural network classifier, training is slow.

In the fuzzy classifier with ellipsoidal regions [10], first for the data belonging to a class we define one fuzzy rule with an ellipsoidal region, by calculating the center and the covariance matrix of the ellipsoidal region for the class. Then we tune the fuzzy rules, i.e., the slopes of the membership functions successively until there is no improvement in the recognition rate of the training data. Then if the number of the data belonging to a class that are misclassified into another class exceeds a prescribed number, we define a new cluster to which those data belong. Then we tune the newly defined fuzzy rules in the similar way as stated above, fixing the already obtained fuzzy rules. We iterate cluster generation and tuning of newly generated fuzzy rules until the number of misclassifications does not exceed the prescribed number. When the input variables are continuous, the fuzzy classifier with ellipsoidal regions shows a recognition rate comparable with the maximum recognition rate of the multilayer neural network classifier for the test data [10].

In this chapter we discuss a new fuzzy classifier with hyperbox regions and its training method. The differences from the classifiers discussed in [2] and [3] are that 1) pyramidal membership functions are used instead of hyperbox membership functions, and 2) locations of the centers as well as the slopes of the membership functions are tuned.

Since a recognition rate is a discrete value and this cannot be expressed as a continuous objective function, in training we usually use the sum of square errors between the target values and the classifier outputs. Thus minimization of the objective function does not necessarily leads to a higher recognition rate [11]. Our tuning algorithm solves this problem for the fuzzy classifier having pyramidal membership functions. We tune the slope and the location of the center of each membership function successively, by counting the increase in the number of misclassifications as well as the increase in the number of correct classifications and analytically determining the slopes and centers that locally maximize the recognition rate. To make this possible we use the pyramidal membership functions with the average operator as will be explained later.

We define one type of hyperboxes and allow the overlap of hyperboxes belonging to different classes. The training of the classifier is as follows. First for each class a hyperbox is defined by calculating the minimum and maximum values in each input variable. Then the center of the hyperbox is taken as the apex of the pyramidal membership function. The location of each hyperbox is tuned successively so that the recognition rate of the training data is maximized. Then, slopes of the membership functions are tuned successively in the similar way as discussed in [9] so that the recognition rate of the training data is maximized. Then if the number of the data belonging to a class that are misclassified into another class exceeds a prescribed number, we define a new cluster to which those data belong. Then we tune the newly defined fuzzy rules in the similar way as stated above, fixing the already obtained fuzzy rules. We iterate cluster generation and tuning of newly generated fuzzy rules until the number of misclassifications does not exceed the prescribed number. Hereafter, we call the classifier the fuzzy classifier with pyramidal membership functions.

In Section 2, we introduce a pyramidal membership function that is suitable for tuning, and in Section 3 we describe a process of fuzzy rule generation. In Sections 4 and 5, we discuss tuning of the hyperbox centers and slopes, respectively.

2 Membership Functions

In [2], [3], the hyperbox membership functions, in which the degree of membership is 1 in the hyperbox, are used. The problem with these membership functions is that tuning is not flexible. Namely, two hyperboxes of different classes cannot overlap. Otherwise, the unclassified region, where the degrees of membership of two classes are 1, will occur. But if we use the triangular membership functions, this does not happen. Thus the possibility of improving the recognition rate is increased. Therefore, in the following we discuss the pyramidal membership function that is an extension of a one-dimensional triangular membership function.

First in the m-dimensional input space, we define the jth hyperbox for class $i (i = 1, \ldots, n)$, A_{ij}, that includes some of the class i training data as follows:

$$A_{ij} = \{\mathbf{x} | v_{ijk} \Box x_k \Box V_{ijk}, k = 1, \ldots, m\} \quad \text{for } j = 1, \ldots, \tag{1}$$

where x_k is the kth element of $\mathbf{x}$, and v_{ijk} and V_{ijk} are, respectively, the minimum and maximum values of the hyperbox A_{ij} with respect to x_k.

Then the center vector $\mathbf{c}_{ij}$ of the hyperbox A_{ij} is given by

$$\mathbf{c}_{ij} = \left(\frac{V_{ij1} + v_{ij1}}{2}, \ldots, \frac{V_{ijm} + v_{ijm}}{2} \right)^t, \tag{2}$$

where t denotes the matrix transpose.

Now we define a membership function for A_{ij}. We assume that the degree of membership at $\mathbf{c}_{ij}$ is 1. Then we define a pyramidal membership function with the apex $\mathbf{c}_{ij}$. First we define, for the input vector $\mathbf{x}$, a one-dimensional tuned weighted distance $h_{ijk}(\mathbf{x})(k = 1, \ldots, m)$ as follows:

$$h_{ijk}(\mathbf{x})^s = \frac{d_{ijk}(\mathbf{x})^s}{\alpha_{ij}}, \tag{3}$$

where s is 1 when the membership function is piecewise linear and 2 when quadratic, α_{ij} is a tuning parameter that determines the slope of the membership function and the initial value is set to be 1, and $d_{ijk}(\mathbf{x})$ is the one-dimensional weighted distance given by

$$d_{ijk}(\mathbf{x}) = \frac{|c_{ijk} - x_k|}{w_{ijk}}, \tag{4}$$

where c_{ijk} is the kth element of $\mathbf{c}_{ij}$ and

$$w_{ijk} = \frac{V_{ijk} - v_{ijk}}{2}. \tag{5}$$

To avoid zero division, if $w_{ijk} < \varepsilon$ where ε is a small positive number, in (4) we replace w_{ijk} with ε.

According to the above definition, the weighted distance at any point on the surface of the hyperbox is the same. This is the same idea with that of the Mahalanobis distance used in the fuzzy classifier with ellipsoidal regions [9], [10].

Then the one-dimensional membership function for input variable $x_k(k = 1, \ldots, m)$, $m_{ijk}(\mathbf{x})$, is defined as follows:

$$m_{ijk}(\mathbf{x}) = \begin{cases} 0 & \text{for} \quad x_k > c_{ijk} + \alpha_{ij} w_{ijk} \\ 1 - h_{ijk}(\mathbf{x})^s & \text{for} \quad c_{ijk} + \alpha_{ij} w_{ijk} \geq x_k \geq c_{ijk} - \alpha_{ij} w_{ijk} \\ 0 & \text{for} \quad x_k < c_{ijk} - \alpha_{ij} w_{ijk}. \end{cases} \tag{6}$$

Equation (6) is not differentiable in $x_k > c_{ijk}$ and $x_k < c_{ijk}$ and tuning of α_{ij} is difficult. Thus to make (6) differentiable except for $x_k = c_{ijk}$, we allow negative degrees of membership:

$$m_{ijk}(\mathbf{x}) = 1 - h_{ijk}(\mathbf{x})^s. \tag{7}$$

Hereafter we use (7) as the membership function. Instead of using (7), we can use $h_{ijk}(\mathbf{x})^s$, in which the minimum $h_{ijk}(\mathbf{x})^s$ corresponds to the largest degree of membership.

The degree of membership for the input variable $\mathbf{x}$, $m_{ij}(\mathbf{x})$, can be defined in two ways using (7). The first one uses the minimum operator:

$$m_{ij}(\mathbf{x}) = \min_{k=1,\ldots,m} m_{ijk}(\mathbf{x}) = 1 - \max_{k=1,\ldots,m} h_{ijk}(\mathbf{x})^s. \tag{8}$$

The second one uses the average operator:

$$m_{ij}(\mathbf{x}) = \frac{1}{m}\sum_{k=1}^{m} m_{ijk}(\mathbf{x}) = 1 - h_{ij}(\mathbf{x})^{s}, \tag{9}$$

where $h_{ij}(\mathbf{x})$ is the tuned weighted distance:

$$h_{ij}(\mathbf{x})^{s} = \frac{1}{m}\sum_{k=1}^{m} h_{ijk}(\mathbf{x})^{s}. \tag{10}$$

The contour surface of (8), on which every location has the same degree of membership, is parallel to the surface of A_{ij}, whereas the contour surface of (9) is not parallel to the surface of A_{ij}.

Now consider the sensitivity of the two membership functions to the slope or location change. First consider the slope change. Let the membership functions with the minimum and average operators be $m'_{ij}(\mathbf{x})$ for α'_{ij}. Then from (3), (8), and (9),

$$m'_{ij}(\mathbf{x}) = 1 + \frac{\alpha_{ij}}{\alpha'_{ij}}\left(m_{ij}(\mathbf{x}) - 1\right). \tag{11}$$

Thus the change of the slope is expressed in a closed form for both membership functions.

Assume that $c_{ijk}(\mathbf{c}_{ij})$ is changed to $c'_{ijk}(\mathbf{c}'_{ij}), k \in \{1,\ldots,m\}$. For the membership function with the minimum operator, if $h_{ijk'}(\mathbf{x}), k' \in \{1,\ldots,m\}$ is the maximum for $\mathbf{c}_{ij}$, there is no guarantee that it is still the maximum for $\mathbf{c}'_{ij}$. Thus it is difficult to express the change of location in a closed form. For the membership function with the average operator, on the other hand,

$$m'_{ij}(\mathbf{x}) - m_{ij}(\mathbf{x}) = \frac{|c_{ijk} - x_{ij}|^{s}}{m\alpha_{ij}w_{ijk}} - \frac{|c'_{ijk} - x_{ij}|^{s}}{m\alpha_{ij}w_{ijk}}. \tag{12}$$

Thus the location change can be expressed in a closed form.

From the above discussions, for the membership functions with the minimum operator, tuning of centers is not easily achieved. Also since tuning of centers is essential to achieve a high recognition rate, in the following we use the membership function with the average operator given by (9) or the tuned weighted distance given by (10).

3 Fuzzy Rule Generation

If we define one hyperbox region for one class, underfitting may easily occur. Thus, the generalization ability may not be good. To avoid this we need to approximate each class region by several hyperbox regions. To do this we can divide the training data for each class into several clusters in advance as discussed in [9], or we can generate hyperboxes dynamically when the recognition rate is not enough as discussed in [10]. Here we use the dynamic rule generation method discussed in [10].

In dynamic rule generation, clustering is postponed after the training is completed. Namely, first for the data belonging to a class we define one fuzzy rule. Then we tune the slopes and locations of the membership functions successively until there is no improvement in the recognition rate of the training data. Then if the number of the data belonging to a class that are misclassified into another class exceeds a prescribed number, N_c, we define a new cluster to which those data belong. Then we tune the newly defined fuzzy rules in the similar way as stated above, fixing the already obtained fuzzy rules. We iterate cluster generation and tuning of newly generated fuzzy rules until the number of the data belonging to a class that are misclassified into another class does not exceed the prescribed number.

The general flow of rule generation is as follows.

1. For class i, we define the hyperbox A_{i1} using the data belonging to the cluster. And for the hyperbox region A_{i1}, we define the following fuzzy rule:
$$R_{ij}: \quad \text{If } \mathbf{x} \text{ is } \mathbf{c}_{ij} \text{ then } \mathbf{x} \text{ belongs to class } i \tag{13}$$
where $j = 1$.
2. Let the maximum allowable number of new misclassifications in tuning α_{ij} be $l_M - 1$. Increase or decrease the values of $\alpha_{ij}(i = 1, \ldots, n, j = 1, \ldots)$ successively so that the recognition rate of the training data is maximized, allowing at most $l_M - 1$ new misclassifications.
3. Let the maximum allowable number of new misclassifications in tuning c_{ijk} be $l_{M_c} - 1$. For each fuzzy rule, tune c_{ijk} for $k = 1, \ldots, m$ successively so that the recognition rate of the training data is maximized, allowing at most $l_{M_c} - 1$ new misclassifications.
4. If the number of the data belonging to a class that are misclassified into another class is larger than or equal to N_c, and these misclassified data are not previously defined as a cluster, define a new cluster that includes only these misclassified data. Then define the hyperbox A_{ij}, and define fuzzy rules given by (13). And go to step 5. Otherwise, terminate the algorithm.
5. Tune the tuning parameters of the newly defined fuzzy rules successively, fixing the already tuned tuning parameters. The tuning procedure is the same as in steps 2 and 3. Go to step 4.

We call the execution of steps 2 and 3 and that of step 5 one epoch of tuning.

4 Tuning of Center Vectors

4.1 Concept

Allowing misclassification, we move the hyperbox A_{ij} so that the recognition rate of the training data is maximized. To simplify matters, we move A_{ij}

along one input axis at a time. First, we explain the concept of hyperbox tuning, using a simple one-dimensional case with two classes shown in Fig. 1. In the figure, one fuzzy rule is defined for each class. Datum 1 is correctly classified into class 2 but data 2 and 3 are misclassified into class 2. Data 2 and 3 are correctly classified if the center of class 1 fuzzy rule is moved into the shaded region, although datum 1 is now misclassified. In this case, since the recognition rate is improved by one datum, we allow misclassification. We can also improve the recognition rate, moving the center of the class 2 hyperbox to the right. Since the membership function is given by (9), the change of the degree of membership for the movement of the center in one axis is easily calculated by (12). Thus tuning of centers is possible for the multi-dimensional input.

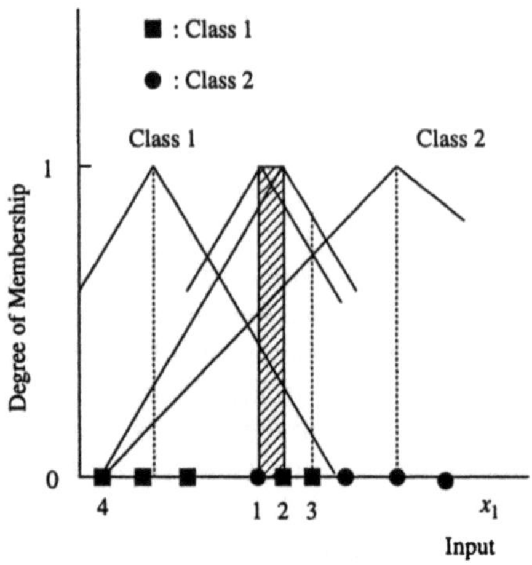

Fig. 1. Concept of tuning. If the center of the membership function for class 1 is moved in the shaded region, datum 1 is misclassified but data 2 and 3 are correctly classified.

Now suppose we tune center $\mathbf{c}_{ij}$ on the kth input axis. To some value we can move c_{ijk} to the right or left without increasing misclassification of class i data. If c_{ijk} moves to x_k, $m_{ij}(\mathbf{x})$ increases until c_{ijk} reaches x_k. And $m_{ij}(\mathbf{x})$ decreases if c_{ijk} moves away from x_k. Thus for the correctly classified input vector $\mathbf{x}$ that belongs to class i, there is an interval $(I_{ijk}(\mathbf{x}), D_{ijk}(\mathbf{x}))$, in which $\mathbf{x}$ is correctly classified when c_{ijk} is moved. And for the correctly classified input vector $\mathbf{x}$ that does not belong to class i, under some condition that is discussed later, there are intervals $(-\infty, D_{ijk}(\mathbf{x}))$ and $(I_{ijk}(\mathbf{x}), \infty)$ in which $\mathbf{x}$ is correctly classified. Thus there are an upper bound $D_{ijk}(1) = \min D_{ijk}(\mathbf{x})(\geq c_{ijk})$ and a lower bound $I_{ijk}(1) = \max I_{ijk}(\mathbf{x})(< c_{ijk})$ that result in no increase in misclassification. Likewise, we have the upper bound $D_{ijk}(l)$ and lower bound $I_{ijk}(l)$ in which $l-1$ correctly classified data are misclassified. Here, suppose that some correctly classified datum that belongs to a class other than class i is misclassified when $c_{ijk} = D_{ijk}(o), 1 \square o < l$. This datum is correctly classified when $c_{ijk} = D_{ijk}(l)$ and $c_{ijk} > I_{ijk}(o)$. This may occur for $I_{ijk}(l)$. But, in calculating $D_{ijk}(l)$ and $I_{ijk}(l)$ we do not consider this to make the calculation simple.

Similarly, if we move c_{ijk} to the right or left, misclassified data may be correctly classified. For the misclassified input vector $\mathbf{x}$ that belongs to class i, there is an interval $(I_{ijk}(\mathbf{x}), D_{ijk}(\mathbf{x}))$ where it is correctly classified. And for the misclassified input vector $\mathbf{x}$ that does not belong to class i, under some condition, there are intervals $(I_{ijk}(\mathbf{x}), \infty)$ and $(-\infty, D_{ijk}(\mathbf{x}))$ where it is correctly classified.

Then using the upper bounds $D_{ijk}(l)$ and the lower bounds $I_{ijk}(l)$ of c_{ijk} that allow $l-1$ new misclassifications and the intervals that resolve misclassification, we can determined the interval of c_{ijk} that makes the highest recognition rate. Figure 2 shows an example. In the upper part of the figure, the six intervals indicate that the misclassified data are correctly classified if c_{ijk} is moved into those intervals. The lower part of the figure shows the net increase of the number of correctly classified data for different c_{ijk}. Suppose c_{ijk} moves from the current position to the right. When c_{ijk} enters into the interval 3, one misclassification is resolved. And if c_{ijk} enters into the interval 5, three misclassifications are resolved. But if c_{ijk} passes over $D_{ijk}(1)$, the one misclassification occurs and the net increase of the correctly classified data becomes two. Thus in this example, if c_{ijk} is moved in the shaded interval, the number of correctly classified data increases by three.

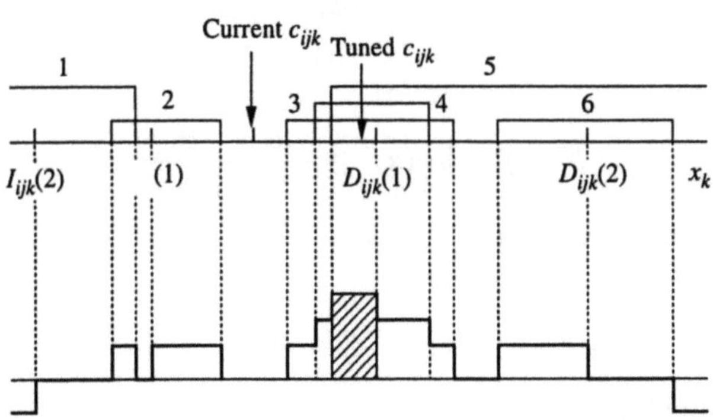

Fig. 2. Movement of c_{ijk}. If the current c_{ijk} is moved to the tuned c_{ijk} in the dashed interval, three misclassified data are correctly classified without causing new misclassification.

In the following we discuss the determination of the upper bound $D_{ijk}(l)$ and the lower bound $I_{ijk}(l)$ that allow $l-1$ new misclassifications, the determination of the intervals that resolve misclassification, and the determination of the center c_{ijk} that achieve the maximum recognition rate allowing at most $l_{M_c} - 1$ new misclassifications.

4.2 Upper and lower bounds of c_{ijk}

We calculate the upper bound $D_{ijk}(l)$ and the lower bound $I_{ijk}(l)$ of c_{ijk} in which $l-1(\geq 0)$ data are newly misclassified. We divide a set of input data into X and Y, where X consists of the data correctly classified using the set of fuzzy rules $\{R_{ij}\}$ and Y consists of the misclassified data. Then, we choose $\mathbf{x}(\in X)$ that belongs to class i, and that satisfies

$$h_{ij}(\mathbf{x}) \Box \min_{k\neq j} h_{ik}(\mathbf{x}). \tag{14}$$

If $\mathbf{x}$ further satisfies

$$h_{ij}(\mathbf{x}) < h_{op}(\mathbf{x}) = \min_{o\neq i, q=1,\ldots} h_{oq}(\mathbf{x}) < \min_{k\neq j} h_{ik}(\mathbf{x}), \tag{15}$$

there are a lower bound $I_{ijk}(\mathbf{x})$ and an upper bound $D_{ijk}(\mathbf{x})$ to keep $\mathbf{x}$ correctly classified.

To calculate these bounds, first we calculate the change in the tuned weighted distance when c_{ijk} is moved. Among the one-dimensional tuned weighted distances $h_{ijl}(\mathbf{x})(l=1,\ldots,m)$, from (4) only $h_{ijk}(\mathbf{x})$ changes. Let $h_{ij}^{k}(\mathbf{x})$ denote the tuned weighted distance of the fuzzy rule R_{ij} for the input vector $\mathbf{x}$, when the center vector $\mathbf{c}_{ij}$ is moved in the kth input axis to c'_{ijk}. Then, the difference $\Delta h_{ij}^{k}(\mathbf{x})^{s} = h_{ij}^{k}(\mathbf{x})^{s} - h_{ij}(\mathbf{x})^{s}$ is given by

$$\begin{aligned} \Delta h_{ij}^{k}(\mathbf{x})^{s} &= h_{ij}^{k}(\mathbf{x})^{s} - h_{ij}(\mathbf{x})^{s} \\ &= \frac{|c'_{ijk} - x_k|^{s}}{m\, \alpha_{ij}\, w_{ijk}} - \frac{|c_{ijk} - x_k|^{s}}{m\, \alpha_{ij}\, w_{ijk}}. \end{aligned} \tag{16}$$

Thus

$$h_{ij}^{k}(\mathbf{x})^{s} - h_{ij}(\mathbf{x})^{s} \geq -\frac{|c_{ijk} - x_k|^{s}}{m\, \alpha_{ij}\, w_{ijk}}, \tag{17}$$

where the equality holds for $c'_{ijk} = x_k$.

At the upper or lower bound,

$$h_{ij}^{k}(\mathbf{x})^{s} = \Delta h_{ij}^{k}(\mathbf{x})^{s} + h_{ij}(\mathbf{x})^{s} = h_{op}(\mathbf{x})^{s}. \tag{18}$$

From(17) and (18),

$$h_{op}(\mathbf{x})^{s} - h_{ij}(\mathbf{x})^{s} \geq -\frac{|c_{ijk} - x_k|^{s}}{m\, \alpha_{ij}\, w_{ijk}}, \tag{19}$$

and from $h_{op}(\mathbf{x})^{s} - h_{ij}(\mathbf{x})^{s} > 0$, the upper and lower bounds always exist and are given by,

$$D_{ijk}(\mathbf{x}) = x_k + \mathrm{sign}(h_{op}(\mathbf{x}) - h_{ij}(\mathbf{x}))a_k, \tag{20}$$

$$I_{ijk}(\mathbf{x}) = x_k - \mathrm{sign}(h_{op}(\mathbf{x}) - h_{ij}(\mathbf{x}))a_k, \tag{21}$$

where

$$\text{sign}(x) = \begin{cases} 1 & \text{for } x \geq 0, \\ 0 & \text{for } x < 0, \end{cases} \tag{22}$$

$$a_k = \left(m\,\alpha_{ij}\,w_{ijk}^s\,(h_{op}(\mathbf{x})^s - h_{ij}(\mathbf{x})^s) + |c_{ijk} - x_k|^s\right)^{1/s}. \tag{23}$$

If (15) is not satisfied, namely

$$h_{ij}(\mathbf{x}) < \min_{k \neq j} h_{ik}(\mathbf{x}) < \min_{o \neq i, p=1,\ldots} h_{op}(\mathbf{x}), \tag{24}$$

c_{ijk} can be moved without making $\mathbf{x}$ become misclassified.

Now we choose $\mathbf{x}(\in X)$ which belongs to class $o(\neq i)$. Let cluster op have the minimum tuned distance $h_{op}(\mathbf{x})$:

$$h_{op} = \min_q h_{oq}(\mathbf{x}). \tag{25}$$

Since the tuned distance $h_{ij}(\mathbf{x})$ is larger than $h_{op}(\mathbf{x})$, if (19) is satisfied, there are $D_{ijk}(\mathbf{x})$ given by (20) and $I_{ijk}(\mathbf{x})$ given by (21).

Now the lower bound $I_{ijk}(1)$, which is defined as the lower bound that does not cause new misclassification, is

$$I_{ijk}(1) = \max_{\mathbf{x} \in X} I_{ijk}(\mathbf{x}). \tag{26}$$

To clarify the discussion, we assume that $I_{ijk}(\mathbf{x})$ is different for different $\mathbf{x}$. Then (26) is satisfied by one $\mathbf{x}$. Similarly, $I_{ijk}(2)$, which is defined as the lower bound that allows one new misclassification, is the second maximum among $I_{ijk}(\mathbf{x})$ and is given by

$$I_{ijk}(2) = \max_{\mathbf{x} \in X, I_{ijk}(\mathbf{x}) \neq I_{ijk}(1)} I_{ijk}(\mathbf{x}). \tag{27}$$

In general,

$$I_{ijk}(l) = \max_{\mathbf{x} \in X, I_{ijk}(\mathbf{x}) \neq I_{ijk}(1), \ldots, I_{ijk}(l-1)} I_{ijk}(\mathbf{x}). \tag{28}$$

As we discussed earlier, when c_{ijk} is moved to $I_{ijk}(l)$, the data which are misclassified when c_{ijk} is moved to $I_{ijk}(o)(1 \,\square\, o < l)$ may become correctly classified. Even when this happens, we do not consider these data in calculation of $I_{ijk}(l)$.

Now the upper bound $D_{ijk}(1)$, which is defined as the upper bound that does not cause any new misclassification, is

$$D_{ijk}(1) = \min_{\mathbf{x} \in X} D_{ijk}(\mathbf{x}). \tag{29}$$

Here we also assume that $D_{ijk}(\mathbf{x})$ is different for different $\mathbf{x}$. Then (29) is satisfied by one $\mathbf{x}$. Similarly, $D_{ijk}(2)$, which is defined as the upper bound that allows one new misclassification, is the second minimum among $D_{ijk}(\mathbf{x})$ and is given by

$$D_{ijk}(2) = \min_{\mathbf{x} \in X, D_{ijk}(\mathbf{x}) \neq D_{ijk}(1)} D_{ijk}(\mathbf{x}). \tag{30}$$

In general,

$$D_{ijk}(l) = \min_{\mathbf{x} \in X, D_{ijk}(\mathbf{x}) \neq D_{ijk}(1), \ldots, D_{ijk}(l-1)} D_{ijk}(\mathbf{x}). \tag{31}$$

As we discussed earlier, when c_{ijk} is moved to $D_{ijk}(l)$, the data which are misclassified when c_{ijk} is moved to $D_{ijk}(o)(1 \Box o < l)$ may become correctly classified. Even when this happens, we do not consider these data in calculation of $D_{ijk}(l)$.

Thus c_{ijk} is bounded by

$$\begin{aligned} \cdots < I_{ijk}(l) < I_{ijk}(l-1) < \cdots < I_{ijk}(1) < c_{ijk} < D_{ijk}(1) \\ < \cdots < D_{ijk}(l-1) < D_{ijk}(l) < \cdots. \end{aligned} \tag{32}$$

If we change c_{ijk} in the range of $(I_{ijk}(1), D_{ijk}(1))$, there are no new misclassification. And if we change c_{ijk} in the range of $[D_{ijk}(l-1), D_{ijk}(l))$, or $(I_{ijk}(l), I_{ijk}(l-1)]$, the $l-1 (l \geq 2)$ correctly classified data are misclassified.

4.3 Resolution of misclassification by changing c_{ijk}

For $\mathbf{x}(\in Y)$ which is misclassified into class i or which belongs to class i but is misclassified into class $o(\neq i)$, we check whether it can be correctly classified by moving c_{ijk}. For the datum $\mathbf{x}$ that belongs to class i but that is misclassified into class o with the minimum tuned distance of $h_{op}(\mathbf{x})$, if (19) is satisfied, there is an interval $(I_{ijk}(\mathbf{x}), D_{ijk}(\mathbf{x}))$ in which $\mathbf{x}$ is correctly classified, where $I_{ijk}(\mathbf{x})$ and $D_{ijk}(\mathbf{x})$ are given by (20) and (21), respectively.

Let $\mathbf{x}$, which belongs to class o, be misclassified into class i. Then similar to the above discussions, we calculate the intervals $(-\infty, D_{ijk}(\mathbf{x}))$ and $(I_{ijk}(\mathbf{x}), \infty)$ that make $\mathbf{x}$ be correctly classified by moving c_{ijk}. First, the minimum tuned distance for class o should be the second minimum among n classes:

$$\min_k h_{ik}(\mathbf{x}) < h_{op}(\mathbf{x}) = \min_{q \neq i, r=1,\ldots} h_{qr}(\mathbf{x}). \tag{33}$$

Second, $h_{ij}(\mathbf{x})$ needs to be the minimum in class i, and the second minimum in class i is larger than the minimum tuned distance in class o:

$$h_{ij}(\mathbf{x}) < h_{op}(\mathbf{x}) < \min_{k \neq j} h_{ik}(\mathbf{x}). \tag{34}$$

Then, since $h_{op}(\mathbf{x}) > h_{ij}(\mathbf{x})$, $D_{ijk}(\mathbf{x})$ and $I_{ijk}(\mathbf{x})$ exist and are given by (20) and (21), respectively.

4.4 Calculation of a net increase of the recognition rate

To calculate the net increase of the recognition rate by moving c_{ijk}, we determine the stepwise function that shows the net increase/decrease of the number of correctly classified data when c_{ijk} is changed as shown in Fig. 2. New misclassification occurs when c_{ijk} is moved in the left direction to $I_{ijk}(1), \ldots, I_{ijk}(l_{M_c} - 1)$, or to $I_{ijk}(l_{M_c})$, and it moves in the right direction to $D_{ijk}(1), \ldots, D_{ijk}(l_{M_c} - 1)$, or $D_{ijk}(l_{M_c})$, and the value of the stepwise function decreases by 1 at these points. Since misclassification is resolved when c_{ijk} is moved in the right direction to $I_{ijk}(\mathbf{x})$ and misclassified again when moved to $D_{ijk}(\mathbf{x})$, and moved in the left direction to $D_{ijk}(\mathbf{x})$ and misclassified again when moved to $I_{ijk}(\mathbf{x})$, the stepwise function increases (or decreases) at these points. Thus to generate the stepwise function we need to find the points at which the stepwise function increases or decreases in the increasing order in the interval $[c_{ijk}, D_{ijk}(l_{M_c})]$ and in the decreasing order in the interval $[I_{ijk}(l_{M_c}), c_{ijk}]$.

Now we sort, in the increasing order, $I_{ijk}(\mathbf{x})$, $D_{ijk}(\mathbf{x})$ and $D_{ijk}(l)$ that belong to the interval $[c_{ijk}, D_{ijk}(l_{M_c})]$ and define the sorted points as $\beta^u_{ijk}(o)(o = 1, \ldots, o_{M_u})$ where o_{M_u} is the number of points included in $[c_{ijk}, D_{ijk}(l_{M_c})]$. Associated with $\beta^u_{ijk}(o)$, we define $\mathrm{Inc}^u(o)$ as follows:

$$\mathrm{Inc}^u(o) = \begin{cases} 1 & \text{for} \quad \beta^u_{ijk}(o) = I_{ijk}(\mathbf{x}), \\ -1 & \text{for} \quad \beta^u_{ijk}(o) = D_{ijk}(l) \text{ or } D_{ijk}(\mathbf{x}). \end{cases} \tag{35}$$

And we sort, in the decreasing order, $D_{ijk}(\mathbf{x}), I_{ijk}(\mathbf{x})$, and $I_{ijk}(l)$ that belong to the interval $[I_{ijk}(l_{M_c}), c_{ijk})$ and define the sorted points as $\beta^l_{ijk}(o)(o = 1, \ldots, o_{M_l})$ where o_{M_l} is the number of points included in $[I_{ijk}(l_{M_c}), c_{ijk})$. Associated with $\beta^l_{ijk}(o)$ we define $\mathrm{Inc}^l(o)$ as follows:

$$\mathrm{Inc}^l(o) = \begin{cases} 1 & \text{for} \quad \beta^l_{ijk}(o) = D_{ijk}(\mathbf{x}), \\ -1 & \text{for} \quad \beta^l_{ijk}(o) = I_{ijk}(l) \text{ or } I_{ijk}(\mathbf{x}). \end{cases} \tag{36}$$

Now we can determine the values of the stepwise function $f^u(o)$ at $\beta^u_{ijk}(o)$ as follows:

```
begin
    f^u(1) = Inc^u(1)
    o = 1
    while  o □ o_{M_u}  do
        f^u(o) = f^u(o − 1) + Inc^u(o)
    end
end
```

Similarly, we can determine the values of the stepwise function $f^l(o)$ at $\beta^l_{ijk}(o)$ as follows:

```
begin
    f^l(1) = Inc^l(1)
    o = 1
```

while $o \Box o_{M_l}$ do
    $f^l(o) = f^l(o-1) + \mathrm{Inc}^l(o)$
  end
end

4.5 Modification of c_{ijk}

For $f^u(o), o = 1, \ldots, o_{M_u}$, we find o^u that satisfies

$$f^u(o^u) = \max_{o=1,\ldots,o_{M_u}} f^u(o). \tag{37}$$

Similarly, for $f^l(o), l = 1, \ldots, o_{M_l}$, we find o^l that satisfies

$$f^l(o^l) = \max_{o=1,\ldots,o_{M_l}} f^l(o). \tag{38}$$

If there are plural o's that satisfy (37) or (38), we choose the smallest o. If $f^u(o^u) \geq f^l(o^l)$, we move c_{ijk} to the right. If we move c_{ijk} in the interval $(\beta^u_{ijk}(o^u), \beta^u_{ijk}(o^u + 1))$ the recognition rate is maximized. Thus we set c_{ijk} in $(\beta^u_{ijk}(o^u), \beta^u_{ijk}(o^u + 1))$ as follows:

$$c_{ijk} = \beta^u_{ijk}(o^u) + \delta^c(\beta^u_{ijk}(o^u + 1) - \beta^u_{ijk}(o^u)), \tag{39}$$

where δ^c satisfies $0 < \delta^c < 1$.

Likewise, if $f^u(o^u) < f^l(o^l)$, we move c_{ijk} to the left. If we move c_{ijk} in the interval $(\beta^l_{ijk}(o^l + 1), \beta^l_{ijk}(o^l))$ the recognition rate is maximized. Thus we set c_{ijk} in $(\beta^l_{ijk}(o^l + 1), \beta^l_{ijk}(o^l))$ as follows:

$$c_{ijk} = \beta^l_{ijk}(o^l) - \delta^c(\beta^l_{ijk}(o^l) - \beta^l_{ijk}(o^l + 1)). \tag{40}$$

The tuning method is a local optimization method and guarantees a monotonic increase in the recognition rate during tuning. By allowing new misclassification during tuning, a relatively good local minimum can be obtained.

5 Tuning of Slopes

Tuning of the tuning parameters α_{ij} is discussed in [9]. Here, we summarize the procedure.

Tuning of α_{ij} is similar to tuning of center vectors, but is simpler. To explain the concept of tuning α_{ij}, we consider a two-class case with one rule for each class as shown in Fig. 3 which is the same initial condition with Fig. 1. Datum 1 is correctly classified into class 2, while data 2 and 3 are misclassified into class 2. If we increase α_{11} or decrease α_{21}, datum 1 is first misclassified, but if we allow datum 1 to be misclassified we can make data 2 and 3 be correctly classified. Figure 3 shows this when α_{11} is increased so that the degree of membership for class 1 lies between the shaded regions. Then by allowing one datum to be misclassified, two data are correctly classified, i.e., the recognition rate is improved by one datum. Comparing this with Fig. 1 (b), the range of α_{ij} that makes the misclassified data 2 and 3 be correctly classified is the half-infinite interval, while the range of c_{ijk} that makes the misclassified data 2 and 3 be correctly classified is the finite interval.

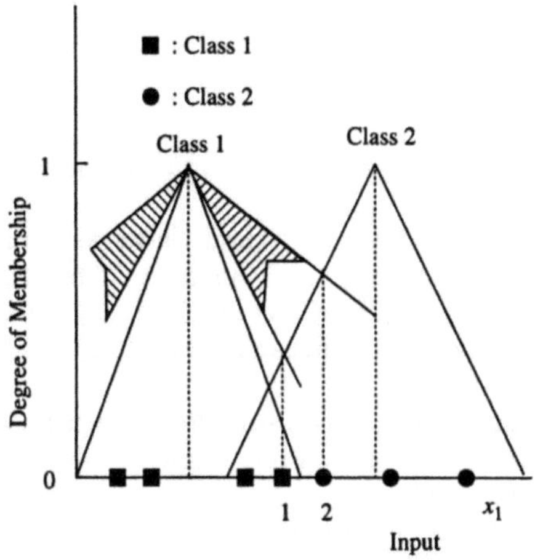

Fig. 3. Concept of tuning α_{ij}. If the slope of the membership function for class 1 is decreased so that the resulting function lies between the shaded regions, datum 1 is misclassified but data 2 and 3 are correctly classified.

Now suppose we tune the tuning parameter α_{ij}. Up to some value we can increase or decrease α_{ij} without causing new misclassification of class i data. Now let $U_{ij}(1)$ and $L_{ij}(1)$ denote the upper and lower bounds that do not cause new misclassification, respectively. Likewise, $U_{ij}(l)$ and $L_{ij}(l)$ denote the upper and lower bounds in which $l-1$ data are newly misclassified, respectively. Then, for instance, if we set a value in the interval $[U_{ij}(1), U_{ij}(2))$ to α_{ij}, one class i datum is newly misclassified.

Similarly, if we increase or decrease α_{ij}, misclassified data may be correctly classified. Let $\beta_{ij}(l)$ denote the upper bound of α_{ij} that is smaller than $U_{ij}(l)$ and that resolves misclassification. And $\gamma_{ij}(l)$ denotes the lower bound of α_{ij} that is larger than $L_{ij}(l)$ and that resolves misclassification.

Then the next task is to find which interval among $(L_{ij}(l), \gamma_{ij}(l))$ and $(\beta_{ij}(l), U_{ij}(l))$ $(l = 1, \ldots)$ gives the maximum recognition rate. To limit the search space, we introduce the maximum l, i.e., l_M. Let $(L_{ij}(l), \gamma_{ij}(l))$ be the interval that gives the maximum recognition rate of the training data among $(L_{ij}(k), \gamma_{ij}(k))$ and $(\beta_{ij}(k), U_{ij}(k))$ for $k = 1, \ldots, l_M$. Then even if we set any value in the interval to α_{ij}, the recognition rate of the training data does not change but the recognition rate of the test data may change. To control the generalization ability, we set α_{ij} as follows:

$$\alpha_{ij} = \beta_{ij}(l) + \delta(U_{ij}(l) - \beta_{ij}(l)), \tag{41}$$

for $(\beta_{ij}(l), U_{ij}(l))$, where δ satisfies $0 < \delta < 1$ and

$$\alpha_{ij} = \gamma_{ij}(l) - \delta(\gamma_{ij}(l) - L_{ij}(l)). \tag{42}$$

for $(L_{ij}(l), \gamma_{ij}(l))$.

References

1. Salzberg, S. (1991) A Nearest Hyperrectangle Learning Method, Machine Learning, **6**, 251-276
2. P. K. Simpson, "Fuzzy Min–Max Neural Networks – Part 1: Classification," *IEEE Trans. Neural Networks*, vol. 3, no. 5, pp. 776–786, 1992.
3. Abe, S. and Lan, M.–S. (1995) A Method for Fuzzy Rules Extraction Directly from Numerical Data and Its Application to Pattern Classification. IEEE Trans. Fuzzy Systems, **3**, 18–28
4. Thawonmas, R. and Abe, S. (1998) Rule Acquisition Based on Hyperbox Representation and Its Applications. Proc. Knowledge-based Intelligent Electronic Systems (KES '98), **1**, 120–125
5. Abe, S. Lan, M.-S. and Thawonmas, R. (1996) Tuning of a Fuzzy Classifier Derived from Data. The International Journal of Approximate Reasoning, **14**, 1–24
6. Uebele, F. Abe, S. and Lan, M.–S. (1995) A Neural Network-Based Fuzzy Classifier. IEEE Trans. Systems, Man, and Cybernetics **25**, 353–361
7. Musavi, M. T. Ahmed, W. Chan, K. H. Faris, K. B. and Hummels, D. M. (1992) On the Training of Radial Basis Function Classifiers. Neural Networks, **5**, 595–603
8. Abe, S. (1996) Neural Networks and Fuzzy Systems: Theory and Applications. Kluwer Academic Publishers, Boston
9. Abe, S. and Thawonmas, R. A (1997) Fuzzy Classifier with Ellipsoidal Regions. IEEE Trans. Fuzzy Systems, **5**, 358–368
10. Abe, S. (1998) Dynamic Cluster Generation for a Fuzzy Classifier with Ellipsoidal Regions. IEEE Trans. Systems, Man, and Cybernetics–Part B, **28**, 869–876
11. Cherkassky, V. and Mulier, F. (1998) Learning from Data: Concepts, Theory, and Methods. John Wiley & Sons, New York

A Comparison of Soft Computing and Traditional Approaches for Risk Classification and Claim Cost Prediction in the Automobile Insurance Industry

Ai Cheo Yeo[1], Kate A. Smith[1], Robert J. Willis[1], and Malcolm Brooks[2]

[1] School of Business Systems, Monash University, Clayton, Victoria 3800, Australia

[2] Australian Associated Motor Insurers Limited

Summary: In the automobile insurance industry, the expected value of claims from insuring risk is not known with certainty. Current methods employed by the industry tend to be rule based. This paper proposes a data driven approach to modelling claim costs by using soft computing techniques. Our approach consists of two stages. The first involves classifying policy holders according to their perceived risk, and the second stage involves modelling the claim costs within each risk group. Three methods are compared for the risk classification stage: k-means clustering, fuzzy c-means clustering and a previously published heuristic method that groups policy holders according to pre-defined factors. The accuracy of the expected claim costs are used to evaluate the three methods. A case study is presented utilising real data, and the results show the benefits of employing a data-driven approach.

Keywords: Soft computing, insurance, claim costs, prediction, k-means clustering, fuzzy c-means clustering, risk classification, heuristic methods

1 Introduction

The environment in which the automobile insurance industry operates is becoming increasingly competitive. Insurance companies strive for a combination of market growth and profitability, and these two goals are at times conflicting. In the process of attaining greater market share, companies may underwrite policy holders with high risk, and profitability may suffer. The decision making process involves setting profitability targets, accurately assessing the estimated risk of individual policy holders, and setting competitive premiums in order to simultaneously retain market share and achieve the targeted profit levels. Over the last four decades, tools such as Management Information Systems (MIS), Decision Support System (DSS), Executive Information Systems (EIS) and On Line Analytical Processing Systems (OLAP) have been used to aid the industry in their decision making. These decision making tools require guidance and expertise from the user, and are used merely to support the expertise of the decision maker.

Recent advances in computer processing and storage technology have enabled businesses to store huge amounts of data in data warehouses, however, and the potential of data-driven approaches to decision making is becoming clear. The insurance industry, like most other competitive industries, is turning to soft computing techniques such as neural networks and fuzzy logic to extract valuable information from its huge databases in order to improve their decision making and capitalise on the investment in business data. The main advantages of soft computing approaches for such decision support lies in their tolerance to imprecise data, and their modelling capabilities in uncertain environments.

This chapter considers risk assessment and prediction of claim cost using k-means and fuzzy c-means clustering. A case study is used to demonstrate the merits of the clustering approach. The accuracy of the predictions are compared to those obtained using a previously published heuristic method (Samson *et al.*, 1987). Section 2 introduces the proposed schemes for classifying risk and predicting claim costs based on this classification. The case study is presented in Section 3, where the clustering methods are compared to a heuristic method in terms of prediction accuracy. Finally, conclusions are drawn and future research is discussed in Section 4.

2 Methodology

Insurance companies classify policy holders into various risk groups based on factors such as territory, demographic variables (such as age, gender and marital status) and other variables (such as use of vehicle, driving record and years of driving experience). These factors are considered predictors of claim costs (Dionne *et al.*, 1992) (Samson, 1986) (Tryfos, 1980). Risk classification has traditionally been achieved using heuristic methods, both within the industry and in academic studies. For example, Samson and Thomas (1987) selected four variables: age of policy holder, area in which the policy holder lives, group rating of the insured automobile and level of no-claim discount; and categorised each variable into three levels of activity. Each policy was placed into one of the 81 (3^4) risk groups. However, the number of factors that can be included is limited for the heuristic method. This is because the number of exposure units in each risk group must be sufficiently large to make claim costs reasonably predictable (Vaughan *et al.*, 1996). To ensure that there is a large number of exposure units in each risk group, the number of risk groups has to be kept small, which in turn affects the number of factors that can be considered. For example adding an additional factor in Samson and Thomas' study would increase the number of risk groups to 243 (3^5) which would significantly reduce the number of exposure units in each risk group.

The classification structure is normally designed to achieve maximum homogeneity within groups and maximum heterogeneity between groups. This can be achieved through clustering, whether crisp or fuzzy. Clustering places objects into groups or clusters based on distances computed on the attributes of the data. The objects in each cluster tend to be similar to each other in some sense, and objects in different clusters tend to be dissimilar. (Johnson, 1998) (Zimmermann, 1996) Clustering allows more factors to be included in risk classification without compromising the size of each risk group compared to the heuristic method.

Several researchers have used clustering techniques for the risk classification stage of the claim cost prediction problem. Williams and Huang (1997) used k-means clustering to identify high claiming policy holders in a motor vehicle insurance portfolio. Derrig and Ostaszewski (1995) used fuzzy c-means clustering for territorial rating of Massachusetts towns. Smith, Willis and Brooks extended the use of clustering by proposing the use of k-means clustering to predict claim costs (Smith *et al.*, 2000). Our work extends this approach by evaluating the k-means clustering model and the fuzzy c-means clustering model as techniques for predicting claims cost, and comparing the results obtained with a heuristic model (Samson *et al.*, 1987) to determine the advantages of a data-driven approach.

Having grouped the policy holders into various risk groups using one of the three methods, the next stage is to predict claim costs for each of the risk groups. For the heuristic method, we have chosen regression, firstly so that we can compare our results to Samson and Thomas (1987), and secondly because regression has been shown in the literature to be preferable to other models that have been used (Chang *et al.*, 1979) (Sant, 1980). As for the two clustering methods, the average claim cost per policy holder within each cluster of the training set was used as the basis for prediction of the test set.

Figure 1 shows the methodology we have used for the risk classification and subsequent prediction of claim costs.

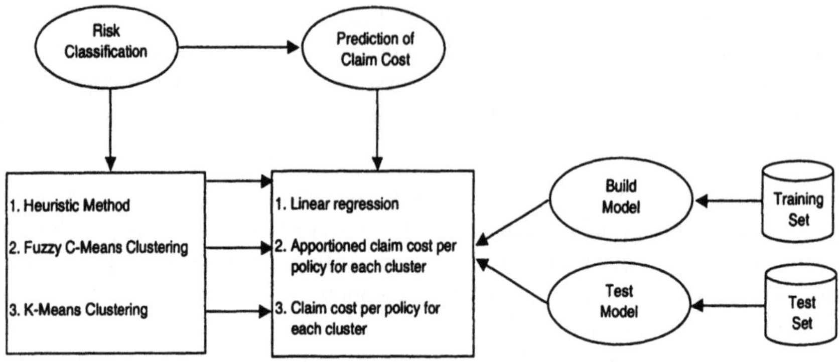

Figure 1: Methodology for Predicting Claim Cost

Table 1: Summary Statistics of Interval Variables

	TRAINING SET		TEST TEST	
Variables	Mean	Standard Deviation	Mean	Standard Deviation
Policy holder's age	42.1	12.8	41.7	14.3
Years on current rating	2.7	2.5	2.7	2.4
Years on rating 1	1.3	2.6	1	2.5
No of years policy held	2.9	2.4	2.7	2.4
Sum insured of vehicle	16,283	11,195	16,602	11,664
Total excess	415	80	409	200
Vehicle age	7.9	4.6	8.3	9.2

3 A Case Study

The data sets used for this research were supplied by an Australian motor insurance company. Two data sets (training set and test set), each consisting of

12-months of comprehensive motor insurance policies and claim information were extracted. The training set consisted of 146,326 policies with due dates from 1 January to 31 December 1998 while the test set consisted of 186,658 policies with due dates from 1 July 1998 to 30 June 1999. The period of overlap was to enable comparison of exposure and claim information over a one-year period. Forty percent of the policies in the test set were new policies. Extraordinary claim events were excluded from the data sets. The training set was used to create the clustering and heuristic models for risk classification, while the test set was used to evaluate the results of the various models. Table 1 and Table 2 show the variables in the training and test sets, and summarise their attributes.

Table 2: No of Classes of Categorical Variables

Variables	No of Classes
Policy holder's gender	2
Area in which vehicle was garaged	29
Rating of policy holder	9
Category of vehicle	26
Vehicle use (Private, Business)	2
Finance (Cash, Financed)	2

3.1 Risk Classification

The two clustering approaches to the risk classification problem (k-means and fuzzy c-means) utilise all of the available information except claim behaviour to find groups of policy holders exhibiting similar characteristics (demographic and historical). The heuristic method of Samson and Thomas (1987) classifies policy holders according to a set of three pre-defined factors. These methods are described in more detail below.

3.1.1 K-Means Clustering Model

The k-means clustering model used to classify policies performs disjoint cluster analysis on the basis of Euclidean distances computed from variables and seeds that are generated and updated by the k-means algorithm (Everitt, 1993). Using the least squares clustering criterion, the sum of the squared distances of observations to the cluster means is minimized.

All thirteen available variables were used for clustering. A minimum cluster size of 1,000 was specified to satisfy the insurability requisite of mass. The initial clustering yielded 6 clusters, with cluster sizes ranging from 1,600 to 58,000. Two more rounds of clustering were done to reduce the cluster sizes to no more than 20,000. A total of 30 risk groups were generated through the three rounds of clustering, each containing between 1,000 and 20,000 policy holders. This process ensures the clustering algorithm finds a balance between the requisite mass and homogeneity criteria.

3.1.2 Fuzzy C-Means Clustering Model

The fuzzy c-means clustering algorithm assigns each policy holder to different clusters to varying degrees specified by a membership grade. The algorithm minimizes an objective function that represents the distance from any given data point to a cluster centre weighted by that data point's membership grade (Bezdek, 1981). To ensure comparability with the k-means clustering, the number of clusters specified was 30. The same thirteen variables used in the k-means clustering were used for the fuzzy c-means clustering.

The training data was clustered using MATLAB Fuzzy Logic Toolbox. A neural network was then trained to learn the fuzzy inference system using the thirteen variables as input and the membership grades of the 30 clusters as output. The number of hidden neurons was fifty, and the hyperbolic tangent activation function was used. Thirty percent of the data was reserved for a validation set, and the R Squared obtained on this set was 0.9608 (with an R squared of 0.97 for the total data set) giving us the confidence to apply the network to the test set.

3.1.3 Heuristic Model

To determine how well the two clustering models group policy holders into various risk groups, a heuristic model based on the approach of Samson and Thomas (1987) was used for comparison. Three factors were used in the heuristic model: age of the policy holders, area in which the vehicles were garaged and the category of the vehicle. Rating area and category of vehicle were sorted by average claim cost per policy holder and age of policy holder was sorted by age. The three variables were then split into 5 classes each. An attempt was made to minimise the difference in average claim cost per policy between the classes and to ensure that each class had at least 10,000 policies, in other words, to comply with the two requirements of mass and homogeneity. A total of 125 (5^3) groups were created.

3.2 Predicting Claim Cost

3.2.1 K-Means Clustering Model

The actual average claim cost per policy for each cluster, found by clustering the training set, was used as the basis for predicting average claim cost per policy for the test set.

3.2.2 Fuzzy C-Means Clustering Model

Claim cost of a policy holder is apportioned to the clusters he belongs to according to his calculated membership grade. For example, if the claim cost of a policy holder was $1,000 and his membership grade of Cluster 1 is 0.8 and that of Cluster 2 is 0.2, $800 will be apportioned to Cluster 1 and $200 will be apportioned to Cluster 2. The claim cost per policy holder in each cluster will be the total cost apportioned to that cluster divided by the total membership value for that cluster. An illustrative example is shown in Table 3.

Table 3: Illustrative Example of Computing Claims Cost Per Policy Holder (Fuzzy Clustering)

Policy Holder	Claim Cost ($)	Membership Grade Cluster1	Membership Grade Cluster2	Apportioning of Claim Cost ($) (Claim Cost * Membership Grade) Cluster1	Apportioning of Claim Cost ($) (Claim Cost * Membership Grade) Cluster2
1	0	0.98	0.02	0	0
2	300	1.00	0.00	300	0
3	0	0.76	0.24	0	0
4	500	0.99	0.01	497	3
5	0	0.00	1.00	0	0
6	0	0.02	0.98	0	0
7	1,000	0.98	0.02	981	19
8	0	0.03	0.97	0	0
9	900	0.16	0.84	144	756
10	0	0.01	0.99	0	0
Total	**2,700**	**4.93**	**5.07**	**1921**	**779**
Claim Cost Per Policy Holder (Total Claim Cost/Total Membership Value)				**389**	**154**

3.2.3 Heuristic Model

A linear regression model, similar to Samson's (Samson *et al.*, 1987) was used to predict average claim cost per policy for the risk groups found by the heuristic model. The linear model we used is shown in the following equation:

$$
\begin{aligned}
y = {} & 123.2a_1 + 13.1459a_2 - 33.6505a_3 - 29.7609a_4 \\
& - 116b_1 - 82.6731b_2 + 7.9831b_3 - 51.317b_4 \\
& - 77.8226c_1 - 65.8959c_2 - 61.8716c_3 - 3.5125c_4 + 402.4
\end{aligned}
$$

where y = claim cost per policy holder in a cell, and each risk factor is represented by two binary variables as shown below:

$a_1 = 1$ for age group 1	$a_2 = 1$ for age group 2
$= 0$ otherwise	$= 0$ otherwise
$a_3 = 1$ for age group 3	$a_4 = 1$ for age group 4
$= 0$ otherwise	$= 0$ otherwise
$b_1 = 1$ for rating area 1	$b_2 = 1$ for rating area 2
$= 0$ otherwise	$= 0$ otherwise
$b_3 = 1$ for rating area 3	$b_4 = 1$ for rating area 4
$= 0$ otherwise	$= 0$ otherwise
$c_1 = 1$ for vehicle category 1	$c_2 = 1$ for vehicle category 2
$= 0$ otherwise	$= 0$ otherwise
$c_3 = 1$ for vehicle category 3	$c_4 = 1$ for vehicle category 4
$= 0$ otherwise	$= 0$ otherwise

3.3 Results

Figures 1, 2 and 3 show the plots of predicted claim cost per policy against the actual claim cost per policy for the k-means clustering model, fuzzy c-means clustering model and heuristic model respectively. From the graphs, it can be seen that the predicted claim cost of the two clustering models are closer to the actual claim cost compared to the heuristic model. However the fuzzy c-means does not appear to be able to discriminate policy holders with high claim cost.

Figure 1: Prediction of Claim Cost (K-Means Clustering)

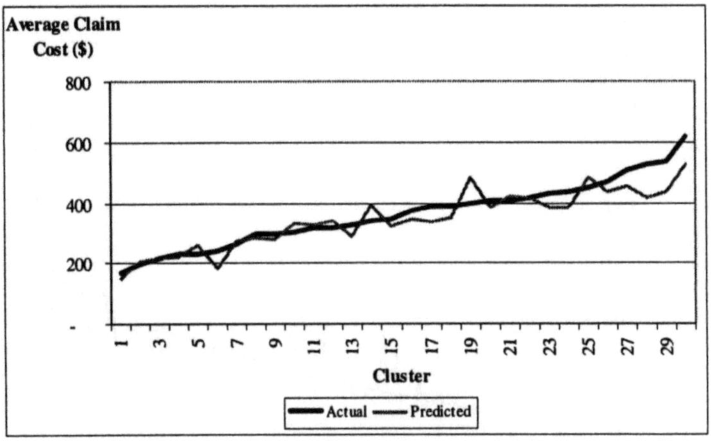

Figure 2: Prediction of Claim Cost (Fuzzy C-Means Clustering)

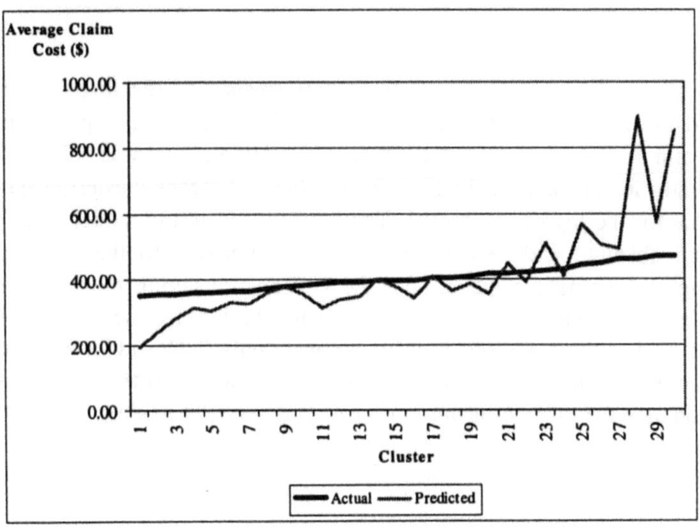

Figure 3: Prediction of Claim Cost (Heuristic)

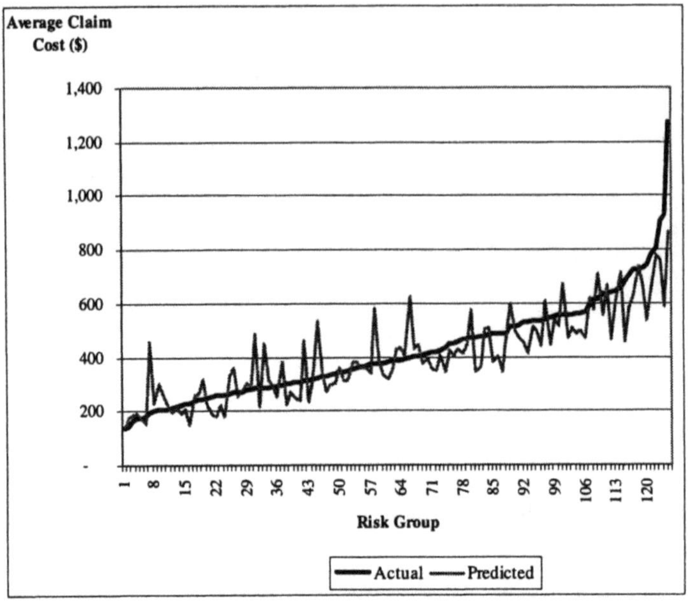

Table 4 shows the various measures of performance of the three models. The weighted mean absolute deviation of predicted claim cost per policy from the actual claim cost per policy for the k-means clustering model was 8.3% which was significantly lower than the 15.63% for the fuzzy c-means clustering model and the 13.3% for the regression model. The k-means clustering model had a smaller maximum absolute and percentage deviation compared to the fuzzy c-means clustering and heuristic models. For the k-means clustering model, 90% of the risk groups had deviation within 20% while for the fuzzy c-means clustering and heuristic models the corresponding figures were only 77% and 72% respectively. In all these measures, the k-means clustering model provided a more accurate prediction than the fuzzy c-means clustering and heuristic models.

Table 4: Measurement of Performance of Models

Measurements	K-Means Clustering Model	Fuzzy C-Means Model	Heuristic Model
Weighted mean absolute deviation	8.30%	15.63%	13.30%
Maximum deviation	$111	$433	$403
Maximum % deviation	23%	93%	129.47%
Deviation within 10%	57%	43%	39%
Deviation within 20%	90%	77%	72%

4 Conclusions

In this paper we have shown that the data-driven k-means clustering approach to risk classification can yield better quality predictions of expected claim costs compared to a previously published heuristic approach (Samson *et al.*, 1987). The clustering approach has the advantage that more variables can be considered, without causing the number of distinct risk groups to increase exponentially, as is the case with the heuristic method. The minimum number of policies for each cluster can be specified and a hierarchical approach has been proposed to enable a balance to be found between the two criteria of mass and homogeneity. The claim cost predictions can be combined with other models that determine the sensitivity of policy holders to premium increments, and set premiums to achieve targeted levels of profitability (Smith *et al.*, 2000). As such, the method proposed here for assessing policy holder risk is one component of a methodology implementing a data-driven decision support tool.

One of the limitations of using clustering to group policy holders however is the interpretation and labelling of the clusters. Business managers would prefer clusters to have clearly defined rules that describe them. At this stage, the superior performance of the clustering approach warrants further research to describe the clusters in terms understandable to the business. Other soft computing methods such as decision trees or fuzzy logic could be combined with the clustering algorithm to find rules and patterns (Gilbert *et al.*, 1998).

Clearly the accuracy of the k-means clustering model will be compromised if factors affecting losses in the future are not similar to those affecting losses in the past. Factors which may affect losses include external influences such the economy and catastrophes, as well as increased competition. When implementing the k-means clustering model as a decision support application therefore, there is a need to periodically check the model to ensure that it is still valid. This may require enhancement of the model to include changes to the data, as well as changes to the environment as discussed above.

In future research, we will address the other components of determining optimal premiums: modelling retention rates and the effect of changes in premim pricing using data driven and soft computing techniques. The results will be used in conjunction with the claims modelling presented in this paper to develop a decision support tool to determine optimal premiums. This data driven approach is quite revolutionary considering the rule-based methods that have driven premium pricing for decades, and is likely to prove to be a critical tool for helping insurance companies satisfy their conflicting goals of increased market share and profitability.

References

Bezdek, J.C. (1981): Pattern recognition with fuzzy objective function algorithms. Plenum Press, New York

Chang, L., Fairley, W.B. (1979): Pricing automobile insurance under multivariate classification of risks: additive versus multiplicative. The Journal of Risk and Insurance 46(1), 75-98

Dionne, G., Vanasse, C. (1992): Automobile Insurance Ratemaking in the Presence of Asymmetrical Information. Journal of Applied Econometrics 7(2), 149-165

Everitt, B.S. (1993): Cluster analysis, 3rd Edition. E. Arnold, London

Gilbert, K., Aluja, T., Cortes, U. (1998): Knowledge discovery with clustering based on rules: Interpreting results. In: Zytkow, J. M., Quafafou, M. (Eds.): Principles of data mining and knowledge discovery : Second European Symposium, PKDD '98, September 1998 proceedings, Nantes, France. Springer, Berlin New York, 81-92

Johnson, D.E. (1998): Applied multivariate methods for data analysts. Duxbury Press, Pacific Grove, Calif

Samson, D. (1986): Designing an automobile insurance classification system. European Journal of Operational Research 27, 235-241

Samson, D., Thomas, H. (1987): Linear models as aids in insurance decision making: the estimation of automobile insurance claims. Journal of Business Research 15, 247-256

Sant, D.T. (1980): Estimating expected losses in auto insurance. The Journal of Risk and Insurance 47(1), 133-151

Smith, K.A., Willis, R.J., Brooks, M. (2000): An analysis of customer retention and insurance claim patterns using data mining: a case study. Journal of the Operational Research Society 51, 1-10

Tryfos, P. (1980): On classification in automobile insurance. Journal of Risk and Insurance 47(2), 331-337

Vaughan, E.J., Vaughan, T.M. (1996): Fundamentals of risk and insurance, 7th Edition. Wiley, New York

Zimmermann, H.J. (1996): Fuzzy set theory and its applications, 3rd Edition. Kluwer Academic Publishers, Boston, Dordecht, London

Evolutionary Rule Generation and its Application to Credit Scoring

Susan E. Bedingfield and Kate A. Smith

School of Business Systems Monash University Clayton, Victoria 3168 Australia

Email: {sue.bedingfield, kate.smith}@infotech.monash.edu.au

1. Introduction

Credit scoring is an important financial application area, concerned with assessing the likely risk of customers defaulting on granted credit. These customers may be bank customers borrowing money, or retail customers being sold goods on a deferred payment scheme. In order to model the relationship between the characteristics of a customer (financial and demographic) and their measured credit risk, data is collected and analysed. The developed model can then be used with new customers to determine their expected behaviour in an effort to automate the decision making process. The task of learning to classify objects in a database according to their attributes or characteristics has been tackled by many approaches over the decades. Statistical approaches such as logistic regression and discriminant analysis have yielded over recent years to soft computing methods such as neural networks, fuzzy logic and genetic algorithms. One of the advantages of soft computing techniques lies in their modeling capabilities in the presence of noisy, imprecise, inaccurate, or missing data. These considerations become specially important in applications such as credit scoring where the data attributes are likely to be incomplete, the classification of the customers as good or bad credit risks may be erroneous, and the costs of misclassification are particularly high.

For credit scoring however, as with many classification problems, we need more than a model that accurately classifies the objects in the database. The discovery of rules to explain the differences between the classes is an important additional outcome required for the classification system to be employed as a practical tool. The need for such explanation is particularly important in the credit scoring industry in order to satisfy legal requirements in many countries. Evolutionary approaches are especially suited to rule generation due to their flexibility when encoding the problem, and the subsequent transparency this creates in the interpretation of the model.

This paper presents a new evolutionary approach to rule generation for credit scoring applications. Classification rules are evolved using an evolutionary algorithm embedded within an information systems framework. Each individual in the genetic algorithm is a rule system, unlike the bit-string representations used by previous researchers that need encoding and decoding [1]. Our evolutionary rule generation system is described in section 2. The chosen fitness functions also contain a variety of novel components designed to address accuracy, coverage, and efficiency criteria. Two different fitness functions are used to combine these components, and are described in Section 3: one which naturally produces conservative rules, and one producing non-conservative rules. Conservative rules are reluctant to classify objects as good credit risks unless the accuracy is high, and thus the coverage of these objects is low and there may be many objects not classified by any rules. Likewise, non-conservative rules tend to classify most objects, so the coverage is high, often at the expense of accuracy. The differences between these two fitness functions is demonstrated statistically in Section 4. The advantage of using the one set of fixed parameters to generate both types of models through fitness function selection is that a linear combination of the two can be used to produce rules that are accurate with good coverage as defined by the user. A single parameter λ can be used to control this process acting much like a utility scale to indicate the user's preferred balance between accuracy and coverage. The user can select the value of λ to produce rules that reflect the best decisions relevant to the misclassification costs, and a list of rules defining poor credit risks can be generated. This approach is demonstrated using a well-benchmarked credit data set in Section 4.

Once a model is developed that reflects the user's preferred balance between accuracy and coverage, we need to focus on the issue of interpretation of the results. Each object in the database can be evaluated to see which of the evolved rules they satisfy, and to which class they belong. A number of scenarios emerge depending on the number of rules an object satisfies and whether any of these are contradictory. Three different methods for analyzing this information are presented in Section 5. Finally, conclusions and future research directions are outlined in Section 6.

2. Evolutionary Rule Generation

In this section we define the information system within which the proposed evolutionary rule generation approach operates. We then discuss the structure of the rule systems, as well as the genetic algorithm operators used to evolve the rules.

2.1 Information system definitions

The definition of an information system originated with Pawlak [2] and provides a structure for the genetic algorithm (GA) to act upon.

Definition 1: An *information system* is a 4-tuple $S = \langle O, A, V, f \rangle$, where O is a finite set of objects, A is a finite set of attributes, V_a is the set of attribute values that $a \in A$ can take,

$$V = \bigcup_{a \in A} V_a \text{ and } f : O \times A \rightarrow V \text{ such that } f(x,a) \in V_a.$$

(f is called an information function).

We will be using a more specific information system called a decision table. This enables us to distinguish between two different types of attributes, condition attributes and decision attributes. For example in a credit scoring database there may be information relating to the circumstances of each applicant such as age, salary etc. which could be classified as condition attributes. There would usually also be information relating to the credit status of each applicant. This could be referred to as a consequence attribute. To accommodate these two types of attributes we introduce a refinement of the information system.

Definition 2: An information system $\langle O, A \cup \{d\}, V, f \rangle$, where d is a function from O to a finite set V_d is called an extension of A by d. In this context, the attributes in A are called the condition attributes and d is called the decision attribute. For example, d might indicate whether a person was a good or bad credit risk.

2.2 Rule structure

Each individual rule is made up of a set of atoms. An atom is a structure of the form (α, β, γ) where α represents an attribute, β is a comparison operator ($\leq$, $\geq$, $=$) and γ a number.

A rule has a structure of the form *(atom1, atom2, ...,atomN)* where $N \leq M$ and $M \geq 2$ is a parameter defining the upper bound on the number of atoms allowed to make a rule. A restriction is placed on the number of atoms allowed to ensure that the rules are easily interpreted and understood by humans. For each rule, $atom_1$, $atom_2$, ...,$atom_{N-1}$ have as the attribute component a condition attribute, and $atom_N$ has as its attribute component a decision attribute.

Note that a particular attribute may appear more than once in an atom, for example when an upper and lower bound for a specific attribute is required. For example, a rule might be:

(age>50, sex = 0, income > 20,000, age <55, married = 1, income < 35,000, credit class = 1)

which can be translated as "if age is between 50 and 55, income is between 20,000 and 35,000 and the person is married, then they are a good credit risk".

A rule system, $\mathfrak{R}$, comprises a finite set of rules {rule1, rule2,...,rulek} where $1 \leq k \leq K$, K being a parameter defining the number of allowable rules in a rule set. The genetic algorithm population consists of P rule systems each containing k individual rules.

2.3 Notation

Before discussing how such rule systems can be evolved using a genetic algorithm, there are certain notations which require explanation. It is assumed that the data is contained in an information system extension $S=<O,A \cup \{d\},V,f>$. If $v \in V_d$, we define O_v to be the subset of O whose decision attribute value is v. The condition attribute values of any $x \in O$ belonging to

$$P \subset \prod_{a \in A} V_a$$

We let $\mathfrak{R}$ represent a rule system acting on S. If $\rho \in \mathfrak{R}$ and $y \in P$, then $y \sim \rho$ means y satisfies the condition attribute values for ρ. If $x \in O$ then $x \in \rho$ means x satisfies the condition and consequence attribute values for ρ, and $x \sim \rho$ means x satisfies the condition attribute values for ρ. (So $x \in \rho \Rightarrow x \sim \rho$). We let d(x) denote the value of the decision attribute associated with x, and we let d(ρ) denote the value that ρ assigns to the decision attribute. If $\rho \in \mathfrak{R}$, then O_ρ is the subset of O whose members satisfy the condition attribute values associated with ρ i.e. $O_\rho = \{x \in O : x \sim \rho\}$ and $O_{\rho, d(\rho)}$ is the subset of O whose members satisfy the condition and decision attribute values associated with ρ, i.e. $O_{\rho,d} = \{x \in O : x \in \rho\}$. (Clearly $O_{\rho,d} \subseteq O_\rho \subseteq O$). Also, we define $O_{d(\rho)}$ to be the subset of O whose d attribute value is the same as the attribute value of ρ, (which we denote by $d(\rho)$), i.e. $O_{d(\rho)} = \{ x \in O : d(x)=d(\rho)\}$. (Note that $O_{\rho,d} \subseteq O_{d(\rho)} \subseteq O$, but we cannot deduce any relationship $O_{d(\rho)}$ and O_ρ).

2.4 Genetic operators

The genetic operators used in this paper are an elitist selection criteria, and crossover and mutation operators.

Crossover between individual rule systems is accomplished randomly based on a fixed crossover probability P_c . The first *C* rules from each rule system are selected and swapped. (The starting point for the swap is also determined as part of the crossover process). *C* may vary from 0 to a number half the size of the smaller rule system. We also include the possibility of the value of *C* not being larger than a fixed amount (say 2) as switching large quantities of rules between rule systems may have a dramatic adverse effect on the functionality of each rule system. The rule systems produced by the crossover are added to the population, thus temporarily increasing its size.

Mutation occurs to the rules within the individual rule bases and takes various forms - random changes to the components of each atom, swapping one or more atoms between two rules, adding a new atom to a specific rule, deleting a rule, adding a new rule. The probabilities of each type of mutation are parameterised (with given probabilities P_{m1}-P_{m5}) in order to establish which are the most useful mutations in terms of optimising the rule systems.

The selection criteria used is elitist - the best *P* rule systems of the current generation are selected to form the next generation.

The initialisation procedure used in this paper was random initialisation; for a given population of size *P* we generate a predefined number of rules for each of the *P* rule systems as follows. An atom is generated by using a random number generator to generate the contents of each component. For example if there are *n* attributes in the information system, each attribute is allocated an integer between 1 and *n*. For the attribute component of each atom, a random integer between 1 and *n* is generated to determine the specific attribute represented by the atom. There are three comparison operators =, >, < which are possible candidates for the second component of each atom and the comparison operator for each specific atom is generated in the same manner. The last component, the value component, is generated in a similar way. In this manner the atoms are generated for each of *k* rules in each of *P* rule systems. The number of atoms belonging to each rule is initially fixed, but allowed to vary as atoms are added and deleted from a given rule. Likewise, the number of rules in a rule system is initially fixed, but allowed to vary as rules are added and deleted during mutation.

3. Fitness Functions

In this section we present the components that comprise the fitness function, as well as two alternative ways of combining these fitness components into fitness functions.

3.1 Fitness function components

There are several measures of fitness of the individual rules in each rule base that we are interested in recording and improving over time. We first consider accuracy measures - both row and column accuracy - of the resulting confusion matrix. The particular information system we are experimenting with has a binary valued decision attribute. This leads to two row accuracy measures C_1 and C_2, and two column accuracy measures C_3 and C_4. C_1 and C_2 effectively measure the coverage by the rule system of the 'good' and 'bad' objects respectively. Whereas C_3 and C_4 measure the accuracy of the 'good' and 'bad' predictions. These four expressions should optimally be equal to 1.

$$C_1 \text{ and } C_2 \text{ are defined as } \frac{Card\left(O_i \cap \bigcup_{d(\rho)=i} O_\rho\right)}{Card(O_i)}$$

$$\text{and } C_3, C_4 \text{ as } \frac{Card\left(O_i \cap \left(\bigcup_{d(\rho)=i} O_\rho\right)\right)}{Card\left(\bigcup_{d(\rho)=i} O_\rho\right)} \quad \text{where} \quad i = 0,1$$

Analogous to the accuracy measures C_1 and C_2 are two measures of inaccuracy C_5 and C_6. These must be included since, for any object in the information system, a given rule system could potentially include at least one rule predicting that the object has a decision attribute value of 0 and at least one rule predicting the object has a decision attribute value of 1, only one of which could be correct. The two inaccuracy measures should optimally be equal to 0. (In practice this may be difficult to achieve if the information system itself contains inconsistencies. This may in turn lead to the generation of inconsistent rules).

C_5, C_6 are defined as $$\frac{Card\left(O_i \cap \left[\bigcup_{d(\rho)\neq i} O_\rho\right]\right)}{Card(O_i)}$$ where $i = 0,1$

We also include a component of the fitness function, C_7, designed to minimise contradictions (incorrect predictions) in the rule systems for each object:

$$C_7 = e^{-\sum_{x\in O}\sum_{\rho\in\Im}\delta'_\rho(x)} \quad \text{where } \Im = \{\rho \in \Re : d(\rho) \neq d(x)\}$$

and $$\delta'_\rho(x) = \begin{cases} 1 & x \in O_\rho \\ 0 & \text{otherwise} \end{cases}$$

The next fitness component, C_8 ,concerns the efficiency of a rule system. In terms of efficiency, it is preferable that one rule gives a correct prediction for a given object rather several rules. C_8 is designed to encourage this behaviour.

$$C_8 = e^{\frac{-\sum_{x\in O}\left(\left(\sum_{\rho\in\Re}\delta_\rho(x)\right)-1\right)^2}{Card(O)}} \quad \text{where } \delta_\rho(x)\begin{cases} 1 \text{ if } & x \in \rho \\ 0 & \text{otherwise} \end{cases}$$

3.2 Combining fitness function components

Thus the total fitness function for the genetic algorithm can be represented as the weighted function of the eight components (C_1-C_8) defined above, and attempts to generate rules over time with maximal coverage, maximal accuracy, minimal contradictions, and high efficiency. All fitness components have optimal values of 1, except C_5 and C_6 which should be 0. The general form for the overall fitness function can take various forms. Two fitness functions in particular, F^1 and F^2, have been reported on [3,4]. F^1 and F^2 are defined as follows:

$$F^1 = \left(\frac{\sum_{i=1}^{4} w_i C_i + \sum_{i=7}^{8} w_i C_i}{1 + \sum_{i=5}^{6} w_i C_i} \right), \quad F^2 = \left(\sum_{i=1}^{4} w_i C_i + \sum_{i=5}^{6} \frac{w_i}{1 + C_i} + \sum_{i=7}^{8} w_i C_i \right),$$

where $w_i \geq 0, i = 1,\ldots,8$ are fitness weights.

4. Comparison of Fitness Functions

To generate and evaluate the rule systems described in this paper we use a well benchmarked data set.

4.1 Data

The data on which the rule system generation is based is the same German credit classification data used by Walker et al. [5] and Graf and Nakhaeizadeh [6]. The data contains 1000 examples of credit applicants, 700 of which are known to be good credit risks, and 300 of which are known to be bad credit risks. The attributes for each example include a combination of continuous and categorical variables such as age, marital status, bank account status, credit history, and the loan amount. There are a total of 20 attributes available with which to model the relationships between attributes and consequence, in this case good or bad credit risk. Since 70% of the credit applicants in the data set are classified as good, the default level of accuracy is 70%, obtained by simply classifying all applicants as good. Similarly, if we decide to classify all applicants as bad, the default level of accuracy is 30%. Thus we are aiming to produce better than 70% accuracy when classifying applicants as good and better than 30% accuracy when classifying applicants as bad. The data was split into two groups for training and testing. The training group contained 80% of the data, 70% of which were selected at random from the 'goods' and 30% of which were selected at random from the 'bads'.

4.2 Generation of rule systems

Rule systems generated by the two fitness functions F^1 and F^2 exhibit quite different behaviours [4]. F^1 tends to generate rule systems with highly accurate rules but lower coverage. Whereas F^2 tends to generate rule systems with higher coverage but less accuracy. This difference in behaviour can be demonstrated by performing an ANOVA on the eight components of the fitness functions F^1 and F^2. Ten experiments were run for 50 generations for each of F^1 and F^2. The various parameters remained the same for each experiment For the final generation each fitness component was averaged over all rule systems. The two averages were then compared using ANOVA tables 1 and 2 demonstrate the result for two of the fitness components

Table 1: F^1 and F^2 comparison for component C_1

Groups	*Count*	*Sum*	*Average*	*Variance*
Fitness function F^1	10	2.815786	0.281579	0.017389
Fitness function F^2	10	9.010643	0.901064	0.003169

ANOVA

Source of Variation	*SS*	*df*	*MS*	*F*	*P-value*	*F crit*
Between Groups	1.918813	1	1.918813	186.6755	6.08E-11	4.413863
Within Groups	0.18502	18	0.010279			
Total	2.103832	19				

Table 2: F1 and F2 comparison for component

Groups	*Count*	*Sum*	*Average*	*Variance*
Fitness function F^1	10	8.90008	0.890008	0.000524
Fitness function F^2	10	7.292911	0.729291	0.000391

ANOVA

Source of Variation	*SS*	*df*	*MS*	*F*	*P-value*	*F crit*
Between Groups	0.12915	1	0.12915	282.2556	1.91E-12	4.413863
Within Groups	0.008236	18	0.000458			
Total	0.137386	19				

Likewise the other fitness components indicated the difference in behaviour of F^1 and F^2.

4.3. Combining fitness functions

By combining F^1 and F^2 using a parameter we can 'force' more moderate behaviour upon the rule systems.

Thus, the fitness function used for the work presented in this paper has the following form:

$F = \lambda F^1 + (1-\lambda)F^2$ where $0 < \lambda < 1$

An illustration of the behaviour of this function is given in Figure 1 which graphs the ratio of the average value component C_1 of to the average value of component C_3 in the final generation of one particular experiment lasting 50 generations. It can be seen that while F^1 dominates, i.e. λ close to 1, the coverage of 'good' objects by the rule system is relatively low compared with the accuracy of the rules. Conversely, when F^2 dominates, or λ is close to 0, the coverage of 'good' objects relative to accuracy is much higher. The value chosen for λ would normally depend on the relative misclassification costs for each of the possible predicted outcomes.

Figure 1: Accuracy vs. coverage ratios for different λ values

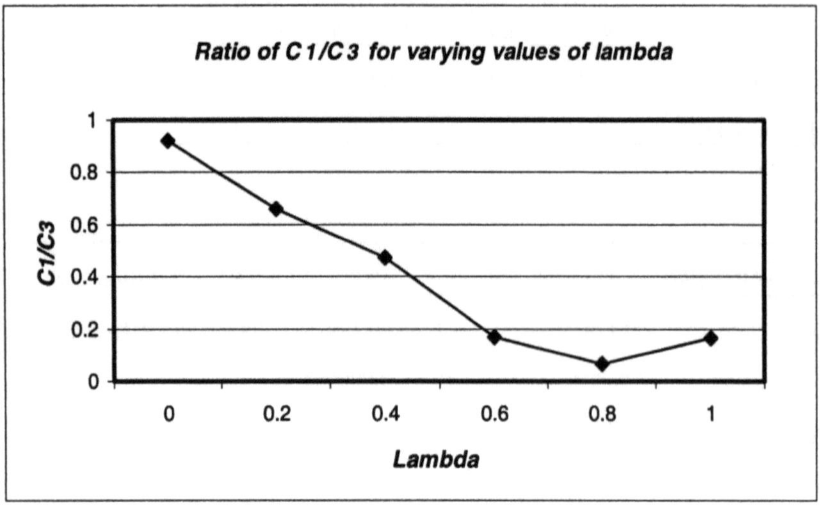

5. Interpretation of a Rule System

The fitness function on which the generation of the rule systems is based is designed to minimize contradictions, misclassification and non classification.

For $x \in \mathfrak{R}$, let $G_x = \{\rho \in \mathfrak{R} : x \sim \rho,\ d(\rho) = \text{'good'}\}$

and $B_x = \{\rho \in \mathfrak{R} : x \sim \rho,\ d(\rho) = \text{'bad'}\}$

Let $p(x)$ denote the predicted value for x

However, it cannot be assumed that the resulting rule systems will be entirely without these problems, partly because the data on which the rules are generated from may itself contain inconsistencies. This section discusses some possible techniques for using the evolved rule systems in practice.

We first consider the possibilities that may occur when a new object x is presented to the rule system:

1. If $G_x \cup B_x = \phi$, then there are no rules whose conditions x satisfies, i.e. a value for $p(x)$ is not provided by $\mathfrak{R}$
2. else if $B_x = \phi$, put $p(x)$ = 'good'
3. else if $G_x = \phi$, put $p(x)$ = 'bad'
4. else $\mathfrak{R}$ provides possible values for $p(x)$ which are contradictory

Case 1. and 4. must be considered further.

Case 1: $\mathfrak{R}$ does not provide a prediction

In this case the approach we have taken is to consider the misclassification costs for each of the possible consequence values. We then select the consequence value with the highest misclassification cost and take that as the prediction. For example in the data that we are using there are two possible outcomes, 'good' or 'bad'. The cost of classifying an actual 'bad' as a 'good' is greater than classifying a 'good' as 'bad', so the default prediction used (in the case where the rule system does not provide a prediction) is 'bad'.

Case 4: $\mathfrak{R}$ provides contradictory predictions:

In this case a prediction needs to be made which takes into account some or all of the following factors:

The accuracies of the rules providing a prediction

The relative misclassification costs

The relative difficulty of generating rules for the various values of the consequence variable (e.g. in the data used there are 700 'goods' and 300 'bads' which makes it more difficult for the rule generation system to generate accurate rules for a 'bad' prediction.)

Method 1: the consensus method.

This method takes into account the relative number of rules which provide a prediction for x for each of the possible consequence values. In effect there may be a number of rules which 'vote' on whether x will be a 'good' or 'bad' risk and we must decide what what the relative 'vote' of 'goods' to 'bads' (or 'bads' to 'goods') is that we will require in order to accept that x is a 'good' risk (or 'bad' risk). For example, we may decide that if 50% of the rules predict that x is a 'bad' risk then we will accept the prediction for x as 'bad'. The acceptability level α that we decide upon will depend on the misclassification costs, so for example if the cost of misclassifying an actual 'bad' as a 'good' is much greater than misclassifying an actual 'good' as a 'bad" then we may only require that at least 10% of the predictions are 'bad' in order to accept the prediction as 'bad'.

More formally,

$Card(B_x) / Card(G_x \cup B_x) > \alpha$ then $p(x) =$ 'bad'

let α be the smallest value such that if

Method 2: the comparison with default method.

We use the following method to arrive at a prediction for a new object x which is presented to the rule system:

Determine ρ_{good} and ρ_{bad} as follows:

$$\text{Accuracy}(\rho_{good}) = \underset{\rho \in G_x}{Max}(accuracy(\rho))$$

$$\text{Accuracy}(\rho_{bad}) = \underset{\rho \in B_x}{Max}(accuracy(\rho))$$

If $\rho_{good} - \alpha_1 > \rho_{bad} - \alpha_0$, put prediction = 'good', else put prediction = 'bad'

where α_1 and α_0 are the default probabilities of making a 'good' / 'bad' prediction, in this case .7 and .3 respectively.

Method 3: the compensating method

Determine ρ_{good} and ρ_{bad} as follows:

$$\text{Accuracy}(\rho_{good}) = \underset{\rho \in G_x}{Max}(accuracy(\rho))$$

$$\text{Accuracy}(\rho_{bad}) = \underset{\rho \in B_x}{Max}(accuracy(\rho))$$

If $\rho_{good} > \omega * \rho_{bad}$, put prediction = 'good', else put prediction = 'bad'

where ω is a weighting factor designed to compensate for the relative difficulty of making a 'bad' prediction.

The best weighting factor ω is determined by fixing a value and making a prediction using the test set, thus obtaining an overall accuracy for the rule system applied to the test set. This procedure is then repeated for various values of ω to determine the optimal value of ω

6. Conclusion

This paper has proposed a new evolutionary approach to rule generation, and evaluated its performance on a credit scoring problem. The approach uses a genetic algorithm to evolve rules designed to improve the classifications according to a variety of measures including coverage and accuracy. Different fitness functions have been shown to produce rule systems exhibiting quite opposing characteristics. Three different methods for interpreting the decisions of the best rule system have been explored and compared.

The overall accuracy achieved on the test set by using the 20 rule systems in the final generation of the evolutionary algorithm varies between rule systems. The results of the rule system giving the most accurate results are provided in Table 3. The table incorporates a comparison of the three methods described which are used in the case when the rule system provides ambiguous or contradictory predictions. (Better results are obtained on the training set, but the results obtained from the test set are the pertinent ones).

Table 3: Comparison of results.

Size of test set	200
Number of predictions provided by rule system	185
Number of correct predictions provided by rule system and Method 1	145
Number of correct predictions provided by rule system and Method 2	143
Number of correct predictions provided by rule system and Method 3	143

Thus the highest accuracy in this case is 72.5% which compares favourably with the default accuracy of 70% and the C4.5 decision tree results of Walker et al [5]. An advantage of using an evolutionary approach is that an evolutionary algorithm has superior optimisation capabilities because of its global search mechanism in comparison with the decision tree approach which uses a greedy search algorithm [1].

Our future research will include an exploration of alternative initialisation strategies including the use of Rough Sets [7]. The results obtained in this paper were obtained using only default values and a thorough search of parameter space may produce improved results. The approach also needs to be evaluated on other data sets and classification problems.

References

[1] Goldberg, D. E., Genetic Algorithms in Search, Optimization, and Machine Learning, Addison-Wesley, Reading, MA, 1989.

[2] Pawlak, Z., "Information Systems, Theoretical Foundations", Information Systems, vol. 6, no. 3, pp. 205-218, 1981.

[3] Bedingfield, S. E. and Smith, K. A., "Evolutionary Rule Generation in an Information System", in Dagli, C. H. et al. (eds.), Intelligent Engineering Systems Through Artificial Neural Networks, ASME Press, NY, volume 9, pp. 485-492, 1999.

[4] Graf, J. and Nakhaeizadeh, G., "Recent Developments in Solving the Credit-Scoring Problem" in Plantamura, V.L., Soucek, B. and Vissagio, G. (eds.), Logistic and Learning for Quality Software, Management and Manufacturing, John Wiley, New York, 1993.

[5] Pawlak, Z., Rough Sets. Theoretical Aspects of Reasoning about Data, Kluwer Academic Publishers, Dordrecht, 1991.

[6] Walker, R. F., Haasdijk, E. W. and Gerrets, M. C., "Credit Evaluation Using a Genetic Algorithm", in Goonatilake, S. and Treleaven, P. (eds.), Intelligent Systems for Finance and Business, Chapter 3, Wiley, Chichester, 1995.

[7] Bedingfield, S. and Smith, K. A., A Comparison of Fitness Functions for Evolutionary Rule Generation, in M. Mohammadian (ed.), Advances in Intelligent Systems: Theory and Applications, IOS Press, 2000, in press.

Social Fuzziology in Action: Acquisition and Making Sense of Social Information

Vladimir Dimitrov and Brendon Stewart

University of Western Sydney
Richmond 2753
Australia
E-mail: {v.dimitrov, br.stewart}@uws.edu.au

1. Introduction

Social Fuzziology [1] studies the fuzziness inherent in human understanding and dealing with social reality.

About everything we do not know for sure in nature, society and life, we usually think, speak or write in a fuzzy way, that is, by using words and expressions, which convey uncertainty, ambiguity and doubt. The truth contained in a fuzzy statement can neither be proved nor disproved, as fuzziness contains both 'truth' and 'non-truth' at the same time.

Fuzziness does not relate to our thinking and speaking and writing only - it permeates our feelings and emotions, our spiritual beliefs and endeavours. We can simultaneously feel happy and sad, attracted and repulsed, strong and weak, determined and hesitant, godly and worldly. This kind of fuzziness does not need words - it 'voices' through innumerable facial expressions, movements of eyes and body, nerve signals and gestures, body position and muscle tone, voice timbre and volume.

Fuzziness can be expressed in our actions - when we act without being sure about what we really do and aim for, or act with information about the goal but ignorant of how to approach it. This is usually the case when we act in a complex or sensitive-to-changes situation. Social reality is full of such situations.

Fuzziness is not something that exists "over there", as a quality of an external object; it is in *our* understanding of both natural and social environment in which we live and constantly create and re-create through *our* thoughts, feelings and actions. But it is also in the way the environment reflects our physical, mental, emotional and spiritual experience and thus constantly creates and re-creates us.

So the sources of fuzziness in our understanding of social reality are in our own self-referential nature - we are simultaneously creators *and* products of this reality.

The fuzziness of our understanding of social reality reflects the fuzziness of our understanding of our own nature. Every time when we succeed in revealing some enigma of our own nature, of our 'inward' individual life, we simultaneously reveal a secret of social reality in which we live, a secret of our 'outward' social life.

While exploring the fuzziness, Social Fuzziology does not intend to eliminate it - this is an impossible task. Social Fuzziology studies fuzziness in order to reveal how to *live* with it so that instead of being seen as a hindrance on the way to our outward realization, it acts as a catalyser, stimulator and reinforcer of the creative urge for this realization.

At the focus of this particular study is application of social fuzziology to the process of acquisition and making sense of social information.

2. Nature of Social Information

Social information mirrors simultaneously what we know *and* what we do not know about our lives in society and nature. What we do not know energizes our strive for knowing - the deeper the process of knowing goes into the enigmas and paradoxes of the individual and social lives, the more intensive the spring of fuzziness in this process. The famous message of Socratis: " the only thing we know for sure, is how little we know" relates to the reproduction of fuzziness in the human knowledge.

Both our ignorance about the unknown and the uncertainty about the known that ever moves, reshapes and changes keep the engine of fuzziness operate in the space of social interactions and stimulate the emergence of different kinds of social information out of these interactions.

The acquisition and making sense of social information crucially differs from the analogical processes in science and engineering. In the latter we collect information in order to reduce the fuzziness imbedded in our knowledge of nature (reflected in various branches of the natural science) or human-made environment (reflected in the engineering science and technology). The approach of reducing fuzziness does not work with social information because society is not separated from us. We are the society - it consists of us, and we also constantly create, destroy and accumulate an infinite amount of social information by sharing our experiences, our thoughts and views, feelings and emotions, beliefs and dreams. At the same time, society crucially affects us in the process of exchange of social information encapsulated in its numerous multimedia incarnations. What we hear, watch, read and write influence our experience, the ways we think, feel, believe, dream, aspire, and thus create or destroy our personalities.

If some aspect of social life appears fuzzy to an individual, say X, the acquisition of more information, specially designed to clarify this aspect (from the point of view of the provider of this information, say Y) can easily worsen its understanding by X. Everything depends of the views, attitudes and intentions of Y. What if Y 's intention is to manipulate people interested to know more about a specific social issue, so that to make them behave in a way that is desirable (or beneficial) to Y?

For example, X may desperately want to know more about the cases revealing life-threatening effects registered when applying approaches of genetic engineering, but instead s/he is bombarded by information full of fuzzy charisma about how good these approaches might be in 'producing' humans with desirable characteristics in future. Behind the charisma, a generous 'non-fuzzy' funding of 'perspective' (that is, providing competitive advantages to the funding bodies) genetic engineering projects continues with an accelerated tempo.

The sociologists use to say that for every well-defined direction of some problem related to a specific social issue, there is an opposite and equally well-defined direction. One can find many examples of presenting and interpreting one and the same piece of social information in totally opposite ways. The politicians are masters of such kind of performances.

3. Misuse of Fuzziness in Social Information

Misuse or abuse of the fuzziness imbedded in human understanding of social reality makes social information susceptible to manipulations. In society manipulations aim at clearing the path for the exercise of power of an oligarchy (or a leader). It is common for social groups holding power (be it economical, political, religion- or science-oriented, etc.) to have a quite clear agenda, but to utilise imbedded fuzziness in a manipulative way as a means of keeping this agenda secret from the largest part of society.

Most of the charismatic speeches and promises of the leaders in the business and politics are designed to evoke a high degree of fuzziness of people's understanding and hence to hide motives and actions entirely opposite to what the manipulative speeches and promises indicate.

A good example was the recent referendum in Australia - whether Australia to be independent republic or to remain monarchy under the British crown. The political manipulations of the Prime Minister and his supporters (mostly with a significant economic power) succeeded to implant such a high degree of fuzziness in people's understanding of the potential results of the referendum that the majority of people were unclear what they were voting for. Even those who sincerely wanted to see Australia as a republic, voted 'no' to the republican alternative, because the phenomenal brainwash made them think that by voting 'no' they were actually supporting a future(?) realization of a better model of republic.

In the present era of economic globalization, social information is in captivity of an ever-strengthening dictator of people's behaviour - The Global Free Market. This dictator is entirely anti-human - it prefers dealing with robot-like individual programmed to consume and work under conditions of a monstrous exploitation. Bourdieu [2] calls the Global Free Market *infernal machine* producing "not only the poverty of an increasingly large segment of the most economically advanced societies, the extraordinary growth in income differences, the progressive disappearance of autonomous universes of cultural production, such as film, publishing, etc. through the intrusive imposition of commercial values, but also and above all two major trends. First is the destruction of all the collective institutions capable of counteracting the effects of the infernal machine, primarily those of the state, repository of all of the universal values associated with the idea of the *public realm*. Second is the imposition everywhere, in the upper spheres of the economy and the state as at the heart of corporations, of that sort of moral Darwinism that, with the cult of the winner, schooled in higher mathematics and bungee jumping, institutes the struggle of all against all and *cynicism* as the norm of all action and behaviour."

Under conditions of *struggle of all against all*, when everyone is clinging to their job and organisation in an atmosphere of insecurity, suffering, and stress, the artificially generated fuzziness serves entirely 'those who count', that is, those with the greatest influence in the global economy. A continuous brainwash keeps consumption-oriented desires of people at the highest possible level and make them believe that they are 'absolutely free' to choose whatever they want when entering the Market [3]. It is up to them to work as hard and stressful as possible in order to enjoy the fruits of the highest possible democracy - the Democracy of the Markets! Voting for politicians is almost senseless - they cannot change anything at all; they are nothing but obedient marionettes whose strings are pulled by unimaginably reach economic actors.

The mass media intensively 'cares' for the process of injecting fuzziness into the human brains to go with an ever-accelerating tempo. Here is what one former chief of staff of the "New York Times," called by his peers "the dean of his profession", used to say to his colleagues: "I am paid weekly for keeping my honest opinions out of the paper I am connected with. Others of you are paid similar salaries for similar things, and any of you who would be so foolish as to write honest opinions would be out on the streets looking for another job. If I allowed my honest opinions to appear in one issue of my paper, before twenty-four hours my occupation would be gone. The business of the Journalist is to destroy truth; To lie outright; To pervert; To vilify; To fawn at the feet of mammon, and to sell his country and his race for his daily bread. You know it and I know it ... We are the tools and vassals for rich men behind the scenes. We are the jumping jacks, they pull the strings and we dance. Our talents, our possibilities and or lives are all the property of other men. We are intellectual prostitutes" (this quotation is from a material distributed through Internet at *http://www.teamlaw.org/*).

The higher the degree of fuzziness of people's understanding of the totalitarian nature of the free market society, the easier the suppression of their possible resistance to The Global Capitalist Establishment.

4. Acquisition and Making Sense of Social Information

While collecting and making sense of social information, it is necessary to discriminate between two types of fuzziness - delusive fuzziness that is deliberately generated to delude people and manipulate their understanding, and fuzziness inherent in human knowing (or knowledge-related fuzziness) that inevitably permeates what humanity considers as known at any stage of its development.

Social Fuzziology helps us to do this discrimination by eliciting the following important moments:

4.1 Social information is created by people.

When discriminating between the delusive and knowledge-related fuzziness, one needs to explore the motivation of those who create or distribute social information. What could be the motives for presenting this information?

For example, the leader of the One Nation Party in Australia, P. Hanson said in the Australian parliament: "Immigration and multiculturalism are issues that, for too long, ordinary Australians have been kept out of any debate by the major parties. I and most Australians want our immigration policy radically reviewed and that of multiculturalism abolished. I believe we are in danger of being swamped by Asians..." [4]

The motive behind this information is quite clear: search for support by those in Australia who cling to racist views and do not share the ideas of building a multicultural society. The knowledge of this motive helps to navigate through the fuzziness generated in the election platforms of this party full of words like 'caring for' and 'thinking seriously about the future of' the people of Australia.

4.2 Society functions through control and interventions.

Therefore, social information reflects (either explicit or implicit) intentions for exercising power. When discriminating between the delusive and the knowledge-related fuzziness, one needs to investigate where the power comes from, towards whom it is directed and what could be the goal it pursues?

For example, while warmly welcoming everybody to come to the Olympic games, Australia does not issue visas to people younger than 40 years from the former communist countries in the Balkan Peninsula, like Bulgaria, Romania, Macedonia, etc. in fear of possible illegal immigration. Of course, this 'non-fuzzy' decision of rejecting the issue of visas is put into a wonderfully served charismatic fuzziness that the Olympic games provide a rare opportunity for the young people from all the planet to come and meet in one of the most beautiful city in the world, that everybody is welcomed to apply for a visit in Australia, that every application will be professionally considered and every individual case will be mindfully decided upon and answered in a due time, etc. The travel agencies, semi-officially informed about this official decision from the Australian embassy in Belgrade, are less charismatic when talking to the young people wishing to visit Olympic Australia: "You are welcome to apply but your visa will be rejected, and once rejected, it will be hard to expect a positive answer on your next application to visit Australia".

What kind of forces is behind this decision, so incompatible with the spirit of the Olympic games? Forces grounded in a racist way of thinking, supported by the Howard's government in Australia. What pursues the fuzziness fabricated around the issue of visas? Keeping high the international image of Australia as a hospitable organizer of the Olympic games. The image is an important thing when doing global economy, so in the name of keeping the image high, the use of the delusive fuzziness is justified. Towards whom the delusive fuzziness is directed? Towards the young people, extremely vulnerable and fragile both economically and socially, who have managed to save money for their trip (with a hard and stressful labour in their low-paid countries). Now they will know that they are treated as a 'low category' people by the Australian authorities - people who are not allowed to sit in the new Olympic stadium in Sydney.

4.3 In society majority tends to exercise power over minorities.

Majority persecutes those who do not obey its rules. The decisions of majority characterise with inertia; it always takes time before the largest part of people changes their opinions and decisions. An individual (and minorities) can spontaneously reflect or adapt to changes or undergo essential transformations. Does this mean that in our time of accelerated changes, majority always creates social decisions, which are wrong, while minorities have the right answers? The following *paradox* manifests the fuzziness of social decision-creation:

The majority, though never right, is not always wrong (most often, its decisions favour the status quo, that is, do not decide anything*), and therefore the minority, though never wrong, is not always right.*

The above paradox reveals the lack of a final arbiter in judging whether the acquired social information is right or wrong - most probably the judgement remains fuzzy: neither right nor wrong, no matter who is the creator of the social information - the 'strong' majority or the 'weak' minority.

4.4 Does it exist any standard by which to decide what is 'wrong' and what is 'right' when dealing with social information?

There is no such a standard. And as far as even this categorical statement cannot be accept dogmatically in social reality - a reality, which is sensitive to perturbations and changes, 'right' and 'wrong' never stop agitating and attracting attention of social researchers. Therefore, the analysis of what is 'wrong' and what is 'right' continues to make sense when discriminating between the two types of fuzziness - the delusive fuzziness qualified as 'wrong', and the knowledge-related fuzziness qualified as 'right', in any specific case of acquisition and making sense of the social information.

4.5 Two generalizations relate to people's ways of living in society.

- Living while remaining immersed *in* the every-day stream of actions.
- Living while trying to transcend the every-day stream of actions by understanding the dynamics that propel it.

Living in the stream of actions means developing routines, customs and habits to cope with the burden of the social control, pressure and stress - a burden, which for many people on the planet becomes unbearable. The competitive struggle for survival in today's typically Orwellian type of social order, ruled entirely by the power of money, absorbs the physical, mental and emotional energy of people and make them incapable for penetrating into the roots of The System. And without exploring the roots, we are *never* able to recognize the delusive fuzziness of the social information, with which The System bombards us, simultaneously creating illusion that our voices matter (by providing well-conducted situations where we are 'generously' allowed to decide upon something on the very surface of the social establishment). *In the free market based global capitalist economy, the only voices which matters are the voices of few unimaginably reach corporations and individuals and their innumerous servile vassals and marionettes - politicians and the media, police and the mafia, militaries and the legal system.*

Being inside the stream of actions, that is, inside the social establishment, we cannot understand it. The System is too complex and interwoven, too turbulent and powerful, that every attempt to stop just for a second or to do something different from what It pushes us to do, is destined to fail - the whirlpools in the stream easily can smash us. There are so many examples of people smashed by The System because of their decisions not to follow the stream. One must get out of the stream, in order to grasp its flow, one must transcend its whirling dynamics in order to see how they act. Then The System immediately loses its potency and all its delusive fuzziness becomes transparent, translucent, incapable to brainwash and insinuate lies. The way of transcending the stream of actions needs an effort - an *extra* effort to stay *awake*, to refer to ourselves and not to what others preach, to reflect ourselves and not others' images and doctrines, to be responsible of ourselves and not dependent on the will of others, to be aware of our infinite potential to create, to connect with our spiritual essence and to grow.

5. Conclusion

The application of social fuzziology for studying the fuzziness of human understanding and dealing with social reality reveals that the process of acquisition and making sense of social information is sensitive to two major types of fuzziness: knowledge-related fuzziness inherent in what people consider as known and delusive fuzziness intensively used in society for manipulation and brainwash conducted by the most powerful actors of today's global economy. Social fuzziology elicits ways for discriminating between the two types of fuzziness through understanding the forces which shape our actions and experience and thus strengthening our capacity to persist in pursuing our inner drive towards self-realization and growth.

References

1. Dimitrov, V. 2000 Fuzziology, *Internet publication*: http://www.uws.edu.au/vip/dimitrov/fuzzysoc.htm
2. Bourdieu, P. 1998 The essence of neoliberalism, *Le Monde Diplomatique*, 12
3. Dimitrov, V. 2000 Complexity, Chaos and Creativity: A Journey beyond Systems Thinking, *Internet publication*: http://www.hawkesbury.uws.edu.au/academic/socialinq/complexity/complexthink.htm< /P>
4. Hanson, P. 1996 (Member for Oxley), First Speech Australian House of Representatives 10 September. Appropriation Bill No. 1 in *Hansards* (Canberra).

MIX
Papier aus verantwortungsvollen Quellen
Paper from responsible sources
FSC® C105338
FSC
www.fsc.org

If you have any concerns about our products,
you can contact us on
ProductSafety@springernature.com

In case Publisher is established outside the EU,
the EU authorized representative is:
Springer Nature Customer Service Center GmbH
Europaplatz 3, 69115 Heidelberg, Germany

Printed by Libri Plureos GmbH
in Hamburg, Germany